Commutative Algebra

GRADUATE STUDIES IN MATHEMATICS 233

Commutative Algebra

Andrea Ferretti

AMS American Mathematical Society
Providence, Rhode Island

2020 *Mathematics Subject Classification.* Primary 13-01, 11R04, 13P99, 11-01, 14A10.

For additional information and updates on this book, visit
www.ams.org/bookpages/gsm-233

Library of Congress Cataloging-in-Publication Data

Names: Ferretti, Andrea, 1981– author.
Title: Commutative algebra / Andrea Ferretti.
Description: Providence, Rhode Island : American Mathematical Society, [2023] | Series: Graduate studies in mathematics, 1065-7339 ; Volume 233 | Includes bibliographical references and index.
Identifiers: LCCN 2023012824 | ISBN 9781470471279 (paperback) | (ebook)
Subjects: LCSH: Commutative algebra. | Commutative rings. | AMS: Commutative algebra – Instructional exposition (textbooks, tutorial papers, etc.). | Number theory – Algebraic number theory: global fields – Algebraic numbers; rings of algebraic integers. | Commutative algebra – Computational aspects and applications – None of the above, but in this section. | Number theory – Instructional exposition (textbooks, tutorial papers, etc.). | Algebraic geometry – Foundations – Varieties and morphisms.
Classification: LCC QA251.3 .F472 2023 | DDC 512/.44–dc23/eng20230711
LC record available at https://lccn.loc.gov/2023012824

∞ The paper used in this book is acid-free and falls within the guidelines
established to ensure permanence and durability.
Visit the AMS home page at `https://www.ams.org/`

10 9 8 7 6 5 4 3 2 1 28 27 26 25 24 23

To Sbambi, who tolerates me every day a little more.

Contents

Preface

I have learned the basics of commutative algebra from the famous book by Atiyah and MacDonald [**AM69**]. The present text was born from the intention to expand the material therein, and give an alternative organization, although at this point it has grown into its own thing.

In writing this book, I have tried to follow a few principles. First, I have tried not to skim on the basics too rapidly. While it is true that some topics—such as Euclidean domains or unique factorization—are often met in a first algebra course, I feel that they belong in a basic text on commutative rings. Second, and more important, I have tried to present the connections with the most important applications of commutative algebra, such as number theory, algebraic geometry, and computational algebra. This approach makes for a less terse style, but I hope that this is repaid by a wider perspective. Finally, while the exercises present many auxiliary topics, the core of the book should make sense on its own and not depend on them, either logically or because important themes are only left to the exercises.

Commutative algebra is at the crossroad between many fertile areas of mathematics, and I hope that this book conveys the various points of view appropriately. In particular, results in commutative algebra are often clarified by their geometric interpretation. In the present book I have not relied on previous knowledge on the topic, but I have opted instead for a chapter that translates the algebra in geometric terms. Number theory is the other root of commutative algebra, and a constant source of inspiration. Many important topics, such as completion, can be better appreciated by first learning a special case of arithmetic relevance—in the case of completion, this is the construction of the field of p-adic numbers.

Computational algebra has clearly seen an explosion since the advent of computers. Chapter 4 is dedicated to it, but other computational topics are scattered in the text, such as the algorithm to compute Smith's normal form in Chapter 1 or the LLL algorithm in the exercises for Chapter 6. Many other topics could be mentioned, but I could only make some small connections with model theory and invariant theory in the exercises. For the former, see [**Sch99**], while a good introduction to invariant theory is [**Dol03**].

Unfortunately, many topics are not covered here, in particular those that depend on homological algebra techniques. Among them, we can mention flatness, spectral sequences, the study of the Koszul complex and regular sequences, Cohen–Macaulay rings, Gorenstein rings, and duality theory. These are introduced in a subsequent volume, called (without much fantasy) "Homological methods in commutative algebra" [**Fer**].

I should remark that this is an introductory text, so I didn't even try to cover the material of more advanced books, such as Mastumura's books [**Mat70**] and [**Mat86**], or Eisenbud [**Eis95**]. To give an idea, the latter was created as a reference for the famous algebraic geometry book [**Har77**], and ended up being far thicker than it. I advise the reader to consult [**Eis95**] for further reading, and remark that it is actually quite enjoyable despite its appearance; my scope here is far more limited than Eisenbud's.

Here is a brief description of the contents of the book; see the introduction to the various chapters for more details on the topics covered therein.

In the first chapter, we introduce the basics notions of commutative algebra, like rings, ideals, and modules. Moreover, we treat a few basic constructions that we will use throughout the book: quotients, localization, and tensor products. A section about Euclidean rings is also present, to show an example of rings that are especially well-behaved. These topics will probably be familiar from a first algebra course, and the knowledgeable reader can just have a quick glance, as we do nothing fancy in this chapter. Finally, we discuss the language of graded rings and modules.

The core of the book starts with Chapter 2, where we introduce some particular finiteness conditions on our rings and modules. We start by considering rings whose ideals are generated by a single element, and then we generalize to the case of *Noetherian* rings, where ideals are generated by finitely many elements. An equivalent condition is that ascending chains of ideals eventually stabilize. This condition is strong enough to produce a lot of result, but no so stringent, so that most of the rings that we will encounter will satisfy this hypothesis. More generally, we will define Noetherian modules by asking that ascending chains of submodules stabilize, and their symmetric counterpart, *Artinian* modules, where ask the same thing for descending chains. We investigate various operations that preserve these properties, proving in particular the famous Hilbert's basis theorem, that guarantees that if A is a Noetherian ring, so is the polynomial ring $A[x]$. Finally, for modules that are both Noetherian and Artinian, we introduce the important numerical invariant of length.

Chapter 3 is about factorization. We start from the simple case of rings that admit a unique factorization for elements. This is a rather special class of rings, so after that we present the theory of primary decomposition of Lasker, which is a generalization of unique factorization that works over arbitrary Noetherian rings and modules. We then specialize this case to study the important case of Dedekind domains, which have a theory of prime factorization for ideals.

In Chapter 4, we focus on the case of polynomial rings, where some explicit computational techniques are available. In particular, we introduce the resultant of two polynomials. Using it, we tackle the problem of elimination, which is about solving polynomial systems in an inductive way. We also use resultants to introduce discriminants and show their basic properties. In the second part of the chapter, we switch to a different approach and introduce Gröbner bases, which are certain special sets of generators of an ideal in a polynomial ring. Again, after having proved their basic properties, and found algorithms to compute them, we use them to study the problem of elimination from a different angle.

Chapter 5 introduces the concept of integral elements. These are the analogue of algebraic elements from field theory in the setting of rings. In fact, much of the theory just mimics what one does for fields. The parallel notion of an algebraic extension of fields is an integral extension of rings. For such an extension $A \subset B$, we present the Cohen–Seidenberg theory that relates prime ideals in A and B in a precise way. We also use these results to give another, more traditional, characterization of Dedekind rings.

Dedekind rings are studied in much more detail in Chapter 6. In particular, we study the properties of factorization of ideals and what happens in integral extensions of Dedekind rings. An important special case here is the class of number rings: these are obtained by taking elements integral over $\mathbb{Z}$ in a finite extension of $\mathbb{Q}$, say of degree n. Such rings have a natural embedding in $\mathbb{R}^n$ that represents them as lattices, and their ideal are sublattices of finite index. One can then use geometric techniques to derive bounds on the index of ideals, and to find elements of small norm in a given ideal. Using these bounds, we prove two important finiteness results in number theory: the fact that the class group of ideals of an algebraic number field is finite, and that the group of units of its ring of integers is finitely generated.

Chapter 7 introduces topological methods. We start by defining absolute values over fields, which are similar to norms in functional analysis. By mimicking the construction of the Cauchy completion of the reals, one can define the completion of a field endowed with an absolute value. In particular, using a suitable absolute value on $\mathbb{Q}$, this construction produces the field of p-adic numbers. A class of absolute values, called nonarchimedean, can be obtained from the more general notion of a valuation, which we study next. Following this, we study the problem of completion from a more algebraic point of view. To do this, we introduce the machinery of direct and inverse limits. The algebraic point of view allows us to define completion of a ring with respect to a topology (called I-adic) determined by an ideal I, a notion that one can also extend to modules. A crucial tool is the Artin–Rees lemma, which relates the I-adic topology of a module and that of its submodules. In the last section, we give yet another generalization of the construction of p-adic numbers, this time using the notion of Witt vectors.

In Chapter 8, while proving few new results, we give the basic definitions about algebraic geometry: affine and projective varieties, Zariski topology, morphisms of varieties and so on. We then show how most of the material covered so far can be used to quickly obtain information about these new objects. We have decided to put this chapter almost at the end of the book, so that most of the text can be read independently of it. Still, one can start reading this chapter even at the beginning, and follow the geometric dictionary while learning new algebraic concepts. The main result from commutative algebra that we introduce in this chapter is Hilbert's Nullstellensatz, which gives a deep link between the points of an algebraic variety and the maximal ideals of its coordinate ring.

Our treatment of algebraic geometry is as elementary as possible; in particular we do not even mention the machinery of schemes. Even in this more limited setting, the interplay between algebra and geometry is veruy fruitful. One can apply results from commutative algebra to some simple rings (usually finitely generated reduced k-algebras) to obtain results which are geometric in nature; conversely the geometry can suggest that a result may be true for some special rings, and often the algebraic result can be proved for a much more general class of rings.

The next chapter is about dimension theory. The concept of dimension is introduced in Chapter 8 from a geometric point of view, but Chapter 9 gives a wide algebraic generalization. We show that a Noetherian local ring has a definite notion of dimension, the main result being that all reasonable definitions suggested by the geometric intuition lead to the same concept. In most of the chapter we work with local or graded rings, and the parallelism here is very strict. The main technical tool to develop the theory is the Hilbert polynomial, which estimates the order of growth of the size of homogeneous components in a graded module, so we begin the chapter by proving its existence and studying its properties.

In Chapter 10 we define regular local rings, which correspond to nonsingular points on a variety. We also study the nonregular case by introducing the concept of multiplicity of a local ring, which is a simple measure of singularity. As it turns out, this is related to the concept of degree of a graded ring. In the geometric case, where a graded ring corresponds to a projective variety, the degree expresses the number of points of intersection with a general linear space of complementary dimension. This chapter develops the theory of multiplicity, and in doing so describes in some detail the structure of local rings, culminating in the celebrated theorem of Cohen, that gives a precise description for complete local Noetherian rings.

The Appendix consists of an exposition of the classical Galois and Kummer theory of field extensions. It aims to give background for the field theoretic results used in the rest of the book (in particular, Galois theory of finite extensions and the notion of separability) but covers more ground than it is strictly needed. In fact, it can be read independently of the rest of the book as a short introduction to field theory. While fields are in many respects simpler than general rings, they also present many new phenomena, and a familiarity with fields is certainly part of the study of commutative algebra in a broad sense.

The prerequisites for reading the book are not many. We assume that the reader is more or less familiar with algebraic objects, and some acquaintance with linear algebra is assumed—for instance, it is useful if the reader has some familiarity with the tensor product construction in the context of vector spaces. Finally, from Chapter 7 we make some use of elementary topology and some notions about metric spaces.

The book is suitable for a semester on algebra at the introductory graduate level. It could also be used to support a shorter course on algebraic number theory, introducing global and local fields, Dedekind rings with their factorization theory and completions.

To help the reader orient themselves, we suggest some possible paths through the book, other than reading it cover to cover.

A standard introduction to commutative algebra, along the lines of [**AM69**], would start from the basics in Chapters 1 to 3, then go through integral extensions in Chapter 5, topological methods in Chapter 7 (going quickly over Sections 7.1 and 7.2), prove the Nullstellensatz in Sections 8.1 and 8.2, then introduce dimension theory in Chapter 9 and at least the basics on regular rings in Section 10.1.

An introduction to algebraic number theory could cover factorization in Chapter 3, the basics of field theory from Appendix A, the first half of Chapter 4 to introduce discriminants, and parts of Chapter 5 to characterize Dedekind rings; then Chapter 6 covers the global theory and Chapter 7 the local theory, especially Sections 7.1, 7.2, and 7.3.

The reader that wants a quick introduction to the methods of computational algebra can just follow Chapter 2, Section 3.1, and Chapter 4, although we advise to complement this with other texts on the matter, such as [**KR00**]. In a similar way, Appendix A could be used, together with other material, in a minicourse on Galois theory.

Finally, a geometrically minded reader could learn the basics in Chapters 1 and 2, then go through Chapter 8 and learn the necessary commutative algebra along the way. Section 8.5 requires learning about completions in Sections 7.4 and 7.5, while the decomposition in irreducible components of Section 8.7 makes use of the primary decomposition of Section 3.2. Finally, Section 8.9 makes use of the theory of Dedekind rings developed in Chapters 5 and 6. The ideas about dimension are then expanded in Chapter 9, while the notion of regularity comes again in Section 10.1. This way of reading the book will require tracking back the prerequisites for some results, but has the advantage of giving geometrical reasons to introduce algebraic constructs.

Examples in the text usually require only trivial verifications. They are part of the core of the text and should not be skipped; some definitions are actually given inside the examples (for instance, the basic operations on ideals). When an example requires more work, it should be considered as an exercise.

On the other hand, exercises vary from simple to hard, and there is (intentionally) no indication to distiguish the level. So try to do as many exercises as you can, and don't feel frustrated if some of them look too hard. Maybe you can come back later, when you are familiar with more techniques.

In general, I have tried to avoid depending on exercises for the main body of the text. The cases where I have done so should be easy verifications. On the other hand, many important and subtle counterexamples are presented as series of exercises.

No contribution in this book is original, except of course the usual amount of errors, that should be attributed only to the author. If you spot some of them, you can send an email to ferrettiandrea@gmail.com.

I hope that you will enjoy reading this book as much as I enjoyed writing it!

If I was able to write this book, it is because Massimo Gobbino and Paolo Tilli, when I was young and did not know better, believed in me and persuaded me to undertake the study of mathematics. This turned out ot be one of the best choices I made, and I have to really thank them for this. Thanks to Roberto Dvornicich, who instilled in me a lasting love for algebra. I take the opportunity to thank the AMS for the editorial support, especially Ina Mette, who believed in the project and followed it with great patience through many years. Most of all, I want to thank my wife Sbambi, who with her love shows me everyday what is really important, and with her patience and understanding has given me the time and peace of mind to do mathematics and finish this book. $\ddot\smile$

Chapter 1

Basics

In this chapter we introduce (or review) the basic notions of commutative algebra, like rings, ideals, and modules. The first three sections cover these basics, then we treat the case of rings admitting a sort of Euclid's algorithm. These are called Euclidean rings and have a very simple structure: in particular we classify (finitely generated) modules over these rings, generalizing the classification of finitely generated abelian groups.

In the last two sections, we introduce the somewhat parallel notions of local and graded rings, even though the parallelism will emerge later, as the theory is developed. The notion of *localization* of a ring is complementary to that of a quotient, and it is a common technique in commutative algebra. In short the idea is to enlarge a ring formally introducing multiplicative inverses of its elements, in the very same way that one passes from $\mathbb{Z}$ to $\mathbb{Q}$.

1.1. Rings and ideals

Not surprisingly we start by defining rings, the objects that will be central in our study. The idea of a ring is that of a set where one is allowed to perform all of the usual operations, except for division. The notion was born to identify the common features shared by the set $\mathbb{Z}$ of integer numbers and things like polynomial rings $k[x_1, \dots, x_n]$, or a bit more exotic examples like $\mathbb{Z}[i]$, the set of complex numbers that have integral real and imaginary part.

Definition 1.1.1. A *ring* A is a set endowed with two (binary) operations, usually denoted $+$ and $\cdot$ and called *addition* and *multiplication*, which are required to satisfy the following axioms. First, the operation $+$ should make

A into an abelian group, that is

(1) addition is *associative*, namely for every $a, b, c \in A$ we have $(a+b)+c = a+(b+c)$;

(2) there exists an element $0 \in A$ such that $0+a = a+0 = a$ for every $a \in A$;

(3) for every $a \in A$ there exists an element $-a \in A$ such that $a+(-a) = 0$;

(4) addition is *commutative*, that is for every $a, b \in A$ we have $a+b = b+a$.

Second, we ask for some properties of the multiplication, explicitly

(5) the operation $\cdot$ is associative too, so for every $a, b, c \in A$ we have $(a \cdot b) \cdot c = a \cdot (b \cdot c)$;

(6) multiplication is *distributive* over addition, that is for every $a, b, c \in A$ we have $a \cdot (b+c) = a \cdot b + a \cdot c$.

Remark 1.1.2. We have slightly abused notation giving a precise name 0 to the neutral element in (2), since a priori it could be the case that other such elements exist. So we check its uniqueness: if another neutral element $0'$ exists, we must have

$$0 = 0 + 0' = 0'$$

by two applications of (2).

In a similar way we can easily check that for a given a there exist at most one element b such that $a + b = 0$, so we can safely call it *the* inverse and denote it by $-a$.

Another standard identity can be derived from the axioms, namely for every $a \in A$ we have $0 \cdot a = a \cdot 0 = 0$. Indeed from $0 = 0 + 0$ and the distributivity axiom (6) it follows that

$$0 \cdot a + 0 \cdot a = 0 \cdot a$$

and the desired identity follows by cancellation of $0 \cdot a$ on both sides, which is allowed since A is a group with respect to addition.

Before giving the obvious examples, together with some slightly less obvious ones, we give some definitions, to restrict the attention to the rings we are actually interested in.

Definition 1.1.3. We say that a ring A is *commutative* if its multiplication is; explicitly

(7) for every $a, b \in A$ we have $a \cdot b = b \cdot a$.

We say that A is a *ring with unit* if there is a unit for multiplication, namely

(8) there exists an element $1 \in A$ such that $1 \cdot a = a \cdot 1 = a$ for every $a \in A$.

Remark 1.1.4. *All the rings that we shall consider in this book are assumed to be commutative with unit.* Of course noncommutative rings are also of interest, but the theory is actually different in many respects. So, starting from the next section, *ring* will be shorthand for *commutative ring with unit.* Notice that some authors include the existence of 1 into the definition of a ring, and use the name *rng* for a ring without unit.

Example 1.1.5.

(a) The set $\mathbb{Z}$ of integer numbers is a ring with the usual operations. So is the set

$$\mathbb{Z}[i] = \{a + ib \mid a, b \in \mathbb{Z}\}$$

of Gaussian integers. Both these examples are commutative rings with unit.

(b) Every field k is in particular a commutative ring with unit.

(c) If A is a commutative ring, then one can form the ring $A[x]$ of polynomials with coefficients in A. Clearly $A[x]$ is again commutative, and has a unit if and only if A has. In particular if k is a field, then $k[x_1, \ldots, x_n]$ is a ring.

(d) Let E be an abelian group, and let

$$\mathrm{End}(E) = \{f \colon E \to E \mid f \text{ is a group homomorphism}\}$$

denote the set of endomorphisms of E. Then $\mathrm{End}(E)$ inherits an abelian group structure from E (in order to add up two functions you just sum their values); moreover one can define multiplication to be composition of endomorphisms. In this way $\mathrm{End}(E)$ becomes a ring with unit, usually noncommutative.

(e) In a similar fashion if V is a vector space, then $\mathrm{End}(V)$ is a noncommutative ring with unit (here we are only considering linear endomorphisms of V).

(f) If X is a topological space, the set $C(X)$ of continuous functions from X to $\mathbb{R}$ becomes a ring under point-wise addition and multiplication.

Similarly if U is a smooth variety (if you don't know what these are, just let U be an open set in $\mathbb{R}^n$) one gets a ring $C^\infty(U)$ of infinitely differentiable functions on U. Both rings are commutative and have the constant function 1 as a unit.

(g) Let $n \in \mathbb{Z}$. Then the set $n\mathbb{Z}$ of multiples of n is a subring of $\mathbb{Z}$. It is commutative, but it has no unit.

(h) If $A_1, \dots, A_n$ are rings, then one can form in the obvious way the direct product

$$A = A_1 \times \cdots \times A_n,$$

with component-wise sum and multiplication. A is clearly a ring; it is commutative if and only if all the A_i are, and it has a unit if and only if all the A_i are rings with unit.

(i) For a fancier example, consider the set A of all functions

$$f \colon \mathbb{Z}_{\geq 1} \to \mathbb{C},$$

that is, sequences of complex numbers. Addition in A is just the usual addition, but for multiplication we take the *Dirichlet convolution* defined by

$$f * g(n) = \sum_{d|n} f(d) g\left(\frac{n}{d}\right).$$

With these operations, A becomes a commutative ring with unit. Everything is clear, except perhaps for the associativity of multiplication, which follows from the fact that

$$\sum_{d|n} f\left(\frac{n}{d}\right) \sum_{e|d} g\left(\frac{d}{e}\right) h(e) = \sum_{abc=n} f(a)g(b)h(c).$$

The unit of this ring is the function

$$f(n) = \begin{cases} 1 \text{ if } n = 1 \\ 0 \text{ if } n \neq 1. \end{cases}$$

Definition 1.1.6. Let A be a ring. We let $A[x]$ be the ring of polynomials in the indeterminate x, having coefficients in A, with the usual operations.

We also define the ring $A[[x]]$ of *formal power series* with coefficients in A. Elements of $A[[x]]$ are formal linear combinations

$$a(x) = a_0 + a_1 x + a_2 x^2 + \cdots$$

with all $a_i \in A$. Note that we allow an infinite number of a_i to be nonzero. There is no request of convergence on the series (actually this does not even make sense unless $A = \mathbb{R}$ or $\mathbb{C}$); still we will usually write $a(0)$ for a_0.

Operations are defined as follows: given

$$a(x) = a_0 + a_1 x + a_2 x^2 + \cdots,$$
$$b(x) = b_0 + b_1 x + b_2 x^2 + \cdots$$

we let

$$(a + b)(x) = (a_0 + b_0) + (a_1 + b_1)x + (a_2 + b_2)x^2 + \cdots$$

while $c(x) = a(x)b(x)$ is defined by the Cauchy product which should be familiar from calculus

$$\begin{aligned} c_0 &= a_0 b_0 \\ c_1 &= a_1 b_0 + a_0 b_1 \\ c_2 &= a_2 b_0 + a_1 b_1 + a_0 b_2 \\ &\vdots \end{aligned}$$

This is just the operation that comes from distributivity and the request that $x^m \cdot x^n = x^{m+n}$.

By construction $A[x]$ is a subring of $A[[x]]$. When it makes sense we will also use the notation $A\{x\}$ for the ring of *convergent* power series.

Let us agree again that *from now on all rings will be commutative with unit.* If B is a ring and $\{A_i\}$ is a collection of subrings of B, the intersection $A = \bigcap_i A_i$ is again a ring. In particular given a set $E \subset B$ there is a smallest ring containing E.

Definition 1.1.7. The smallest ring containing the set E is said to be the ring *generated* by E. If A is a subring of B and $b \in B$, the symbol $A[b]$ denotes the ring generated by A and b; this is easily seen to be the set of polynomial expressions in b with coefficients in A.

We remark that the notation introduced above is consistent with the notation $A[x]$ for the polynomial ring with coefficients in A in the indeterminate x. Indeed the latter can be seen as the smallest ring containing A and the indeterminate x.

Inside the class of rings we identify a smaller subclass:

Definition 1.1.8. Let A be a ring. We say that A is an *integral domain*—or simply that A is *integral*—if $ab = 0$ implies that a or b is 0. Unlike the identity $a \cdot 0 = 0$, this property does not follow from the ring axioms.

Example 1.1.9.

(a) The rings $\mathbb{Z}$, $\mathbb{Z}[i]$ are integral domains.

(b) Every field is an integral domain.

(c) Every subring of an integral domain is integral; in particular every subring of a field is an integral domain.

(d) The ring $C(X)$ of continuous functions on a topological space X is usually not an integral domain. Similarly for the ring $C^\infty(U)$, where U is a manifold (or an open set in $\mathbb{R}^n$).

(e) If A is an integral domain, $A[x]$ is again an integral domain: this is easily seen considering the monomials of highest (or lowest) degree in a product.

Definition 1.1.10. An element $a \in A$ is called *invertible* if there exists $b \in A$ such that $ab = 1$. The element b is then uniquely determined and denoted by a^{-1}. The set of invertible elements of A is denoted by A^*. Sometimes an invertible element is also called a *unit.*

Example 1.1.11. An element $a(x) \in A[[x]]$ is invertible if and only if $a(0)$ is invertible in A. Indeed the equations

$$\begin{aligned} 1 &= a_0 b_0 \\ 0 &= a_1 b_0 + a_0 b_1 \\ 0 &= a_2 b_0 + a_1 b_1 + a_0 b_2 \\ &\vdots \end{aligned}$$

can be solved for the b_i inductively, provided $a_0 \in A^*$.

In particular if $A = k$ is a field, $a(x)$ is invertible if and only if $a(0) \neq 0$; it follows that every $a(x) \in k[[x]]$ can be written

$$a(x) = x^r b(x)$$

for some $b(x) \in k[[x]]$ invertible; this makes the algebra of power series much easier than that of polynomials.

Definition 1.1.12. An element $a \in A$ is called a *zero divisor* if there exists $b \neq 0$ such that $ab = 0$. 0 itself is considered to be a zero divisor. Of course nontrivial zero divisors are present only if A is not an integral domain. If $a^n = 0$ for some n, a is called *nilpotent*; of course this is stronger than being a zero divisor.

We next introduce the other players: the ideals.

Definition 1.1.13. Let $I \subset A$ be an *additive subgroup.* We say that I is an *ideal* if for every $i \in I$ and $a \in A$ we have $ai \in I$. Any ideal different from A itself is called *proper.*

The proper ideal I is called *prime* if $ab \in I$ implies that either $a \in I$ or $b \in I$. It is called *maximal* if it is not properly contained in any other (proper) ideal.

Remark 1.1.14. For any subset $E \subset A$ there is a smallest ideal containing E, namely the intersection of all ideals containing E. This is denoted by (E). When $E = \{a_1, \dots, a_n\}$ we simply denote it by $(a_1, \dots, a_n)$. Explicitly, the ideal $(a_1, \dots, a_n)$ is the set of elements $a \in A$ which can be written in the form

$$a = x_1 a_1 + \cdots + x_n a_n.$$

To see this, one only has to check that the set thus defined is an ideal; of course it is the smallest ideal containing $a_1, \ldots, a_n$.

Definition 1.1.15. If $I = (a)$ is generated by a single element, we say that I is *principal*. When (a) is a prime ideal we say that a is a *prime element*.

Remark 1.1.16. Any proper ideal is contained in a maximal one. This follows directly from Zorn's lemma, together with the fact that the union of an ascending chain of proper ideals is an ideal (it is not the whole ring because it does not contain 1).

Proposition 1.1.17. *Any maximal ideal $\mathcal{M}$ is prime.*

Proof. Let $a \notin \mathcal{M}$ and $ab \in \mathcal{M}$; as $\mathcal{M}$ is maximal $(\mathcal{M}, a)$ must be the whole ring, so we can write

$$1 = m + xa$$

for some $m \in \mathcal{M}$ and $x \in A$. Multiplying by b we get

$$b = bm + xab \in \mathcal{M}. \qquad \square$$

Example 1.1.18.

(a) $\mathbb{Z}$ has the principal ideals (n) for every integer n; these are the only ideals because they are the only additive subgroups. This will be generalized in Proposition 1.5.5. The ideal (n) is prime if and only if $n = 0$ or n is a prime of $\mathbb{Z}$.

(b) The ideal (5) is not prime in $\mathbb{Z}[i]$; indeed $5 = (2+i)(2-i)$. So a prime element of a ring need not be prime in a bigger ring.

(c) The ideal (x, y) of $k[x, y]$ is not principal. (Why?)

(d) An ideal I is the whole ring if and only if it contains 1, if and only if it contains any invertible element. In particular the only ideals of a field k are 0 and k itself. Conversely, any ring having just the trivial ideals is a field.

(e) The ideal (0) is prime if and only if A is an integral domain.

(f) For any ring A the set of nilpotent elements is an ideal $\mathcal{N}(A)$, called the *nilradical* of A. Recall that an element $a \in A$ is *nilpotent* if $a^n = 0$ for some n. We shall prove in Proposition 1.6.10 that it is the intersection of all prime ideals of A.

(g) We define the *Jacobson radical* $\mathcal{J}(A)$ as the intersection of all maximal ideals of A. We then have $\mathcal{N}(A) \subset \mathcal{J}(A)$.

(h) If I, J are ideals of A, then we can define the ideals

$$I + J = \{i + j \mid i \in I,\, j \in J\} = (I, J),$$
$$I \cdot J = \big(\{i \cdot j \mid i \in I,\, j \in J\}\big).$$

Note that we always have $I \cdot J \subset I \cap J$.

(i) If I, J are ideals of A, then we can define the ideal

$$(I : J) := \{a \in A \mid aJ \subset I\}$$

(check that this is indeed an ideal!). When $J = (x)$ is principal, we will simply write $(I : x)$ to mean $(I : J)$, and similarly when I is principal. This operation may behave slightly differently from what the notation suggests. For instance in $\mathbb{Z}$ we have

$$(8 : 4) = (2) \text{ but } (8 : 5) = (8).$$

We easily see that $I \subset (I : J)$ in any case.

(j) Let $I \subset A$ be any ideal. Then the set

$$\sqrt{I} = \{a \in A \mid a^n \in I \text{ for some } n\}$$

is an ideal, called the *radical* of I. The only nontrivial check is that if $a, b \in \sqrt{I}$, then $a + b \in \sqrt{I}$. Let n be big enough, so that $a^n, b^n \in I$. Then

$$(a+b)^{2n} = \sum_{i=0}^{2n} \binom{2n}{i} a^i b^{2n-i} \in I$$

because each addend is in I. Note that by definition we have $\sqrt{0} = \mathcal{N}(A)$.

The ideal I is called *radical* when $I = \sqrt{I}$.

The Jacobson radical admits the following characterization.

Proposition 1.1.19. *Let A be a ring. Then $a \in \mathcal{J}(A)$ if and only if $1 - ab$ is invertible for all $b \in A$.*

Proof. Let $a \in \mathcal{J}(A)$; then $1 - ab$ is not contained in any maximal ideal, so $(1 - ab) = A$ and finally $1 - ab$ is a unit.

Vice versa, let $\mathcal{M}$ be a maximal ideal. If $a \notin \mathcal{M}$, then $(a, \mathcal{M}) = A$, so we can write

$$1 = ab + m,$$

for some $b \in A$ and $m \in \mathcal{M}$; but then $1 - ab \in \mathcal{M}$ is not invertible. □

The defining property of prime ideals with respect to elements extends in some way to ideals.

Proposition 1.1.20.

(i) *Let $P_1, \ldots, P_n \subset A$ be prime ideals and $I \subset A$ be any ideal. If*

$$I \subset P_1 \cup \cdots \cup P_n,$$

then $I \subset P_k$ for some k.

(ii) *Let* $I_1, \dots, I_n \subset A$ *be ideals and* $P \subset A$ *be a prime ideal. If*

$$P \supset I_1 \cap \cdots \cap I_n,$$

then $P \supset I_k$ *for some* k.

Proof.

(i) By induction on n, we can assume that I is not contained in any union of $n-1$ of these primes, so we can take

$$a_i \in I \setminus \bigcup_{j \neq i} P_j.$$

Then we must have $a_i \in P_i$. Let

$$a = \sum_{i=1}^{n} a_1 \cdot \cdots \cdot \widehat{a_i} \cdot \cdots \cdot a_n,$$

where $\widehat{a_i}$ means that a_i is omitted. Then all addends but one lie in P_i, so $a \notin P_i$, for every i. This is a contradiction because $a \in I$.

(ii) Assume that each $I_k \not\subset P$. Take $a_k \in I_k \setminus P$; then

$$a = a_1 \cdot a_2 \cdot \cdots \cdot a_n \notin P,$$

which contradicts the fact that $a \in I_k$ for each k. □

1.2. Quotients

After one introduces rings, the next natural steps is to define the class of maps between them.

Definition 1.2.1. Let A, B be rings. A *ring homomorphism*, or simply homomorphism, between them is a map

$$f\colon A \to B$$

such that for every $a, b \in A$ we have

$$f(a+b) = f(a) + f(b) \quad \text{and} \quad f(ab) = f(a)f(b).$$

We will always require that our homomorphisms are *unital*, that is $f(1) = 1$.

Definition 1.2.2. Given a homomorphism $f\colon A \to B$ we define its kernel

$$\ker f := f^{-1}(0) = \{a \in A \mid f(a) = 0\}.$$

It is a straightforward check that $\ker f$ is an ideal of A.

Remark 1.2.3. More generally for any ideal I of B, $f^{-1}(I)$ is an ideal of A, and if I is prime, then $f^{-1}(I)$ is prime too (check these facts!).

Remark 1.2.4. If $f\colon A \to B$ is a homomorphism, then $f(A)$ is only a subring of B, not necessarily an ideal. (Can you produce an example where $f(A)$ fails to be an ideal?)

As special cases we have

Definition 1.2.5. A homomorphism $f\colon A \to B$ is called an *isomorphism* when is it injective and surjective, so it admits an inverse function $g\colon B \to A$. In this case g is necessarily a homomorphism, since f is.

When $A = B$ we will say that f is an *automorphism* of A.

Example 1.2.6.

(a) Multiplication by n is *not* a ring homomorphism $\mathbb{Z} \to \mathbb{Z}$. Indeed

$$nh \cdot nk \neq n \cdot hk$$

for nonzero h, k, as soon as $n \neq 1$.

(b) If $a \in A$ satisfies $a^2 = a$ (in which case a is called *idempotent*), multiplication by a does satisfy the first two conditions of a ring homomorphism $A \to A$, yet it is not unital.

(c) For any ring A, there is a unique homomorphism $\phi\colon \mathbb{Z} \to A$. Its kernel $\ker\phi$ is generated by a nonnegative number n, possibly 0 when ϕ is injective. We call n the *characteristic* of A, denoted $\mathrm{char}(A)$.

(d) Let $a = (a_1, \ldots, a_n) \in k^n$, and consider the evaluation map

$$ev_a\colon k[x_1, \ldots, x_n] \to k$$

defined by $ev_a(f) = f(a_1, \ldots, a_n)$. This is a homomorphism of rings; indeed polynomial rings are constructed exactly in such a way that this holds.

The same example works for any ring A in place of k.

(e) In the same way for every topological space X and for every $x \in X$ we have the valuation homomorphism

$$ev_x\colon C(X) \to \mathbb{R}$$

defined by $ev_x(f) = f(x)$.

(f) The conjugation map

$$c\colon \mathbb{Z}[i] \to \mathbb{Z}[i]$$

is a ring automorphism.

Similarly, let $\omega \in \mathbb{C}$ be a primitive third root of unity, and take $A = \mathbb{Z}[\omega]$. There is a unique automorphism $A \to A$ exchanging the third roots of unity ω and ω^2.

(g) Let $f\colon k^n \to k^n$ be a polynomial function. Then composition with f yields a homomorphism

$$c_f\colon k[x_1,\ldots,x_n] \to k[x_1,\ldots,x_n]$$

given by $c_f(g) = g \circ f$. This is an automorphism when f admits a polynomial inverse, for instance if f is an invertible linear function.

An example of such a function outside the realm of linear maps is $f(x,y) = (x+y^2, y)$.

There is an automatic way to produce a homomorphism, actually one that will produce a homomorphism with an assigned kernel.

Definition 1.2.7. Given a ring A and an ideal I we define a ring A/I as follows. As a set, A/I is A modulo the equivalence relation $a \sim b$ if there is some $i \in I$ such that $a = b + i$. The equivalence class of a in A/I is denoted $\overline{a}$, or also $\overline{a}_I$ when we want to emphasize the dependence on I.

The operation are defined on representatives:

$$\overline{a} + \overline{b} := \overline{a+b} \qquad \text{and} \qquad \overline{a} \cdot \overline{b} := \overline{ab}.$$

These are well defined: for instance if $a = a' + i$ we have

$$\overline{ab} = \overline{(a'+i)b} = \overline{a'b + ib} = \overline{a'b}$$

since $ib \in I$.

The set A/I, endowed with the above ring structure is called the *quotient* of A by I.

As a notation we sometimes write

$$a \equiv b \pmod{I}$$

and say that a and b are *congruent modulo* I whenever $\overline{a}_I = \overline{b}_I$, that is, $a - b \in I$.

By construction we have a surjective homomorphism, which we usually call *projection*,

$$\pi_I\colon A \to A/I$$

sending a to $\overline{a}$. We see at once that $\ker \pi_I = I$. In a precise sense these cover all the examples of homomorphisms.

Proposition 1.2.8. *Let $f\colon A \to B$ be a homomorphism, and let $I = \ker f$. Then we have an injective homomorphism $\widehat{f}\colon A/I \to B$ such that $\widehat{f}(\overline{a}) = f(a)$. Moreover $\widehat{f}$ is surjective (hence an isomorphism) if and only if f is surjective.*

Proof. One checks that $\widehat{f}$ given as above is well defined. Then $\ker \widehat{f} = I/I = 0$, so $\widehat{f}$ is injective. By definition, it has the same image of f. □

Proposition 1.2.9. *The preimage map $J \to \pi_I^{-1}(J)$ gives a bijective correspondence between ideals of A/I and ideals of A containing I. This correspondence preserves prime and maximal ideals.*

Proof. For every ideal J of A/I, $\pi_I^{-1}(J)$ is an ideal of A. The inverse correspondence sends the ideal $K \supset I$ to $K/I = \pi(K)$. This is an additive subgroup of A/I (this is true for any homomorphism of groups). To check that it is an ideal we take $\overline{a} \in A/I$ and $\overline{k} \in K/I$ and note that

$$\overline{a} \cdot \overline{k} = \overline{ak} \in K/I.$$

If J is prime then $\pi^{-1}(J)$ is prime; this is true for any homomorphism. Vice versa, let $P \supset I$ be a prime of A and assume that

$$\overline{a} \cdot \overline{b} = \overline{ab} \in P/I.$$

Then $ab \in P+I = P$, so either $a \in P$ or $b \in P$. It follows that P/I is prime.

Finally, the correspondence preserves inclusions, hence it preserve maximality of ideals. □

Since 0 is prime in A if and only if A is an integral domain, and is maximal if and only if A is a field, we deduce the following,

Corollary 1.2.10. *The ideal $I \subset A$ is prime if and only if A/I is an integral domain. It is maximal if and only if A/I is a field.*

From the corollary we can deduce again that maximal ideals are prime.

Example 1.2.11. Consider the homomorphism v_a of Example 1.2.6. It is a surjective homomorphism whose kernel is

$$I_a := \ker v_a = \{f \in k[x_1, \ldots, x_n] \mid f(a_1, \ldots, a_n) = 0\}.$$

From the isomoprhism

$$k[x_1, \ldots, x_n]/I_a \cong k$$

we deduce that I_a is a maximal ideal.

We end the section with the classical

Theorem 1.2.12 (Chinese remainder theorem). *Let A be a ring, $I, J \subset A$ two ideals which are* coprime, *in the sense that $I + J = A$. Then we have a canonical isomorphism*

$$f\colon A/(I \cap J) \to A/I \times A/J$$

such that $f(\overline{a}_{I \cap J}) = (\overline{a}_I, \overline{a}_J)$.

Proof. Define a homomorphism

$$g\colon A \to A/I \times A/J$$

by $g(a) = (\overline{a}_I, \overline{a}_J)$. If we prove that g is surjective, the thesis follows from Proposition 1.2.8, since $\ker g = I \cap J$.

Then for any $\overline{b} \in A/I$ and $\overline{c} \in A/J$ we want to find $x \in A$ such that

$$b = x + i$$
$$c = x + j$$

for some $i \in I$, $j \in J$. This can be solved for x provided

$$b - c = i - j.$$

The existence of suitable elements i, j then follows from the hypothesis $I + J = A$. □

1.3. Modules

Formally, modules over a ring are defined exactly like vector spaces over a field.

Definition 1.3.1. Let A be a ring, M a set endowed with a (binary) operation $+$, called addition, and a map

$$A \times M \longrightarrow M$$

$$(a, m) \longrightarrow a \cdot m,$$

called (scalar) multiplication. We will call M an *A-module*, or simply a module, if the operations satisfy the following axioms. First, the operation $+$ should make M into an abelian group, that is

(1) addition is *associative*, namely for every $m, n, p \in M$ we have $(m + n) + p = m + (n + p)$;

(2) there exists an element $0 \in M$ such that $0 + m = m + 0 = m$ for every $m \in M$;

(3) for every $m \in M$ there exists an element $-m \in M$ such that $m + (-m) = 0$;

(4) addition is *commutative*, that is for every $m, n \in M$ we have $m+n = n + m$.

Second we ask for some properties of the multiplication, explicitly

(5) the operation $\cdot$ is associative, in the sense that for every $a, b \in A$ and $m \in M$ we have

$$(a \cdot b) \cdot m = a \cdot (b \cdot m)$$

(note that the two mutiplications involved are actually different operations);

(6) multiplication is *distributive* over addition, that is for every $a \in A$ and $m, n \in M$ we have $a \cdot (m+n) = a \cdot m + a \cdot n$.

Definition 1.3.2. A *submodule* of M is just a subset $N \subset M$ which is closed under the operations, so that it inherits the structure of an A-module itself.

Remark 1.3.3. Uniqueness of the neutral element and of the additive inverse are proved exaectly in the same way as for rings, so there is no ambiguity in using the symbols 0 and $-m$.

Remark 1.3.4. Let M be an abelian group. Giving M the structure of an A-module is the same as giving a homomorphism of rings $A \to \operatorname{End}(M)$.

Example 1.3.5.

(a) When $A = k$ is a field an A-module is just a vector space over k. In general A-modules have a much more subtle structure, as the next examples show.

(b) A^n is a module over A with component-wise multiplication.

(c) The submodules of A are just its ideals.

(d) Let $I \subset A$ be an ideal. Then A/I is an A-module.

(e) Every abelian group G has a unique structure of $\mathbb{Z}$-module. This is because there is exactly one homomorphism $\mathbb{Z} \to \operatorname{End}(G)$ sending 1 to the identity. Explicitly we have

$$n \cdot g = \underbrace{g + \cdots + g}_{n \text{ times}}.$$

(f) For the reader familiar with vector bundles, let X be a topological space, $E \to M$ a vector bundle. Let $C(X, E)$ be the set of continous sections of E; then $C(X, E)$ is a module over $C(X)$.

A similar example can be obtained when $X = M$ is a manifold, E is a differentiable vector bundle and only C^∞ functions and sections are considered.

(g) Let M be a manifold (if you prefer just take M an open set in $\mathbb{R}^n$). Then $C(M)$ is a module over $C^\infty(M)$.

(h) Let V be a k-vector space and choose any linear endomorphism $L \in \operatorname{End}(V)$. Then V becomes a module over $k[x]$; multiplication by x is defined by

$$x \cdot v = L(v).$$

(i) Let M be an A-module, $E \subset M$. There is a smallest submodule of M containing E; this is called the submodule *generated* by E, and denoted $\langle E \rangle_A$.

(j) Let M be an A-module, and denote $M[x]$ the set of all formal finite linear combinations

$$m_n x^n + \cdots + m_1 x + m_0$$

with $m_n, \ldots, m_0 \in M$; then $M[x]$ is an $A[x]$-module in the obvious way.

(k) Let M be an A-module, $I \subset A$ an ideal. We define $I \cdot M$ as the submodule of M generated by all products $i \cdot m$ for $i \in I$ and $m \in M$, in symbols

$$I \cdot M = \langle i \cdot m \mid i \in I,\, m \in M \rangle_A .$$

This agrees with our previous definition of product when M is an ideal of A.

(l) If $M, N \subset R$ are submodules, the sum

$$M + N := \{m + n \mid m \in M,\, n \in N\}$$

is a submodule of R; it coincides with $\langle M, N \rangle_A$.

(m) If $M \subset N$ is a submodule, the set

$$(M : N) := \{a \in A \mid aN \subset M\}$$

is an ideal of A. This is sometimes denoted $(M :_A N)$ when the ring is unclear.

Remark 1.3.6. Let M be an A-module, $I \subset A$ an ideal. Then M has an induced structure of A/I-module if and only if $I \cdot M = 0$.

In particular if $I \cdot M = 0$, the A-submodules and A/I submodules of M are the same. When $M = A/I$ we conclude that ideals of A/I are the same of A-submodules of A/I.

This fact will be used in many places, often implicitly.

Next we introduce some notions analogous to those already seen for rings.

Definition 1.3.7. Let M, N be A-modules. A *homomorphism* between M and N is a map

$$f \colon M \to N$$

such that

$$f(m+n) = f(m) + f(n) \qquad \text{and} \qquad f(am) = af(m)$$

for all $m, n \in M$ and $a \in A$. We will also say that f is a A-*linear* map.

When $M = N$ we say that f is an *endomorphism* of M. When f is both injective and surjective, so that there is an inverse A-linear map, we say that f is an *isomorphism*.

Definition 1.3.8. Given a homomorphism $f\colon M \to N$ we define its kernel

$$\ker f := f^{-1}(0) = \{m \in M \mid f(m) = 0\}$$

and its image $\operatorname{im} f := f(M) \subset N$. Both $\ker f$ and $\operatorname{im} f$ are submodules, of M and N respectively.

Not surprisingly every submodule of M arises as the kernel of suitable homomorphism, which is the projection to the quotient.

Definition 1.3.9. Let $N \subset M$ be A-modules. We define the *quotient A-module* M/N as follows. As a set M/N is formed by equivalence classes of elements of M modulo the equivalence relation $l \sim m$ if there is some $n \in N$ such that $l = m + n$. The equivalence class of m is denoted $\overline{m}$ or sometimes $\overline{m}_N$.

Operations on M/N are defined on representatives:

$$\overline{m} + \overline{n} := \overline{m + n} \qquad \text{and} \qquad a \cdot \overline{m} := \overline{am}.$$

The verification that these are well defined is identical to the one we made for rings.

As a notation we sometimes write

$$m \equiv n \pmod{N}$$

and say that m and n are *congruent modulo N* whenever $\overline{m}_N = \overline{n}_N$, that is, $m - n \in N$.

We have a surjective homomorphism, which we usually call *projection*,

$$\pi_N\colon M \to M/N$$

sending m to $\overline{m}$. By construction $\ker \pi_N = N$. As in the case of rings we have (with the same proof)

Proposition 1.3.10. *Let M, R be A-modules, $f\colon M \to R$ a homomorphism, and let $N = \ker f$. Then we have an injective homomorphism $\widehat{f}\colon M/N \to R$ such that $\widehat{f}(\overline{m}) = f(m)$. Moreover $\widehat{f}$ is surjective (hence an isomorphism) if and only if f is surjective.*

With the same notation, we have:

Proposition 1.3.11. *The preimage map $R \to f^{-1}(R)$ gives a bijective correspondence between submodules of R and submodules of M containing N.*

Definition 1.3.12. Given a homomorphism $f\colon M \to N$ we define the *cokernel* of f as

$$\operatorname{coker} f := N/\operatorname{im} f.$$

In a sense which will be clearer later on, this is the specular notion of the kernel.

Usually we use diagrams to keep track of kernels and cokernels.

Definition 1.3.13. Consider a sequence (finite or infinite) of A-modules with maps between them:

$$M_\bullet\colon \qquad \cdots \longrightarrow M_{n-1} \xrightarrow{f_{n-1}} M_n \xrightarrow{f_n} M_{n+1} \longrightarrow \cdots .$$

If $f_n \circ f_{n-1} = 0$ for every n we say that $M_\bullet$ is a *complex* of A-modules. This means that

$$\operatorname{im} f_{n-1} \subset \ker f_n.$$

If we have

$$\operatorname{im} f_{n-1} = \ker f_n$$

we say that $M_\bullet$ is an *exact sequence*. A three-term exact sequence like

$$0 \longrightarrow M \longrightarrow N \longrightarrow P \longrightarrow 0$$

is called a *short exact sequence*. In this case the first map is injective and the last one is surjective.

Remark 1.3.14. If

$$0 \longrightarrow M \xrightarrow{f} N \xrightarrow{g} P \longrightarrow 0$$

is a short exact sequence, then M is isomorphic to $\ker g$ and P to $\operatorname{coker} f$.

Vice versa, any quotient M/N fits into an exact sequence

$$0 \longrightarrow N \longrightarrow M \longrightarrow M/N \longrightarrow 0 .$$

Remark 1.3.15. Any long exact sequence can be split into short exact sequences like

$$\begin{gathered} 0 \longrightarrow \ker f_n \longrightarrow M_n \longrightarrow \ker f_{n+1} \longrightarrow 0 \\ 0 \longrightarrow \ker f_{n+1} \longrightarrow M_{n+1} \longrightarrow \ker f_{n+2} \longrightarrow 0 ; \\ \vdots \end{gathered}$$

this allows us to reduce many statements about arbitrary exact sequences to the case of short ones.

Example 1.3.16.

(a) For any ideal I of A we have the short exact sequence *of A-modules*

$$0 \longrightarrow I \longrightarrow A \longrightarrow A/I \longrightarrow 0\,;$$

notice that the modules I and A/I have quite a different role—in particular inherits A/I a ring structure, while I does not.

(b) Multiplication by n defines a homomorphism *of $\mathbb{Z}$-modules* $\mathbb{Z} \to \mathbb{Z}$, which gives rise to the exact sequence

$$0 \longrightarrow \mathbb{Z} \longrightarrow \mathbb{Z} \longrightarrow \mathbb{Z}/n\mathbb{Z} \longrightarrow 0\,;$$

this is not a homomorphism of rings since we require the latter to be unital.

(c) A homomorphism from A^m to A^n is defined by a $m \times n$ matrix with coefficients in A, as in the field case.

(d) The set of homomorphism from M to N is in the obvious way an A-module, denoted $\mathrm{Hom}(M,N)$, or else $\mathrm{Hom}_A(M,N)$ if the ring is not clear from the context. We also use the notation $\mathrm{End}(M) := \mathrm{Hom}(M,M)$.

(e) Let $U \subset \mathbb{C}$ be an open set and regard $M = C^\infty(U,\mathbb{C})$ as a module over $A = \mathcal{O}(U)$. Then the derivative $d/d\overline{z}$ is an endomorphism of M, while d/dz is not A-linear.

After all these preliminary definitions, we actually begin to prove something about modules.

Definition 1.3.17. An A-module M is said to be *finitely generated* if there is a finite number of elements $m_1, \ldots, m_r \in M$ such that

$$M = \langle m_1, \ldots, m_r \rangle_A\,.$$

Equivalently, every element of M is a linear combination of the elements $m_1, \ldots, m_r$ with coefficients in A.

Remark 1.3.18. Every finitely generated module is a quotient of A^r for some r, and vice versa.

The following may be the most widely used result in commutative algebra.

Theorem 1.3.19 (Nakayama's lemma). *Let A be a ring, $J = \mathcal{J}(A)$ its Jacobson radical. Let M be a finitely generated module such that $J \cdot M = M$; then $M = 0$.*

Proof. Assuming $M \neq 0$, let $m_1, \dots, m_r$ be a *minimal* set of generators of M. Since $m_r \in J \cdot M$ we can write

$$m_r = a_1 m_1 + \cdots + a_r m_r,$$

with $a_1, \dots, a_r \in J$. Rewrite this as

$$(1 - a_r) m_r = a_1 m_1 + \cdots + a_{r-1} m_{r-1}.$$

According to Proposition 1.1.19 we can invert $1 - a_r$ to write m_r as a linear combination of the other m_i, contradicting the minimality. □

The above result and its corollaries are often used when A is a *local ring*, that is A has only one maximal ideal $\mathcal{M}$. In this case of course we just have $J = \mathcal{M}$.

Applying Nakayama's lemma to the quotient M/N we get the seemingly stronger form.

Corollary 1.3.20. *Let A be a ring, $J = \mathcal{J}(A)$ its Jacobson radical. Let M be a finitely generated module, $N \subset M$ a submodule, such that $M = N + J \cdot M$; then $M = N$.*

There is a notion of finite generation for a different kind of objects, which we now introduce.

Definition 1.3.21. Let $A \subset B$ be rings. Then on B we have both the structure of ring and that of A-module. We say that B is a A-*algebra*. More generally we speak of a A-algebra when we have a (not necessarily injective) homomorphism $f \colon A \to B$. A homomorphism between two A-algebras is a homomorphism of rings which is also A-linear.

Remark 1.3.22. When $A = k$ is a field, every homomorphism $k \to B$ is injective, so the distiction above does not apply.

If B is an A-algebra and $I \subset A$ an ideal we will write $I \cdot B$ to denote the ideal generated by I in B. This is sometimes called the *extension* of I in B.

Definition 1.3.23. Let B be an A-algebra, $E \subset B$ a subset. We say that E generates B if the smallest sub-A-algebra of B containing E is B itself. If we can choose a finite set E that generates B, we say that B is *finitely generated*.

Remark 1.3.24. Of course, if B is finitely generated as an A-module, it is also finitely generated as an A-algebra, but the converse is in general false.

Remark 1.3.25. If B is a finitely generated A-algebra, there is a surjective homomorphism

$$A[x_1, \dots, x_n] \to B,$$

and conversely.

Example 1.3.26.

(a) Just as any abelian group is a $\mathbb{Z}$-module, every ring B is a $\mathbb{Z}$-algebra, via the only homomorphism $\mathbb{Z} \to B$ which sends 1 to 1.

(b) For any ring A we can consider the ring $A[x_1, \ldots, x_n]$. This is finitely generated as an A-algebra, but not as an A-module.

(c) $\mathbb{Q}$ is not finitely generated as $\mathbb{Z}$-algebra. If it was, only a finite number of prime factors could appear in the denominators of rational numbers.

(d) Every quotient A/I is a A-algebra via the canonical projection, and is of course finitely generated.

1.4. More constructions with modules

Next we describe more operations that we can perform on A-modules.

Definition 1.4.1. Let $\{M_i\}_{i \in I}$ be a (not necessarily finite) collection of A-modules. We define the *direct product* of the M_i to be

$$\prod M_i := \{(m_i)_{i \in I} \mid m_i \in M_i\}$$

and their *direct sum* as the submodule

$$\bigoplus M_i := \{(m_i)_{i \in I} \mid m_i \in M_i,\ m_i \neq 0 \text{ for finitely many } i\}.$$

Operations on $\prod M_i$ and $\bigoplus M_i$ are defined component-wise.

We say that M is *free* if it is isomorphic to a direct sum of copies of A.

Remark 1.4.2. Given a collection of A-module homomorphisms

$$f_i \colon N \to M_i,$$

there is a unique homomorphism

$$f \colon N \to \prod M_i$$

such that $f_i = \pi_i \circ f$, where π_i is the projection on the factor M_i.

Symmetrically given homomorphisms

$$g_i \colon M_i \to N$$

there is a unique homomorphism

$$g \colon \bigoplus M_i \to N$$

which agrees with g_i on each summand.

Of course, when the index set is finite direct product and direct sum agree, and enjoy both properties.

Remark 1.4.3. Let M be any A-module, $E = \{e_i\}$ a set of generators. Take any set $S = \{s_i\}$ in bijective correspondence with E, and consider the direct sum of copies of A, one for each s_i. This is a free module with basis $\{s_i\}$, call it F. We have a surjective homomorphism $F \to M$ sending s_i to e_i. So every A-module is a quotient of a free one.

Example 1.4.4.

(a) Every vector space over a field k has a basis, hence it is free.

(b) Regard $A[x]$ as an A-module; then it is the direct sum of a denumerable quantity of copies of A. Instead $A[[x]]$ is their direct product.

(c) Let M, N, R be A-modules. If $M = N \oplus R$ we have an exact sequence

$$0 \longrightarrow N \longrightarrow M \longrightarrow R \longrightarrow 0\,.$$

Conversely, given an exact sequence

$$0 \longrightarrow N \xrightarrow{\ i\ } M \xrightarrow{\ p\ } R \longrightarrow 0,$$

M is the direct sum of N and R (in such a way that i and p are the natural inclusion and projection) if and only if there is an A-linear map

$$s\colon R \to M$$

such that $p \circ s = \mathrm{id}_R$. In this case we say that the exact sequence *splits*.

(d) The condition above is not always verified; for instance we have an exact sequence of Z-modules

$$0 \longrightarrow \mathbb{Z}/2\mathbb{Z} \xrightarrow{\ \alpha\ } \mathbb{Z}/4\mathbb{Z} \longrightarrow \mathbb{Z}/2\mathbb{Z} \longrightarrow 0,$$

where $\alpha(\overline{n}) = \overline{2n}$, which does not split. Indeed $\mathbb{Z}/4\mathbb{Z}$ has an element of order 4, while $\mathbb{Z}/2\mathbb{Z} \oplus \mathbb{Z}/2\mathbb{Z}$ does not.

(e) For another example of an exact sequence which is not split take $A = k[x]/(x^2)$, where k is a field. Then we have the exact sequence

$$0 \longrightarrow k \xrightarrow{\ \alpha\ } A \longrightarrow k \longrightarrow 0,$$

where $\alpha(t) = tx$. Can you see why it does not split?

We go on with some definitions.

Definition 1.4.5. Let M be an A-module, $m \in M$. The *annihilator* of m is the ideal

$$\mathrm{Ann}(m) = \{a \in A \mid a \cdot m = 0\}.$$

When $\operatorname{Ann}(m) \neq 0$ we say that m is a *torsion* element. More generally, if $N \subset M$, we denote

$$\operatorname{Ann}(N) = \{a \in A \mid a \cdot n = 0 \text{ for all } n \in N\}.$$

Now assume that A is an integral domain. The *torsion submodule* $\mathrm{T}(M)$ of M is the set of torsion elements. To check that this is a submodule, let m, n be torsion, so that $am = bn = 0$ for some nonzero $a, b \in A$. Then

$$ab(m+n) = 0 \qquad \text{and} \qquad a \cdot cm = 0$$

for all $c \in A$, so both $m + n$ and cm are torsion (here we are using that $ab \neq 0$).

We say that M is *torsion-free* when $\mathrm{T}(M) = 0$; equivalently multiplication by any nonzero $a \in A$ defines an injective homomorphism $M \to M$.

Remark 1.4.6. A free module is always torsion-free; the converse holds only for some particular classes of rings.

Proposition 1.4.7. *Let A be an integral domain, M an A-module. Then $M/\mathrm{T}(M)$ is torsion-free.*

Proof. Assume $a\overline{m} = 0$ in $M/\mathrm{T}(M)$; then am is torsion, so is m. □

The reader may already know the next construction in the context of vector spaces.

Definition 1.4.8. Let M, N be A-modules. We define their *tensor product* $M \otimes N$ as follows. First we consider the free module with basis a symbol $m \circledcirc n$ for every pair $m \in M$, $n \in N$, call it $M \circledcirc N$.

The tensor product $M \otimes N$ is the quotient of $M \circledcirc N$ by the submodule generated by the relations

$$\begin{aligned}
(m+m') \circledcirc n - m \circledcirc n - m' \circledcirc n \\
(am) \circledcirc n - a \cdot (m \circledcirc n) \\
m \circledcirc (n+n') - m \circledcirc n - m \circledcirc n' \\
m \circledcirc (an) - a \cdot (m \circledcirc n)
\end{aligned}$$

for every choice of $m, m' \in M$, $n, n' \in N$ and $a \in A$.

The equivalence class of $m \circledcirc n$ is denoted $m \otimes n$, so elements of $M \otimes N$ are formal linear combinations

$$\sum m_i \otimes n_i.$$

In particular, if the elements $\{m_i\}$ generate M and the $\{n_i\}$ generate N, the products $\{m_i \otimes n_j\}$ generate $M \otimes N$.

Remark 1.4.9. By construction we have a map

$$M \times N \to M \otimes N$$

which is A-linear in each variable; we shall say that such a map is A-bilinear. Namely it is the map sending (m, n) to $m \otimes n$.

Moreover given any A-module R with a bilinear map

$$M \times N \to R$$

there is a unique A-*linear* map $M \otimes N \to R$ making the diagram

$$\begin{array}{ccc} M \times N & \longrightarrow & M \otimes N \\ & \searrow & \downarrow \\ & & R \end{array}$$

commute. This is all immediate from the definition.

This property is the *raison d'être* of the tensor product. It allows us to reduce the study of bilinear maps to that A-linear ones, at the cost of changing the domain. By iterating this construction, one can also use tensor products to investigate *multilinear maps*, that is, maps

$$M_1 \times \cdots \times M_n \to R$$

which are A-linear on each factor. These are in bijective correspondence with A-linear maps

$$M_1 \otimes \cdots \otimes M_n \to R.$$

Remark 1.4.10. Let M, N, R be A-modules. From another point of view, to specify a bilinear map $M \times N \to R$ is the same as to give a linear map $M \to \mathrm{Hom}(N, R)$. This observation yields a bijective correspondence

$$\mathrm{Hom}(M \otimes N, R) \leftrightarrow \mathrm{Hom}(M, \mathrm{Hom}(N, R)),$$

which is easily seen to be an isomorphism of A-modules. For those who know some categorical nonsense, this means that $\cdot \otimes N$ and $\mathrm{Hom}(N, \cdot)$ are *adjoint functors*.

The above remark can be used to prove the following

Proposition 1.4.11. *Let*

$$0 \longrightarrow M_1 \xrightarrow{\alpha} M_2 \xrightarrow{\beta} M_3 \longrightarrow 0$$

be a short exact sequence of A-modules and N another A-module. Then the sequences

$$0 \longrightarrow \mathrm{Hom}(N, M_1) \xrightarrow{\alpha_*} \mathrm{Hom}(N, M_2) \xrightarrow{\beta_*} \mathrm{Hom}(N, M_3),$$

$$0 \longrightarrow \mathrm{Hom}(M_3, N) \xrightarrow{\beta^*} \mathrm{Hom}(M_2, N) \xrightarrow{\alpha^*} \mathrm{Hom}(M_1, N),$$

and

$$M_1 \otimes N \xrightarrow{\alpha \otimes \mathrm{id}_N} M_2 \otimes N \xrightarrow{\beta \otimes \mathrm{id}_N} M_3 \otimes N \longrightarrow 0$$

are exact.

We should say where the maps in the various exact sequences come from. Given a homomorphism

$$\alpha \colon M_1 \to M_2$$

we obtain a corresponding homomorphism

$$\alpha_* \colon \mathrm{Hom}(N, M_1) \to \mathrm{Hom}(N, M_2)$$

defined by $\alpha_*(f) = \alpha \circ f$ for $f \colon N \to M_1$. The homomorphism α^* is defined by composition in the other direction.

We also have a homomorphism

$$\alpha \otimes \mathrm{id}_N \colon M_1 \otimes N \to M_2 \otimes N$$

sending $m \otimes n$ to $\alpha(m) \otimes n$. More generally given another homomorphism

$$\beta \colon N_1 \to N_2$$

there is an induced homomorphism

$$\alpha \otimes \beta \colon M_1 \otimes N_1 \to M_2 \otimes N_2$$

defined by

$$\alpha \otimes \beta(m \otimes n) = \alpha(m) \otimes \beta(n);$$

indeed the map $M_1 \times N_1 \to M_2 \otimes N_2$ which sends (m, n) to $\alpha(m) \otimes \beta(n)$ is bilinear.

Proof. Let $f \in \ker \alpha_*$, so $f \colon N \to M_1$ and $\alpha \circ f = 0$. Since α is injective we have $f = 0$, showing that α_* is injective.

Clearly

$$\beta_* \circ \alpha_* = (\beta \circ \alpha)_* = 0,$$

so we need only to prove that $\ker \beta_* \subset \operatorname{im} \alpha_*$. Let $f \in \ker \beta_*$, so $f \colon N \to M_2$ and $\beta \circ f = 0$. Then f takes values in $\ker \beta = \operatorname{im} \alpha$, so it can be regarded as a map $N \to M_1$. This shows that $f \in \operatorname{im} \alpha_*$, proving that the $\mathrm{Hom}(N, \cdot)$ sequence is exact.

The proof of exactness of the $\mathrm{Hom}(\cdot, N)$ sequence is similar.

By Remark 1.4.10, for every A-module R we have

$$\mathrm{Hom}(M_i \otimes N, R) \cong \mathrm{Hom}(N, \mathrm{Hom}(M_i, R)).$$

By what we have already proved we obtain the exact sequence

$$0 \longrightarrow \mathrm{Hom}(L_3, R) \longrightarrow \mathrm{Hom}(L_2, R) \longrightarrow \mathrm{Hom}(L_1, R),$$

where $L_i := M_i \otimes N$. So we end the proof if we show that this implies the exactness of

$$L_1 \longrightarrow L_2 \longrightarrow L_3 \longrightarrow 0.$$

This is really more complicated to write up than to prove by yourself, so we leave it as an exercise. □

Remark 1.4.12. The above proof using Remark 1.4.10 is much more straightforward than it would be to prove exactness at $M_2 \otimes N$ directly—try it and see!

Remark 1.4.13. The extent to which the above exact sequences fail to be exact is the subject of homological algebra. We will not treat this topic here, but see for instance [**Rot79**] or [**Wei95**] if you are interested.

We can use the tensor product to prove that the rank of a free A-module is well defined.

Proposition 1.4.14. *Assume that $A^m \cong A^n$. Then $n = m$.*

Proof. Let $\mathcal{M}$ be any maximal ideal of A and consider the field $k = A/\mathcal{M}$. Then

$$A^m \otimes A/\mathcal{M} = A^m \otimes k \cong k^m,$$

and the same for A^n. So we get $k^m \cong k^n$ as A-modules, and a fortiori as k-vector spaces, and we obtain the conclusion by the usual linear algebra. □

Remark 1.4.15. If A is an integral domain we can argue in the same way, using the field of fractions of A (see Section 1.6) instead.

Remark 1.4.16. The proposition works equally for the case of infinite rank, but not for the noncommutative case; see Exercise 3.

Definition 1.4.17. Let M be a finitely generated free A-module, so $M \cong A^n$ for some n. We call n the *rank* of M. It is well defined by the above proposition.

1.5. Euclidean rings

We now introduce the most well-behaved class of rings, those which admit an analogue of the Euclidean algorithm. This section is just an example of application of the preceding concepts in a simple situation.

Definition 1.5.1. Let A be an integral domain. We say that A is *Euclidean* if it admits a function

$$N\colon A \setminus \{0\} \to \mathbb{N},$$

called a *norm*, such that:

(1) for all nonzero $a, b \in A$, $N(a) \leq N(ab)$;

(2) for each $a, b \in A$, $b \neq 0$, there exists $q, r \in A$ such that

$$a = qb + r$$

and either $r = 0$ or $N(r) < N(b)$.

In short Euclidean rings are those where we can perform the division with remainder, in such a way that the remainder is "smaller" than the divisor.

Remark 1.5.2. Unlike the Euclidean algorithm for the integers, we do not ask that the decomposition $a = qb + r$ is unique, even up to invertible elements.

Remark 1.5.3. The invertible elements of A are exactly those with minimal norm. Indeed the first condition tells us that $N(1) \leq N(b)$ for every nonzero $b \in A$. So $m = N(1)$ is the minimal possible norm. If $a \in A$ is a unit, then $N(a) \leq N(aa^{-1}) = m$, so $N(a) = m$ by minimality.

Conversely, assume that $N(a) = m$. Then the division property tells us that there are $q, r \in A$ with

$$1 = qa + r;$$

since we cannot have $N(r) < N(a)$ we must have $r = 0$, that is, a is invertible.

Example 1.5.4.

(a) The ring $\mathbb{Z}$ is Euclidean; the norm is just the usual absolute value: $N(n) = |n|$.

(b) For every field k, the ring $k[x]$ is Euclidean. For the norm we can take $N(f) = \deg(f)$; the division is just the usual division between polynomials.

(c) Now a more subtle example: the ring $\mathbb{Z}[i]$ is Euclidean; here the norm is the squared modulus for complex numbers, that is

$$N(a + ib) = a^2 + b^2.$$

Since multiplicativity is clear, let us verify that we can perform the division. Let $z, w \in \mathbb{Z}[i]$ and consider the usual quotient $u = z/w \in \mathbb{C}$. If $u \in \mathbb{Z}[i]$ we are done; otherwise u lies in the interior of some 1×1 square with integer coordinates. Let q be the nearest vertex of this square, and define r by

$$r = z - qw.$$

Then $N(r) = N((u - q)w)$ is less than $N(w)$ since the modulus of $u - q$ is less than 1.

Note that in this example there may be other vertices that work, so we really don't have any kind of uniqueness.

(d) The ring $k[[x]]$, for k a field, is Euclidean too. For a power series $a(x)$ we let $N(a)$ be the degree of the first nonzero monomial; that is, if

$$a(x) = x^t a_1(x)$$

with $a_1(0) \neq 0$, we let $N(a) = t$.

The first condition is immediate. For the existence of the division take any two power series $a(x), b(x)$ with $b \neq 0$. If $N(a) \geq N(b)$ write

$$\begin{aligned} a(x) &= x^{N(a)} a_1(x) \\ b(x) &= x^{N(b)} b_1(x), \end{aligned}$$

then we can divide exactly a by b taking

$$q(x) = x^{N(a)-N(b)} a_1(x) b_1(x)^{-1}, \qquad r(x) = 0.$$

If $N(a) < N(b)$ just take $q = 0$ and $r = a$.

If A is Euclidean we can perform the *Euclidean algorithm*. That is, given $a, b \in A$ we perform the repeated divisions

$$\begin{aligned} a &= q_1 b + r_1 \\ b &= q_2 r_1 + r_2 \\ r_1 &= q_3 r_2 + r_3 \\ &\vdots \end{aligned}$$

until some $r_n = 0$. This must happen in a finite number of steps because

$$N(b) > N(r_1) > N(r_2) > \cdots$$

is a decreasing sequence of natural numbers. So we have a last step

$$r_{n-2} = q_n r_{n-1}.$$

By induction we see that r_{n-1} divides both a and b. Moreover, inverting the steps one at a time we find

$$\begin{aligned} r_{n-1} &= r_{n-3} - q_{n-1} r_{n2} \\ &= r_{n-3} - q_{n-1}(r_{n-4} - q_{n-2} r_{n3}) = \cdots = xa + yb \end{aligned}$$

for some $x, y \in A$. So if c divides both a and b, it must divide r_{n-1}. In this case we say that r_{n-1} is a *greatest common divisor* between a and b. The relation $r_{n-1} = xa + yb$ is called *Bézout's identity*.

Proposition 1.5.5. *Let A be a Euclidean ring. Then every ideal of A is principal.*

Proof. Let $I \subset A$ be an ideal, and take any nonzero $a \in I$ with minimal norm, say $m = N(a)$. If $b \in I$, write

$$b = qa + r;$$

since we cannot have $N(r) < m$ we must have $r = 0$, that is $b \in (a)$. $\square$

We see from the preceding proposition that ideals of an Euclidean ring have a very simple structure. This is true also for more general modules.

Theorem 1.5.6. *Let A be Euclidean and let F be a finitely generated free module over A. If $N \subset F$ is a submodule, then N is free. More precisely, there exists a basis $\{f_1, \ldots, f_n\}$ of F and elements $a_1, \ldots, a_k$ of A, $k \leq n$, such that $\{a_1 f_1, \ldots, a_k f_k\}$ is a basis of N.*

Proof. We give an algorithmic proof. Starting from any basis of $F \cong A^n$, we can write an element $f \in F$ as a column vector with entries in A. Let $n_1, \ldots, n_s$ be generators for N, and write the respective column vectors as a $s \times n$ matrix. We will allow ourselves to perform some elementary operation on the matrix.

First, we can shuffle rows (this amounts to a permutation of the vectors of the basis of F) or columns (a permutation of the generators of N). We can also take two rows r_1 and r_2 and substitute r_1 with

$$r_1' = r_1 + ar_2$$

for any $a \in A$: this is an invertible change of basis for F. Similarly, we can perform the corresponding column operation.

We claim that a repeated application of the elementary operations takes the matrix in the form

$$\begin{pmatrix} a_1 & 0 & & \cdots & & 0 \\ 0 & a_2 & & \cdots & & 0 \\ \vdots & & \ddots & & & \vdots \\ 0 & \cdots & & a_k & \cdots & 0 \\ 0 & & & \cdots & & 0 \\ \vdots & & & & & \vdots \\ 0 & & & \cdots & & 0 \end{pmatrix}, \tag{1.5.1}$$

where $a_1 | a_2 | \cdots | a_k$. From this, the thesis follows.

Let $x_{1,1}$ and $x_{2,1}$ be the first two elements of the first column. Using row operations and the Euclidean algorithm we can substitute $x_{1,1}$ with the greatest common divisor of the two. We repeat the process until we have one element on the first column that divides all the others. We put it in first position, and subtract a suitable multiple of the first row from the others, so that all the elements in the first column below the first are 0. So we get

$$\begin{pmatrix} x_1 & * & \cdots & * \\ 0 & * & \cdots & * \\ \vdots & & & \vdots \\ 0 & * & \cdots & * \end{pmatrix}.$$

Now we do the same for the first row. This may ruin our first column, anyway we obtain

$$\begin{pmatrix} x_2 & 0 & \cdots & 0 \\ * & * & \cdots & * \\ \vdots & & & \vdots \\ * & * & \cdots & * \end{pmatrix}.$$

We turn back to the first column and so on. Since x_1 is multiple of x_2, which is multiple of x_3 and so on, the process will stop. When it does we are left with

$$\begin{pmatrix} a_1 & 0 & \cdots & 0 \\ 0 & * & \cdots & * \\ \vdots & & & \vdots \\ 0 & * & \cdots & * \end{pmatrix}.$$

At this point we can conclude by induction on the number of columns. □

A matrix in the form of (1.5.1) (with $a_1|a_2|\cdots|a_k$) is said to be in *Smith normal form*. Essentially, what we have proved is that a matrix with entries in an Euclidean ring can be brought in Smith normal form with a sequence of elementary row and column operations.

Corollary 1.5.7 (Structure of modules over Euclidean rings)**.** *Let M be a finitely generated module over the Euclidean ring A. Then there exist $a_1, \ldots, a_k \in A$ such that*

$$M \cong A^r \times A/(a_1) \times \cdots \times A/(a_r).$$

Proof. Since M is finitely generated we have a surjective homomorphism $\phi\colon F \to M$, where F is finitely generated free. Apply the proposition with $N = \ker \phi$. □

Example 1.5.8. Since a $\mathbb{Z}$-module is just an abelian group, we recover the usual structure theorem for finitely generated abelian groups.

Corollary 1.5.9. *Let G be a finitely generated free abelian group of rank n, $H < G$ a subgroup. Then H is finitely generated of rank $m \leq n$. Moreover, if $m = n$, the quotient is finite with $|\det M|$ elements, where M is any matrix* *expressing the generators of H in coordinates with respect to generators of G, that is,*

$$h_i = \sum_{j=1}^{n} M_{i,j} g_j,$$

for generators $g_1, \dots, g_n$ of G and $h_1, \dots, h_n$ of H.

Proof. A special case of the above when $A = \mathbb{Z}$. Fix generators $g_1, \dots, g_n$ of G and identify G with $\mathbb{Z}^n$. A finite set of elements in H can be represented with a matrix. By taking this matrix in Smith normal form, we don't change the generated subgroup, which shows that H has at most rank m. If moreover $m = n$, the moves that bring the matrix in Smith normal form preserve the determinant. When H is generated by the columns of a matrix in Smith normal form, the thesis is clear. □

These results will be generalized for principal ideal domains and for Dedekind rings in the following sections.

1.6. Localization

In this section we introduce the process of enlarging a ring A by the use of "fractions" with entries in A. Things would be simpler if A was an integral domain, but for our purpose we need the more general case. For reasons which will become apparent when translated in geometric context, this process is called *localization.*

Definition 1.6.1. Let A be a ring, $S \subset A$ a subset. We say that S is a *multiplicative set* if $1 \in S$ and for every $s, t \in S$ we have $st \in S$.

Example 1.6.2.

(a) For every nonzero $a \in A$ the set
$$S = \{1, a, a^2, \dots\}$$
is a multiplicative set.

(b) If $P \subset A$ is a prime ideal, then $S = A \setminus P$ is a multiplicative set. More generally if $\{P_i\}$ is a family of prime ideals of A, we have the multiplicative set $S = A \setminus \bigcup_i P_i$.

(c) The set of nonzero divisors of A is a multiplicative set. So is the set of invertible elements.

(d) If $I \subset A$ is an ideal, $S = 1 + I$ is a multiplicative set.

(e) Take $A = k[x_1, \dots x_n]$ (k a field) and let $V \subset k^n$ be any subset. Then
$$S = \{f \in k[x_1, \dots, x_n] \mid f(x) \neq 0 \text{ for all } x \in V\}$$
is a multiplicative set.

(f) Let k be a field and $A = k[x]$; then

$$S = \{f \in k[x] \mid f \text{ does not have any root in } k\}$$

is a multiplicative set.

Definition 1.6.3. Given a ring A with a multiplicative set S we define the *localization* of A at S as the ring $S^{-1}A$ obtained as follows.

Elements of $S^{-1}A$ are couples $(a, s) \in A \times S$ modulo the following equivalence relation. Two couples (a, s) and (b, t) are equivalent if there exists $u \in S$ such that

$$atu = bsu.$$

Apart from the presence of u this is the usual cross relation for fractions. We cannot avoid u unless A is an integral domain.

Not surprisingly we denote the class of equivalence of (a, s) by a/s.

We define addition and multiplication in $S^{-1}A$ by the familiar rules

$$a/s + b/t = (at + bs)/(st), \qquad a/s \cdot b/t = (ab)/(st).$$

It is a simple exercise to check that the operations are well defined, so $S^{-1}A$ becomes a commutative ring, with unit $1/1$.

By construction, we have a natural homomorphism

$$\iota\colon A \to S^{-1}A$$

sending a to $a/1$. In general this is not injective: indeed

$$\ker \iota = \{a \in A \mid \text{there exists } u \in S \text{ such that } au = 0\}.$$

So ι is injective precisely when S does not contain any zero divisor; in this case we shall allow ourselves to denote $\iota(a) = a/1$ simply by a. When we want to recall S in the notation we shall write $\iota = \iota_S$, or even $\iota = \iota_P$ if $S = A \setminus P$. For more about the kernel of ι_P, see 7.5.26.

Corresponding to some cases in Example 1.6.2, we have a special notation for $S^{-1}A$. When $S = \{a^n, n \geq 0\}$ we denote $S^{-1}A = A_a$.

When $S = A \setminus P$, P a prime a ideal, we denote $S^{-1}A = A_P$; this is the most important case of localization.

Finally when S is the set of nonzero divisors, we denote $S^{-1}A = \mathcal{F}(A)$, and call it the *total ring of fractions* of A. If A is an integral domain then $S = A \setminus \{0\}$, and $\mathcal{F}(A)$ is a field, which we shall call the *field of fractions* of A.

Remark 1.6.4. Any localization of an integral domain A is contained in the field $\mathcal{F}(A)$, hence it is again an integral domain.

The following universal property comes for free from the definition:

Proposition 1.6.5. *Let $f: A \to B$ be a homomorphism of rings, $S \subset A$ a multiplicative set such that $f(S) \subset B^*$. Then there exists a unique homomorphism $\widehat{f}: S^{-1}A \to B$ such that $f = \widehat{f} \circ \iota$.*

So $S^{-1}A$ together with the map ι is universal for such homomorphisms.

Ideals behave well under localization; in particular we have the following proposition, which should be compared to Proposition 1.2.9.

Proposition 1.6.6. *Let A be a ring, $S \subset A$ a multiplicative set.*

(i) *For any ideal $I \subset S^{-1}A$, I is generated by $\iota(\iota^{-1}(I))$; in particular every ideal of $S^{-1}A$ is the extension of an ideal of A.*

(ii) *For any ideal $I \subset A$ we have*

$$\iota^{-1}(I \cdot S^{-1}A) = \bigcup_{s \in S} (I : s).$$

(iii) *The ideal $I \cdot S^{-1}A$ is the whole $S^{-1}A$ if and only if $I \cap S \neq \emptyset$.*

For instance, let $P \subset A$ be a prime ideal and take $S = A \setminus P$. According to the proposition, $S^{-1}P$ is the only maximal ideal of A_P, which is then a local ring:

Definition 1.6.7. A ring A is called *local* if it has only one maximal ideal. It is called *semilocal* if it has finitely many maximal ideals. A local ring A with maximal ideal $\mathcal{M}$ will sometimes be denoted simply by the pair $(A, \mathcal{M})$.

Proof of Proposition 1.6.6.

(i) Let

$$J := \iota^{-1}(I) \cdot S^{-1}A.$$

The inclusion $J \subset I$ is obvious. For the converse, let $a/s \in I$; then $a \in \iota^{-1}(I)$, so $a/s \in J$.

(ii) Let

$$J := \iota^{-1}(I \cdot S^{-1}A)$$

We first prove that $(I : s) \subset J$ for any $s \in S$. Indeed let $a \in (I : s)$, so $as \in I$; then

$$a \in \iota^{-1}\left(\frac{as}{s}\right) \subset J.$$

For the other inclusion let $a \in J$; this means that $a/1 = b/s$ for some $b \in I$, $s \in S$. So there is $u \in S$ such that

$$asu = bu \in I.$$

This shows that $a \in (I : su)$.

(iii) The ideal $I \cdot S^{-1}A$ is trivial if and only if it contains 1. This is equivalent to $1 = i/s$ for some $i \in I$ and $s \in S$, which means that $iu = su$ for some $u \in S$. Finally, this means that $iu = su \in I \cap S$. □

Unfortunately, we don't have a bijective correspondence between ideals of $S^{-1}A$ and ideals of A not meeting S.

Example 1.6.8. In the ring $\mathbb{Z}_{(2)}$, the ideals (2) and (6) of $\mathbb{Z}$ extend to the same ideal, since 3 becomes invertible after localization.

The situation is better for prime ideals:

Corollary 1.6.9. *Let A be a ring, $S \subset A$ a multiplicative set. There is a bijective correspondence between prime ideals of $S^{-1}A$ and prime ideals of A which don't meet S.*

Proof. It remains to show that if $P \subset A$ is prime, then $P = \iota^{-1}(P \cdot S^{-1}A)$. According to Proposition 1.6.6 we have

$$\iota^{-1}(P \cdot S^{-1}A) = \bigcup_{s \in S} (P : s)$$

and $(P : s) = P$ for all $s \in S$, since P is prime and $s \notin P$. □

We can use our new tool to prove

Proposition 1.6.10. *The nilradical $\mathcal{N}(A)$ of A is the intersection of all prime ideals of A.*

Recall from Example 1.1.18 that $\mathcal{N}(A)$ is the ideal formed by nilpotent elements of A.

Proof. Let P be a prime ideal and let $a \in \mathcal{N}(A)$, then some power $a^n = 0 \in P$, so $a \in P$. This proves one inclusion.

For the other, let a be contained in every prime ideal. Then A_a is ring without prime ideals, that is, A_a must be the 0 ring. (Why?) Then $\iota(1) = 0$ means that $1 \cdot a^n = 0$ for some n, so a is nilpotent. □

By considering the quotient A/I we obtain

Corollary 1.6.11. *Let I be an ideal of A. Then $\sqrt{I}$ is the intersection of all prime ideals containing I.*

It is useful to know that localization and quotients commute, in the following sense.

Proposition 1.6.12. *Let S be a multiplicative set, $I \subset A$ an ideal such that $I \cap S = \emptyset$. Then there is a natural isomorphism between $S^{-1}A/S^{-1}I$ and $T^{-1}(A/I)$, where T is the image of S inside A/I.*

Proof. It is clear that T is a multiplicative set. The composition of the natural maps

$$A \to S^{-1}A \to S^{-1}A/S^{-1}I$$

has kernel $\iota^{-1}(S^{-1}I) = I$ by Proposition 1.6.6. So we have an injective homomorphism

$$A/I \to S^{-1}A/S^{-1}I.$$

Since this sends T into invertible elements, the universal property of localization yields a homomorphism

$$\phi\colon T^{-1}(A/I) \to S^{-1}A/S^{-1}I.$$

Vice versa, consider the composition

$$A \to A/I \to T^{-1}(A/I).$$

This sends S into invertible elements, so we get a homomorphism

$$S^{-1}A \to T^{-1}(A/I),$$

whose kernel contains $S^{-1}I$. Hence we have a map in the other direction

$$\psi\colon S^{-1}A/S^{-1}I \to T^{-1}(A/I).$$

It is immediate to check that ϕ and ψ are mutual inverses. □

Corollary 1.6.13. *Let $P \subset A$ be a prime. Then the fields of fractions of the integral domain A/P is canonically isomorphic to A_P/PA_P, the quotient of the local ring A_P by its maximal ideal.*

Definition 1.6.14. Let $P \subset A$ be a prime ideal. The field

$$k(P) = \mathcal{F}(A/P) = A_P/PA_P$$

is called the *residue field* of A at P.

Modules can be localized in the same way as rings.

Definition 1.6.15. Let A be a ring, S a multiplicative set and M an A-module. Then we define the $S^{-1}(A)$-module $S^{-1}M$ as follows. Elements of $S^{-1}M$ are couples $(m, s) \in M \times S$, modulo an equivalence relation. The couples (m, s) and (n, t) are equivalent when there exists $u \in S$ such that

$$mtu = nsu.$$

The equivalence class of (m, s) is denoted by m/s.

As in the ring case, we have a homomorphism *of A-modules*

$$\iota_M : M \to S^{-1}M$$

given by $\iota_M(m) = m/1$.

The same proof of Proposition 1.6.6 yields:

Proposition 1.6.16. *Let M be an A-module.*

(i) *For any submodule $N \subset S^{-1}M$, N is generated by $\iota(\iota^{-1}(N))$; in particular every submodule of $S^{-1}M$ is the extension of a submodule of M.*

(ii) *For any submodule $N \subset M$ we have*

$$\iota^{-1}(S^{-1}N) = \bigcup_{s \in S} (N : s).$$

Here $S^{-1}N$ denotes the extension of N in $S^{-1}M$, that is, the $S^{-1}A$ submodule generated by N. The submodule $(N : s)$ is defined by

$$(N : s) := \{m \in M \mid ms \in N\}.$$

Finally we have the following useful results.

Proposition 1.6.17. *Localization of modules preserves exact sequences.*

Proof. Let

$$\cdots \to M_{n-1} \xrightarrow{\alpha_{n-1}} M_n \xrightarrow{\alpha_n} M_{n+1} \to \cdots$$

be an exact sequence of A- modules; then we have an induced sequence of $S^{-1}A$-modules

$$\cdots \to S^{-1}M_{n-1} \xrightarrow{\alpha_{n-1}} S^{-1}M_n \xrightarrow{\alpha_n} S^{-1}M_{n+1} \to \cdots .$$

Let us check exactness at M_n.

First for any $m/s \in M_{n-1}$ we have

$$\alpha_n(\alpha_{n-1}(m/s)) = \alpha_n(\alpha_{n-1}(m))/s = 0.$$

Vice versa, let $m/s \in M_n$ and assume that $\alpha_n(m/s) = \alpha_n(m)/s = 0$. Then there exists $u \in S$ such that

$$\alpha_n(um) = u\alpha_n(m) = 0.$$

So we have $um = \alpha_{n-1}(m')$ for some $m' \in M_{n-1}$ and finally $m/s = \alpha_{n-1}(m'/(su))$. □

Proposition 1.6.18. *Let M be an A-module, $m \in M$. Then $m = 0$ if and only if $\iota_P(m) = 0 \in M_P$ for every maximal (or prime) ideal $P \subset A$.*

Proof. One implication is obvious. For the other assume that $\iota_P(m) = 0$ for all maximal ideals $P \subset A$. This means that for every such P we find some $s \notin P$ such that $sm = 0$. In other words the ideal $\mathrm{Ann}(m)$ is not contained in any maximal ideal, so it must be the whole A. □

This has some useful corollaries

Corollary 1.6.19. *Let M be an A-module; then $M = 0$ if and only if $M_P = 0$ for every maximal (or prime) ideal $P \subset A$.*

Corollary 1.6.20. *Let $f\colon M \to N$ be an A-module homomorphism. Then f is injective (resp., surjective) if and only if the induced homomorphism $f_P\colon M_P \to N_P$ is injective (resp., surjective) for every maximal (or prime) ideal $P \subset A$.*

Proof. Use the fact that localization preserves exact sequences and apply the above corollary to $\ker(f)$ (resp., $\mathrm{coker}(f)$). □

Properties like the above, which hold true for a ring (or a module, or a homomorphism...) if and only if they hold for the localization at every prime ideal are called *local properties.* We will see more examples in the next chapters.

1.7. Graded rings and modules

We begin with some routine definitions.

Definition 1.7.1. An *abelian monoid* is a set G endowed with a binary operation $+$, called *addition*, such that

(1) addition is *associative*, namely for every $g, h, l \in G$ we have $(g + h) + l = g + (h + l)$;
(2) there exists an element $0 \in G$ such that $0 + g = g + 0 = g$ for every $g \in G$;
(3) addition is *commutative*, that is for every $g, h \in G$ we have $g + h = h + g$.

In down to earth terms, an abelian monoid is just an abelian group without the requirement of the existence of inverse elements.

Remark 1.7.2. Since inverses don't exist in G, one does not have a cancellation law!

Example 1.7.3.

(a) Any abelian group is an abelian monoid.

(b) The set $\mathbb{N}$ of natural numbers is a monoid which is not a group; it will be our main example. More generally $\mathbb{N}^k$ is a monoid.

(c) We can add to $\mathbb{N}$ a symbol ∞ and give $\mathbb{N} \cup \{\infty\}$ the structure of an abelian monoid by declaring that

$$n + \infty = \infty + \infty = \infty$$

for every $n \in \mathbb{N}$.

(d) An example quite different from the monoids we will consider is A^*, which is an abelian monoid under multiplication for every ring A.

The reason why we have introduced monoids is the following

Definition 1.7.4. Let A be a ring. We say that A is *graded* over the monoid G if we have a decomposition

$$A = \bigoplus_{g \in G} A_g$$

as *abelian groups* such that

$$ab \subset A_{g+h} \text{ for } a \in A_g,\ b \in A_h.$$

Elements of some A_g will be said *homogeneous.* Every element can be written uniquely as a finite sum of nonzero homogeneous elements, which are called it homogeneous components.

If we speak of a graded ring without further specification we always mean graded over $\mathbb{N}$.

Remark 1.7.5. If A is a graded ring, A_0 is a subring and A has the structure of A_0-algebra.

In the case where A is graded over $\mathbb{N}$, the subset

$$A_+ := \bigoplus_{n \geq 1} A_n$$

is an ideal of A, which—for geometric reasons which we shall see later (Exercise 2 in Chapter 8)—is sometimes called the *irrelevant ideal.*

Example 1.7.6.

(a) The most fundamental example of a graded ring is $k[x_1, \ldots, x_n]$. Usually it is graded over $\mathbb{N}$ by considering the total degree, but it has also the structure of a $\mathbb{N}^n$-graded ring by considering a separate degree for each variable.

(b) For any topological space X the cohomology ring $H^*(X)$ is graded over $\mathbb{N}$. The complex K-theory ring $K^*(X)$ is graded over $\mathbb{Z}/2\mathbb{Z}$ by Bott periodicity.

(c) Let A be any ring, $I \subset A$ an ideal. The *graded ring associated to I* is

$$\mathrm{Gr}_I(A) := \bigoplus_{k=0}^{\infty} I^k/I^{k+1} = A/I \oplus I/I^2 \oplus \cdots .$$

Check that the operations are well defined!

In the case where $(A, \mathcal{M})$ is a local ring, we simply write $\mathrm{Gr}(A)$ for $\mathrm{Gr}_{\mathcal{M}}(A)$. This association is at the source of a deep connection between the structure of graded and local rings, which will emerge when we study the concept of dimension in Chapter 9.

In the context of graded rings there are some ideals and modules which are particularly well-behaved.

Definition 1.7.7. Let A be a G-graded ring. The ideal $I \subset A$ is called *homogeneous* if

$$I = \bigoplus_{g \in G} I_g,$$

where $I_g := I \cap A_g$. This amounts to saying that if $a \in I$, the homogeneous components of a are still in I.

Note that if I and J are homogeneous, so is $I \cap J$.

Proposition 1.7.8. *Let I be an ideal in a graded ring. Then I is homogeneous if and only if it generated by homogeneous elements.*

Proof. Assume I is homogeneous. Take any set of generators of I; then the set of their homogeneous components still generates I.

Vice versa assume $I = (E)$, where E is composed of homogeneous elements. Let $a \in I$ and write

$$a = e_1 f_1 + \cdots + e_k f_k$$

for some $e_i \in E$ and $f_i \in A$. Taking the homogeneous component of both members and taking in account that the e_i are homogeneous shows that each homogeneous component of a is again a combination of the e_i, so it lies in I. □

Corollary 1.7.9. *If I and J are homogeneous ideals, so are $I + J$ and $I \cdot J$.*

Definition 1.7.10. Let A be a ring graded over G, M an A-module. We say that M is *graded* if we have a decomposition

$$M = \bigoplus_{g \in G} M_g$$

as *abelian groups* such that

$$am \in M_{g+h} \text{ for } a \in A_g,\ m \in M_h.$$

As for rings we have the definition of homogeneous elements, homogeneous components, and so on. A *graded submodule* $N \subset M$ is a submodule such that

$$N = \bigoplus_{g \in G} N_g,$$

where $N_g := N \cap M_g$.

As before, if A is any ring, $I \subset A$ an ideal and M an A-module we can define the *associated graded module*

$$\mathrm{Gr}_I(M) := \bigoplus_{k=0}^{\infty} I^k M / I^{k+1} M = M/IM \oplus IM/I^2M \oplus \cdots ;$$

this is a module over $\mathrm{Gr}_I(A)$. When $(A, \mathcal{M})$ is local we will write $\mathrm{Gr}(M)$ for $\mathrm{Gr}_{\mathcal{M}}(M)$.

Definition 1.7.11. More generally, we call a nested sequence of submodules of $M \supset M_1 \supset \cdots \supset M_n \supset \cdots$ a *filtration* and we define the associated graded module

$$\mathrm{Gr}(M) := \bigoplus_{k=0}^{\infty} M_n / M_{n+1}.$$

Definition 1.7.12. Let $f\colon M \to N$ be a homomorphism of G-graded A-modules. We say that f is *graded* of degree h if $f(M_g) \subset N_{g+h}$.

In the most common case $h = 0$, so we simply ask that $f(M_g) \subset N_g$; that is, f preserves the degree of homogeneous elements. In this case we will simply speak of a graded homomorphism, without mentioning the degree.

Example 1.7.13.

(a) Write

$$\mathrm{Hom}(M, N)_g := \{f\colon M \to N \mid f \text{ is homogeneous of degree } g\}.$$

Then we have the submodule

$$\mathrm{Hom}(M, N)_{gr} := \bigoplus_{g \in G} \mathrm{Hom}(M, N)_g$$

of $\mathrm{Hom}(M, N)$, which is a graded module.

(b) If M and N are graded over $\mathbb{N}$, the tensor product is also graded. Indeed we define

$$(M \otimes N)_n := \bigoplus_{p+q=n} M_p \otimes N_q.$$

Then one can check for exercise that

$$M \otimes N = \bigoplus_{n \in \mathbb{N}} (M \otimes N)_n.$$

When $f\colon M \to N$ is a graded homomorphism, the image $f(N)$ is a graded submodule of M. If $N \subset M$ is a graded submodule, the quotient M/N has a natural structure of graded module, since

$$M/N = \bigoplus_{g \in G} M_g / N_g.$$

In particular the quotient of a graded ring for a homogeneous ideal is again a graded ring.

Remark 1.7.14. This does *not* mean that $\overline{m} \in (M/N)_g$ if and only $m \in M_g$! A homogeneous class can have various inhomogeneous representatives, and even homogeneous representatives of different degrees. The only case where the preceding assertion is true is when $N = N_0$.

The following result is often useful.

Proposition 1.7.15. *Let A be a graded ring (over $\mathbb{N}$). If the ideal A_+ is finitely generated, A is finitely generated as A_0-algebra.*

Proof. Let $A_+ = (a_1, \ldots, a_n)$; since A_+ is homogeneous we can take the a_i to be homogeneous. We shall show that

$$A = A_0[a_1, \ldots, a_n].$$

Let $B := A_0[a_1, \ldots, a_n]$; we show that $A_k \subset B_k$ by induction, the case $k = 0$ being trivial.

If $k > 0$, $A_k \subset A_+$, so given $a \in A_k$ we can write

$$a = b_1 a_1 + \cdots + b_n a_n$$

for some $b_1, \ldots, b_n \in A$. We can choose the b_i homogeneous, since the components in degree different from k will cancel out; moreover the b_i will have degree less than k. By induction $b_1, \ldots, b_n \in B$, so we conclude that $a \in B$. □

1.8. Exercises

1. Any (not necessarily commutative, not necessarily with unit) ring is a subring of a ring with unit. (Give the abelian group $\mathbb{Z} \oplus A$ a suitable ring structure.)

2. Let A be a (not necessarily commutative) ring with unit. Prove that A is a subring of $\mathrm{End}(E)$ for some abelian group E. (Consider the action of A on itself by multiplication.)

3. Let V be a vector space of infinite countable dimension, $A = \mathrm{End}(V)$ the (noncommutative) ring of linear endomorphisms of V. Show that for any $m, n \geq 1$, A^n and A^m are isomorphic as modules over A. Hence, Proposition 1.4.14 fails in the noncommutative case.

4. Give an example of a module—which is not finitely generated—that does not satisfy the conclusion of Nakayama's lemma.

5. Let A be a ring such that for all $a \in A$ there exists $n > 1$ (depending on a) for which $a^n = a$. Prove that every prime ideal of A is maximal.

6. Show that a finite integral domain is a field.

7. Let A be the ring of Example 1.1.5 i.

(i) Prove that an element $f \in A$ is invertible if and only if $f(1) \neq 0$.

(ii) Let μ be the inverse of the constant function 1 (μ is usually called the *Möbius function*). Let $r(n)$ be the number of primes dividing n. Prove that

$$\mu(n) = \begin{cases} (-1)^{r(n)} & \text{if } n \text{ is squarefree} \\ 0 & \text{if } n \text{ is not squarefree.} \end{cases}$$

(iii) Prove the *Möbius inversion formula*: if

$$g(n) = \sum_{d|n} f(d),$$

then

$$f(n) = \sum_{d|n} g(d)\mu\left(\frac{n}{d}\right).$$

8. The only maximal ideals of $A = C([0,1])$ are those of the form

$$I_x = \{f \in A \mid f(x) = 0\}$$

for some $x \in [0,1]$.

9. Formulate and prove the Chinese remainder theorem (Theorem 1.2.12) for more than two ideals.

10. Show that if the ideals I and J are coprime then $I \cap J = I \cdot J$.

11. Let $I \subset A = k[x_1, \ldots, x_n]$ be an ideal generated by monomials; we shall say that I is a *monomial ideal.* Let $f \in I$; prove that any monomial appearing in f is in I. Assume

$$I = (x^{\alpha_i}),$$

where each α_i is a multiindex. A monomial x^β is in I if and only if there is some α_i such that x^{α_i} divides x^β.

12. Let

$$\begin{array}{ccccccccc} 0 & \longrightarrow & M_1 & \longrightarrow & M_2 & \longrightarrow & M_3 & \longrightarrow & 0 \\ & & \downarrow{\scriptstyle f_1} & & \downarrow{\scriptstyle f_2} & & \downarrow{\scriptstyle f_3} & & \\ 0 & \longrightarrow & N_1 & \longrightarrow & N_2 & \longrightarrow & N_3 & \longrightarrow & 0 \end{array}$$

be a map of exact sequences of A-modules, in the sense that all displayed maps are A-linear, the rows are exact sequences and the diagram commutes.

Show that we have a long exact sequence

$$\begin{array}{l} 0 \longrightarrow \ker f_1 \longrightarrow \ker f_2 \longrightarrow \ker f_3 \\ \longrightarrow \operatorname{coker} f_1 \longrightarrow \operatorname{coker} f_2 \longrightarrow \operatorname{coker} f_3 \longrightarrow 0. \end{array}$$

This result is sometimes called the *snake lemma.*

13. Let P be a finitely generated A-module. Show that the following conditions are equivalent:

(i) there exists an A-module N such that

$$P \oplus N \cong A^n$$

for some n;

(ii) for all homomorphisms of A-modules

$$f \colon S \to T, \qquad g \colon P \to T$$

with f surjective there exists $h \colon P \to S$ such that $g = f \circ h$.

In this case M is called *projective.* In categorical style, the second condition means that one can always complete the dashed arrow in the diagram

$$\begin{array}{ccccc} & & P & & \\ & \overset{h}{\swarrow} & \downarrow g & & \\ S & \xrightarrow{f} & T & \longrightarrow & 0. \end{array}$$

14. Show that a finitely generated torsion-free module over $A = k[x]$ is free.

15. For a converse, show that the ideal (x, y) is a torsion-free module over $A = k[x, y]$, but it is not free.

16. This exercise uses some topology. Let X be a topological space, E a vector bundle on X. The continuous sections of E are a module over the ring $A = C(X)$.

Now take $X = S^2$, the 2-sphere, and let $E = TS^2$ be its tangent bundle. Prove that the associated module M is not free, but $M \oplus A$ is free.

17. We work out an algebraic version of the above exercise. Let

$$A = \mathbb{R}[x_1, x_2, x_3]/(x_1^2 + x_2^2 + x_3^2 - 1).$$

Inside A^3 consider the submodules

$$\begin{aligned} M &= \{(f_1, f_2, f_3) \mid x_1 f_1 + x_2 f_2 + x_3 f_3 = 0\} \\ N &= \{(x_1 f, x_2 f, x_3 f) \mid f \in A\}. \end{aligned}$$

(i) Prove (possibly algebraically) that $N \cong A$ and that

$$M \oplus N \cong A^3;$$

still M is not free.

(ii) Let $\mathcal{M} \subset A$ be the ideal generated by $x_1 - 1, x_2, x_3$. Prove that $\mathcal{M}$ is maximal, in particular prime, and that $M_{\mathcal{M}}$ is free over $A_{\mathcal{M}}$.

18. This exercise uses a little field theory. Let k'/k be a finite separable extension of fields (this means that for all $\alpha \in k'$ the minimum polynomial of α over k has distinct roots). Prove that $k' = k[\alpha]$ for some $\alpha \in k'$.

(Reduce to the case $k' = k(\alpha, \beta)$, and prove that $k' = k(\gamma)$ where $\gamma = \alpha + t\beta$ for a suitable $t \in k$. For this let f and g be the minimum polynomial of α and β respectively. What is the greatest common divisor of $f(\gamma - tx)$ and $g(x)$? Use the Euclidean algorithm to prove that this greatest common divisor is still in $k(\gamma)[x]$.)

19. Prove that the ring $A = \mathbb{Z}\left[\frac{1+\sqrt{-19}}{2}\right]$ is not Euclidean. We shall see later (Exercise 5 of Chapter 6) that nevertheless, it has the property that all its ideals are principal. (First describe A^*. Assuming that A has a norm function, let $a \in A \setminus A^*$ with minimum norm; prove that $A/(a)$ is formed by the classes of 0 and the invertible elements. How big can $A/(a)$ be? Derive a contradiction.)

20. Let k be a field of characteristic 0, and let $A = k[x, y]$. Find generators for $(x^2, y) \cap (x, y^2)$.

21. Let A be a ring, S a multiplicative set, and consider the family

$$F = \{I \subset A \text{ ideal} \mid I \cap S = \emptyset\}.$$

Let $I \in F$ be a maximal element; prove that I is prime.

22. Let A be an integral domain, so that all localizations A_P at prime ideals can be seen as subgrings of $k = \mathcal{F}(A)$. Show that

$$A = \bigcap_{\mathcal{M} \text{ maximal}} A_{\mathcal{M}} = \bigcap_{P \text{ prime}} A_P.$$

23. [**vDdB**] Let k be a field, and $K := k(x_1, x_2, \dots)$ the field of rational functions in infinitely many indeterminates. Consider the rings $A := K[y_1, y_2, \dots]$ and $B := A_{y_1} = A[y_1^{-1}]$. Show that A is a localization of B, and conversely, but A is not isomorphic to B. (Notice that $A^* \cup \{0\}$ is closed under addition.)

24. Show that the ring $\mathbb{C}\{x_1, \dots, x_n\}$ of power series in n variables which have a positive radius of convergence is local.

25. Let A be a ring graded over $\mathbb{Z}$, and assume that A does not have any nonzero prime homogeneous ideals. Then A_0 is a field and either $A = A_0$ or $A \cong A_0[x, x^{-1}]$.

26. Consider the ring $A = k[x, x^{-1}]$, where $\deg x^n = n$. Any graded module over A is free. (Imitate the corresponding proof for vector spaces.)

27. Let $A = \bigoplus_{m\in\mathbb{N}} A_m$ be a ring graded over $\mathbb{N}$ with A_0 a field, and take a *truncation* $B = \bigoplus_{m\in\mathbb{N}} A_{mk}$. Assume that A is finitely generated as an A_0-algebra and it is an integral domain. Then B is generated in degree 1 for k large enough.

28. Let $P \subset A$ be a homogeneous ideal. Then P is prime if and only if for every *homogeneous* $a, b \in A$ such that $ab \in P$ we have $a \in P$ or $b \in P$.

29. The radical of a homogeneous ideal is again homogeneous.

30. Let $A = \bigoplus_{n=0}^{\infty} \mathbb{Z}$ and $A = \prod_{n=0}^{\infty} \mathbb{Z}$. Prove that $\operatorname{Hom}(A, \mathbb{Z}) \cong B$ and $\operatorname{Hom}(B, \mathbb{Z}) \cong A$. (The first one is straightforward. For the second one, show that the map $\operatorname{Hom}(B, \mathbb{Z}) \to B$ sending f to $(f(e_i))$ is injective with image A. For this, it will be useful to notice that every integer x can be written in the form $x = a_n 2^n + b_n 3^n$, for any $n \in \mathbb{N}$.)

Chapter 2

Finiteness Conditions

In this chapter we introduce the most common finiteness conditions one can put on rings or modules, the so-called chain conditions. Basically, we ask that our modules don't have infinite (on the right or on the left) chains of submodules. In the Section 2.2, we study these two conditions on modules; here the theory is essentially symmetric (although the two conditions are not equivalent).

For rings, this symmetry breaks up. First we study the class of Noetherian rings, that is, rings that don't have infinite *ascending* chains of ideals. This is by far the most common class of rings one encounters in commutative algebra. The condition turns out to be equivalent to the request that ideals are finitely generated; indeed, to give some motivation, we start by studying the case where all ideals are actually principal.

Then we study Artinian rings, that is, rings that don't have infinite *descending* chains of ideals. This condition turns out to be much more restrictive, and indeed we show that an Artinian ring is Noetherian as well. Moreover we can give a quite precise structure theorem for this class of rings.

Finally, we turn to modules which are both Artinian and Noetherian: for these modules we have a notion of length, which generalizes the concept of dimension for vector spaces. We also prove the classical Jordan–Hölder theorem, which compares different series of composition for the same module.

2.1. Principal ideal domains

Much of the complication in studying rings more general than say $\mathbb{Z}$ or $k[x]$ comes from the fact that ideals can have a nontrivial structure. From this point of view, the most fortunate case is that of rings whose ideals are all generated by a single element.

Definition 2.1.1. Let A be an integral domain. We say that A is a *principal ideal domain* if all the ideals of A are principal.

Remark 2.1.2. If $I \subset A$ is any ideal of a principal ideal domain, then $I \cong A$ as A-modules. Indeed let $I = (a)$; then the map

$$\begin{aligned} A &\longrightarrow I \\ x &\longmapsto ax \end{aligned}$$

is both injective and surjective.

Example 2.1.3. We shall not give many examples. We have seen that Euclidean rings are principal ideal domains, in particular $\mathbb{Z}$, $\mathbb{Z}[i]$, $k[x]$ and $k[[x]]$, for k a field.

On the other hand, it is not easy to distinguish the class of principal ideal domains from that of Euclidean rings. The simplest example of a principal ideal domain which is not Euclidean is $\mathbb{Z}\left[\frac{1+\sqrt{-19}}{2}\right]$, see Exercise 19 of Chapter 1, and Exercise 5 of Chapter 3.

In this section we shall not say much about principal ideal domains, the main result being Theorem 3.1.5, which states that in a principal ideal domain we have unique factorization. We start with a simple but important fact.

Proposition 2.1.4. *Let A be a principal ideal domain. Then every* nonzero *prime ideal of A is maximal.*

Proof. Let $P = (p)$ be a nonzero prime, and take $a \notin P$; we only need to show that $(a, p) = A$.

Let $(a, p) = (b)$, so $p = bh$ for some $h \in A$. Note that $b \notin P$, since $a \in (b)$, so we have $h \in P$. This means that

$$p = bh = bkp$$

for some $k \in A$, from which we deduce $bk = 1$, since $p \neq 0$ and A is a domain. Hence b is invertible and $(a, p) = A$. □

We can give a generalization of Theorem 1.5.6, characterizing modules over a principal ideal domain. Since the proof is not much different, we leave this as Exercise 2. Here we give a different argument, which yields a weaker result, but works equally well for the infinite case. The reader who is not familiar with ordinals and transfinite induction may try to translate the following in a proof by ordinary induction (in the finite case).

The idea is to proceed by induction, starting from the base case $F = A$, which is just Remark 2.1.2. This proof can be found in [**Rot79**].

Theorem 2.1.5. *Let A be a principal ideal domain and let F be a free module over A. If $N \subset F$ is a submodule, then N is free.*

Proof. Let

$$F = \bigoplus_{s \in S} A_s,$$

where S is any set and A_s is a copy of A; call f_s a generator of A_s. Choose a well-ordering $\leq$ of S, and put

$$F_{\leq s} := \bigoplus_{t \leq s} F_t$$
$$F_{<s} := \bigoplus_{t<s} F_t.$$

We also let $N_{\leq s} := N \cap F_{\leq s}$ and similarly for $N_{<s}$.

Then the quotient $N_{\leq s}/N_{<s}$ is a submodule of A_s, hence it is either 0 or it is generated by a single element $a_s f_s$. In the latter case, choose a vector $n_s \in N_{\leq s}$ having such a component in A_s.

Call S' the set for which the quotient is not 0. We want to show that $\{n_s\}_{s \in S'}$ is a basis of N.

First, assume there is a linear combination

$$b_1 n_{s_1} + \cdots + b_k n_{s_k} = 0$$

with b_j not all 0. We can assume k is minimal, and we can arrange the order so that s_k is the biggest index. Then projecting into A_{s_k} gives

$$b_k a_{s_k} f_{s_k} = 0,$$

thereby showing that $b_k = 0$, a contradiction. Hence the vectors are linearly independent.

Assume by contradiction that these vectors do not generate. Every $n \in N$ lies in $N_{\leq s}$ for some s. Since S is well-ordered we can choose a minimal s such that there exists $n \in N_{\leq s}$ which is not in the span of the $\{n_i\}$.

Let af_s be the component of n in A_s; then $a_s|a$, say $a = ba_s$, so

$$n - bn_s \in N_{<s}.$$

By minimality of s, $n - bn_s$ is in the span of the $\{n_j\}$, and so n is. □

We cannot hope to get the stronger result that there exists a basis $\{f_i\}$ of F and elements $\{a_i\}$ of A, such that $\{a_i f_i\}$ is a basis of N. Since any A-module is the quotient of a free module, this would imply that every A-module would be the direct sum of addends, each one isomorphic to $A/(a)$ for some $a \in A$. This is false, as shown by the following.

Example 2.1.6. We take the ring $A = \mathbb{Z}$, so A-modules are just abelian groups. The additive group $\mathbb{Q}$ is divisible, that is, for every $x \in \mathbb{Q}$ and $m \in \mathbb{Z}$ we can find $y \in \mathbb{Q}$ such that $my = x$. So $\mathbb{Q}$ cannot be isomorphic to the direct sum of cyclic groups.

Still the conclusion is true for finitely generated free modules, as you are asked to prove in Exercise 2, so we obtain:

Theorem 2.1.7. *Every finitely generated module over a principal ideal domain A is isomorphic to*

$$A^r \oplus A/(a_1) \oplus \cdots \oplus A/(a_k)$$

for some $r \in \mathbb{N}$ and a suitable choice of $a_1, \ldots, a_k \in A$.

2.2. Artinian and Noetherian modules

One can try to generalize the notion of a principal ideal domain considering rings whose ideals are finitely generated. In even greater generality one could consider modules having the property that all their submodules are finitely generated. We start with the following characterization.

Proposition 2.2.1. *Let M be an A-module. Then the following properties are equivalent:*

(i) *every submodule of M is finitely generated;*

(ii) *every ascending chain of submodules*

$$N_1 \subset N_2 \subset N_3 \subset \cdots$$

is eventually stationary.

Proof. Assume the first property, and let $N := \bigcup N_i$—this is a submodule of M since the N_i are nested. Since N is generated by a finite number of elements, these all appear in some N_k; after k, the inclusions become equalities.

Conversely, assume the second property, and let $N \subset M$ be a submodule. Take $n_1 \in N$; if $N = (n_1)$ we are done. Otherwise take $n_2 \in N \setminus (n_1)$, and so on. We cannot have an infinite chain

$$(n_1) \subsetneq (n_1, n_2) \subsetneq (n_1, n_2, n_3) \subsetneq \cdots,$$

hence for some k we have $N = (n_1, \ldots, n_k)$. □

Definition 2.2.2. Condition ii) of Proposition 2.2.1 is called the *ascending chain condition*. We can also formulate a *descending chain condition*:

every descending chain of submodules

$$N_1 \supset N_2 \supset N_3 \supset \cdots$$

is eventually stationary.

Definition 2.2.3. An A-module M is called *Noetherian* (resp., *Artinian*) if it satisfies the ascending (resp., descending) chain condition.

The ring A is said to be Noetherian or Artinian if it is as a module over itself, that is, it satisfies the appropriate chain condition for ideals.

Remark 2.2.4. The characterization by finite generation has no analogue for Artinian modules.

Although the definitions look rather similar, we will see that being Artinian is a very special condition, at least for rings, while the Noetherian hypothesis is at the basis of most interesting results in commutative algebra.

Example 2.2.5.

(a) Every finite A-module is trivially both Artinian and Noetherian.

(b) Every principal ideal domain is a Noetherian ring, in particular $k[x]$. Note that any ring of the form $A[x]$ cannot be Artinian, because of the descending sequence of ideals

$$(x) \supset (x^2) \supset (x^3) \supset \cdots .$$

(c) The ring $k[x_1, x_2, x_3, \dots]$ is neither Artinian nor Noetherian.

(d) Consider the abelian group μ_n formed by all nth roots of 1, say inside $\mathbb{C}^*$. For $p \in \mathbb{N}$ a prime, we let

$$\mu_{p^\infty} := \bigcup_{n=1}^{\infty} \mu_{p^n}.$$

Then μ_{p^∞} is an Artinian $\mathbb{Z}$-module, but it is not Noetherian.

The fact that it is not Noetherian is easily seen by the ascending chain of subgroups

$$\mu_p \subset \mu_{p^2} \subset \mu_{p^3} \subset \cdots .$$

To prove that it is Artinian, note that any subgroup of μ_{p^∞} which contains roots of 1 of arbitrarily high order must be the whole μ_{p^∞}. So any proper subgroup contains roots of unity of bounded order, hence it is finite (actually it must be one of the μ_{p^n}).

(e) The ring $A = C([0,1])$ of continous, real valued functions on the unit interval, is not Noetherian. Indeed, let

$$I_k := \{f \in A \mid f(x) = 0 \text{ on } [0, 1/k]\};$$

this is a strictly ascending sequence of ideals.

In fact the Noetherian hypothesis is rather strong: we will see that it holds often for rings coming from algebraic objects, like polynomial rings, but it is seldom (if ever) verified for rings of functions on topological objects. In a vague sense, it is the condition that distinguishes the good properties of polynomial-like rings from the intractable zoo of arbitrary rings.

There is a useful reformulation of the chain conditions. To avoid duplication, we formulate the result for a general partially ordered set; we will then apply it to the set of submodules of a given module M, ordered by ascending or descending inclusion.

Proposition 2.2.6. *Let $(S, \leq)$ be a partially ordered set. The following conditions are equivalent:*

(i) *every ascending chain*

$$s_1 \leq s_2 \leq s_3 \leq \cdots$$

is eventually stationary;

(ii) *every subset of S has a maximal element.*

Proof. Assume ii) and take an ascending chain

$$s_1 \leq s_2 \leq s_3 \leq \cdots ;$$

if s_k is maximal in the set $\{s_i\}$, the sequence becomes stationary after k.

Conversely, assume i), and let $T \subset S$ be any subset. Choose $s_1 \in T$; if s_1 is maximal, we are done. If not there is some $s_2 > s_1$ in T. If this is not maximal, proceed to find $s_3 > s_2$ and so on. Since we cannot have any infinite strictly ascending sequence, we must eventually find a maximal element of T. □

So we see that every family of submodules of a Noetherian (resp., Artinian) module has a maximal (resp., minimal) element; this is an equivalent characterization.

We finish this section by studying the behavior of Noetherian and Artinian modules under quotients and localization. We start with a simple observation.

Remark 2.2.7. If M is Noetherian (resp., Artinian) and $N \subset M$, both N and M/N are Noetherian (resp., Artinian). This is just because submodules of N are also submodules of M, and submodules of M/N correspond to submodules of M by Proposition 1.3.11.

The following is a useful converse.

Proposition 2.2.8. *Let $N \subset M$. If both N and M/N are Noetherian (resp., Artinian) then M is.*

Proof. Let $\{M_i\}$ be any ascending (resp., descending) sequence of submodules. Denote by $\pi_N\colon M \to M/N$ the projection. Then both $M_i \cap N$ and $\pi_N(M_i)$ must eventually be constant.

The conclusion follows from the simple remark that for any submodules $M', M'' \subset M$ the equalities

$$M' \cap N = M'' \cap N \qquad \text{and} \qquad \pi_N(M') = \pi_N(M'')$$

imply that $M' = M''$, *provided* $M' \subset M''$.

The proof is straightforward: if $m'' \in M''$, find some $m' \in M'$ such that $m' \equiv m''$ modulo N. Then $m'' - m' \in N \cap M'' \subset M'$, so $m'' \in M'$. □

Remark 2.2.9. The request that $M' \subset M''$ in the proof is necessary, otherwise the conclusion of the remark is false, even for vector spaces!!

Corollary 2.2.10. *If $M_1, \ldots, M_k$ are Noetherian (resp., Artinian) then $M_1 \oplus \cdots \oplus M_k$ is.*

Corollary 2.2.11. *If A is Noetherian (resp., Artinian) and M is a finitely generated A-module, then M is Noetherian (resp., Artinian).*

Proof. Indeed M is the quotient of some A^k, which is Noetherian (resp., Artinian) by Corollary 2.2.10. □

We then treat the case of localization. Let M be an A-module, $S \subset A$ a multiplicative set. Then $S^{-1}M$ can be regarded both as A-module and $S^{-1}A$-module. Clearly every $S^{-1}A$-submodule is a fortiori an A-submodule. It follows that if M is Noetherian (resplectively Artinian) as A-module, it is as $S^{-1}A$-module. However the converse is false, so it is important to distinguish which is the base ring when we speak about $S^{-1}M$.

Proposition 2.2.12. *Let M be a Noetherian (resp., Artinian) A-module, $S \subset A$ a multiplicative set. Then $S^{-1}M$ is Noetherian (resp., Artinian) as $S^{-1}A$-module.*

Proof. By Proposition 1.6.16 every submodule of $S^{-1}M$ is the extension of a submodule of M. Hence an infinite ascending (or descending) chain of submodules of $S^{-1}M$ induces an infinite chain of submodules of M. □

In particular the proposition says that every localization of a Noetherian (resp., Artinian) ring is Noetherian (resp., Artinian).

2.3. Noetherian rings

In this section, we study the Noetherian condition for rings. As it turns out, the apparent symmetry that exists between Artinian and Noetherian modules breaks up for rings. The condition that a ring is Noetherian is pretty general—in fact it is a standard finiteness condition that will underlie many of our results. On the opposite side, Artinian rings are pretty special: we will eventually characterize them as Noetherian rings of dimension 0.

The crucial result that distinguishes the two is Hilbert's basis theorem, which states that if A is a Noetherian ring, the ring $A[x]$ is Noetherian as well. This is today a pretty routine result (we will see many other properties that can be carried from A to the polynomial ring $A[x]$), but when it was first devised it caused a diatribe about effective and abstract methods in algebra.

Around the beginning of the twentieth century, many algebraist were concerned with rings of invariants, which are obtained by taking an action of a finite or linear group G on $k[x_1, \ldots, x_n]$ and considering the ring of elements that are invariant under G. Typical examples are the determinant or the Pfaffian, seen as polynomials over the entries of a matrix. Many results of that period concerned finding explicit finite set generators for such rings, that were seen as a sort of universal invariants.

By using Hilbert's basis theorem, one can prove that many rings constructed in this fashion are Noetherian (see for instance Exercise 7 of Chapter 5), and in particular their ideals are finitely generated. This nonconstructive approach is in sharp constrast with the explicit computations in vogue at the time, and was regarded as highly controversial. It was the start of a revolution in algebra that eventually lead to abstract algebra as we know today. The key condition was then abstracted by Noether, and today we call this property by her name. In particular, she proved that the Noetherian property is enough to carry over the theory of primary decomposition, that was originally developed by Lasker for polynomial rings and their quotients. This is the approach that we will follow in Section 3.2.

After this historical parenthesis, we begin this section by proving a partial converse of Proposition 2.2.12.

Proposition 2.3.1. *Let A be a ring and take elements $a_1, \ldots, a_n \in A$ such that $(a_1, \ldots, a_n) = A$. If every localization A_{a_j} is Noetherian, A is itself Noetherian.*

Proof. Let $A_j = A_{a_j}$, and let

$$\iota_j \colon A \to A_j$$

be the localization map. We claim that for any ideal $I \subset A$ we have

$$I = \bigcap_{j=1}^{n} \iota_j^{-1}(I \cdot A_j).$$

One inclusion is obvious.

For the other, take an element b in the right-hand side intersection. Then for each j we can write

$$\iota_j(b) = b_j/a_j^{n_j},$$

so in A we have

$$a_j^{m_j}(a_j^{n_j}b - b_j) = 0.$$

By increasing the n_j and the m_j and changing the b_j accordingly, it is not restrictive to assume that all n_j are equal to a fixed n and all m_j to a fixed m.

So $a_j^{m+n}b \in I$ for all j. The powers a_j^{m+n} generate the whole ring (why?), so we can write

$$1 = \sum_{j=1}^{n} c_j a_j^{m+n};$$

multiplying both sides by b we deduce that $b \in I$.

Now take any ascending chain $\{I_k\}$ of ideals of A. Every chain $\{I_k \cdot A_j\}_{k\in\mathbb{N}}$ becomes stationary; since there are finitely many localizations A_j, the chain $\{I_k\}$ becomes stationary by the claim. □

Remark 2.3.2. The hypothesis of the proposition easily implies that every localization A_P at a prime ideal is Noetherian. So the result can be seen as something just a bit weaker than saying that being Noetherian is a local property (which is in fact false, see Exercise 16).

Both the rings A_a and A_P have a geometric meaning, and the relation between the usual notion of local property and the variant used in the proposition is made clear in the theory of schemes, see [**Har77**, Prop. *II*.3.2].

The main result about Noetherian rings is Hilbert's basis theorem. This is a cornerstone of commutative algebra, so we will give two proofs.

Theorem 2.3.3 (Hilbert's basis theorem). *Let A be a Noetherian ring. Then $A[x]$ is Noetherian.*

Remark 2.3.4. The analogue for Artinian rings is definitely false!

Before giving the proofs, it will be convenient to establish some notation.

Definition 2.3.5. Let A be a ring, $f \in A[x]$ a polynomial. The *leading coefficient* of f is the coefficient of the highest-order term in f; it is denoted by $\mathrm{LC}(f)$. If $I \subset A[x]$ is any ideal, we let

$$\mathrm{LC}(I) := \{\mathrm{LC}(f) \mid f \in I\};$$

it is straightforward to check that $\mathrm{LC}(I)$ is in fact an ideal of A.

First Proof. Let $I \subset A[x]$ be any ideal. Since $\mathrm{LC}(I)$ is finitely generated we can choose $f_1, \ldots, f_k \in I$ such that

$$\mathrm{LC}(I) = (\mathrm{LC}(f_1), \ldots, \mathrm{LC}(f_k)).$$

Let n be the maximum degree of all f_i, and take any $f \in I$. If $\deg f \geq n$, we can write

$$f = g_1 f_1 + \cdots + g_k f_k + r,$$

where r has degree smaller than f. Just choose the g_i so to cancel the leading term of f.

Repeating this step, we see that I is generated by $f_1, \ldots, f_k$ and by the intersection $I \cap A[x]_{<n}$, where $A[x]_{<n}$ denotes the polynomials of degree less than n.

The A-module $A[x]_{<n}$ is finitely generated, hence Noetherian by Corollary 2.2.11, so $I \cap A[x]_{<n}$ is finitely generated as an A-module, and a fortiori as an $A[x]$-module. We conclude that I is finitely generated as an $A[x]$-module, which is the thesis. □

Second Proof. This second proof does not use the notion of module. Let $I \subset A[x]$ be an ideal, and choose $f_1 \in I$ to be an element of I of least degree. If $I = (f_1)$ we are done, otherwise choose f_2 of minimal degree in $I \setminus (f_1)$. If $I \neq (f_1, f_2)$, repeat this process to choose f_3 and so on.

We only need to show that this process eventually stops. Let k be the first integer such that $\mathrm{LC}(f_1), \ldots, \mathrm{LC}(f_k)$ generate $\mathrm{LC}(I)$. We claim that $I = (f_1, \ldots, f_k)$. Indeed take any $f \in I$; if $\deg f < \deg f_k$ we must have $f \in (f_1, \ldots, f_{k-1})$ by the choice of f_k.

Otherwise $\deg f \geq \deg f_k$. As in the first proof we can write

$$f = g_1 f_1 + \cdots + g_k f_k + r,$$

where r has degree smaller than f; moreover r is in $(f_1, \ldots, f_k)$ if and only if f is. By repeating this step enough times we find that $f \in (f_1, \ldots, f_k)$. □

The Hilbert's basis theorem has some corollaries which can be seen as generalizations of the theorem itself. First, if A is Noetherian, then $A[x_1, \ldots, x_n]$ is again Noetherian. More generally every finitely generated A-algebra is a quotient of some $A[x_1, \ldots, x_n]$, so we get

Corollary 2.3.6. *Let A be a Noetherian ring, B a finitely generated A-algebra. Then B is Noetherian.*

We conclude this section with some facts on ideals of Noetherian rings. These results will be generalized by the theory of primary decomposition in Section 3.2, but we will need them in next section to compare the Artinian and Noetherian property.

Proposition 2.3.7. *Any ring A has minimal prime ideals. More precisely any prime $P \subset A$ contains a minimal prime ideal.*

Proof. The intersection of a descending chain of prime ideals is a prime ideal; now apply Zorn's lemma. □

By considering the quotient A/I we see that indeed there are minimal primes in the family of primes containing a fixed ideal I. These will be simply called the *minimal primes* of I. For Noetherian rings we also have:

Proposition 2.3.8. *Let A be a Noetherian ring, $I \subset A$ an ideal. Then there are* finitely many *minimal primes of I.*

Proof. Assume that this is not true, and let $\mathcal{F}$ be the family of ideals which don't satisfy the thesis. Let $I \in \mathcal{F}$ be maximal; by taking the quotient A/I we can assume that $I = 0$.

In particular 0 is not prime, so there are nonzero $a, b \in A$ such that $ab = 0$. By maximality, (a) and (b) have finitely many minimal primes. Any minimal prime of (0) either contains a or b, hence it is a minimal prime of (a) or (b). So 0 has finitely many minimal primes, a contradiction. □

Corollary 2.3.9. *In a Noetherian ring, the nilradical is a finite intersection of prime ideals.*

Proposition 2.3.10. *Let A be a Noetherian ring, $I \subset A$ an ideal, $J = \sqrt{I}$. Then there exists m such that $J^m \subset I$.*

Proof. Let

$$J = (a_1, \ldots, a_n)$$

and take k big enough so that $a_i^k \in I$ for every i. Let $m = kn$; then every monomial of degree m in $a_1, \ldots, a_n$ contains at least one a_i at the power k, hence is in I. □

In particular we note that the nilradical of a Noetherian ring is nilpotent, that is, $\mathcal{N}(A)^m = 0$ for some m.

2.4. Artinian rings

We now turn to the study of Artinian rings. Unlike the Noetherian case, we will be able to obtain a very precise picture. We start with the following easy fact.

Proposition 2.4.1. *Let A be an integral domain. If A is Artinian, then it is a field.*

Proof. Let $a \in A$ be nonzero and consider the descending sequence

$$(a) \supset (a^2) \supset (a^3) \supset \cdots .$$

The sequence stabilizes, hence $(a^{m+1}) = (a^m)$ for m big enough. In particular $a^m = ba^{m+1}$ for some $b \in A$; cancelling a^m we get $ab = 1$, so a is invertible. □

Corollary 2.4.2. *Let A be an Artinian ring. Every prime ideal of A is maximal.*

This is already a sign that the structure of Artinian rings is rather simple. We then investigate maximal ideals.

Proposition 2.4.3. *Let A be an Artinian ring. Then A has only finitely many maximal ideals.*

Proof. Assume we have an infinite sequence of distinct maximal ideals $\mathcal{M}_k$, and consider the descending chain

$$\mathcal{M}_1 \supset \mathcal{M}_1 \cap \mathcal{M}_2 \supset \mathcal{M}_1 \cap \mathcal{M}_2 \cap \mathcal{M}_3 \supset \cdots .$$

The chain stabilizes, hence for some n

$$\mathcal{M}_{n+1} \supset \bigcap_{i=1}^{n} \mathcal{M}_i.$$

By Proposition 1.1.20 ii) we get that $\mathcal{M}_{n+1}$ contains some $\mathcal{M}_k$, hence they coincide, a contradiction. □

Combining Corollary 2.4.2 and Proposition 2.4.3, we see that in an Artinian ring A we have $\mathcal{J}(A) = \mathcal{N}(A)$, and moreover the nilradical is a finite intersection of maximal ideals.

Lemma 2.4.4. *Let A be an Artinian ring. Then $\mathcal{N}(A)^k = 0$ for some k.*

Proof. Consider the descending chain $\{\mathcal{N}(A)^k\}$; by the Artinian property we have $\mathcal{N}(A)^{k+1} = \mathcal{N}(A)^k$ for k big enough. We claim that then $\mathcal{N}(A)^k = 0$.

If this is not the case, consider the set of ideals I such that $I \cdot \mathcal{N}(A)^k \neq 0$; since A is Artinian, we can choose I minimal in this set. In particular $I = (a)$ for some $a \in I$ such that $a \cdot \mathcal{N}(A)^k \neq 0$. We have

$$a \cdot \mathcal{N}(A)^{k+1} = a \cdot \mathcal{N}(A)^k \neq 0,$$

hence $a \cdot \mathcal{N}(A)$ is one of the ideals considered above. By minimality we have $(a) = a \cdot \mathcal{N}(A)$, so $a = a \cdot n$ for some $n \in \mathcal{N}(A)$.

But then

$$a = an = an^2 = \cdots = 0,$$

since n is nilpotent, a contradiction. □

The above lemma allows us to reduce the theory of arbitrary Artinian rings to the local case.

Theorem 2.4.5. *Let A be an Artinian ring, $\mathcal{M}_1, \dots, \mathcal{M}_k$ its maximal ideals. Then we have*

$$A \cong A/\mathcal{M}_1^{m_1} \times \cdots \times A/\mathcal{M}_1^{m_k}$$

for suitable integers $m_1, \dots, m_k$. Moreover each ring $A/\mathcal{M}_i^{m_i}$ is local.

Proof. By Lemma 2.4.4 we see that

$$\mathcal{M}_1^{m_1} \cdots \mathcal{M}_1^{m_k} = 0$$

for some integers $m_1, \dots, m_k$.

The only maximal ideal which contains a power of $\mathcal{M}_i$ is $\mathcal{M}_i$ itself, by Proposition 1.1.20 ii). So for $i \neq j$ the ideal $\mathcal{M}_i^{m_i} + \mathcal{M}_j^{m_j}$ is not contained in any maximal ideal, hence $\mathcal{M}_i^{m_i} + \mathcal{M}_j^{m_j} = A$.

We can then apply the Chinese remainder theorem 1.2.12, or better the refined version given in Exercises 9 and 10 of Chapter 1, to obtain the first part of the thesis.

The ideals of $A/\mathcal{M}_i^{m_i}$ correspond to ideals of A which contain $\mathcal{M}_i^{m_i}$, hence the only maximal ideal of the quotient is the image of $\mathcal{M}_i$. □

Corollary 2.4.6. *With the notations of the previous theorem, we have*

$$A/\mathcal{M}_i^{m_i} \cong A_{\mathcal{M}_i}.$$

Proof. Let $\pi \colon A \to A/\mathcal{M}_i^{m_i}$ and $\iota \colon A \to A_{\mathcal{M}_i}$ be the natural maps. Since $A/\mathcal{M}_i^{m_i}$ is local with maximal ideal $\pi(\mathcal{M}_i)$, the projection of any element outside $\mathcal{M}_i$ is invertible in $A/\mathcal{M}_i^{m_i}$. This yields a map

$$A_{\mathcal{M}_i} \to A/\mathcal{M}_i^{m_i}.$$

To give the inverse map we only need to check that $\iota(a) = 0$ for $a \in \mathcal{M}_i^{m_i}$. From the previous proof we have

$$\mathcal{M}_1^{m_1} \cdots \mathcal{M}_1^{m_k} = 0.$$

Note that

$$I = \mathcal{M}_1^{m_1} \cdots \widehat{\mathcal{M}_i^{m_i}} \cdots \mathcal{M}_1^{m_k} \not\subset \mathcal{M}_i,$$

where $\widehat{\mathcal{M}_i^{m_i}}$ means that this factor has been omitted. This follows from Proposition 1.1.20 ii). Take any $x \in I \setminus \mathcal{M}_i$, and let $a \in \mathcal{M}_i^{m_i}$. Then $ax = 0$, whence $\iota(a) = 0$. □

Remark 2.4.7. In the last corollary, we found a localization map which coincides with a quotient map. In particular, we see that a localization map can look drastically different from the inclusion in a bigger ring. We are now rewarded for having developed localization for rings more general than integral domains.

Corollary 2.4.8. *Every Artinian ring is the direct product of its localization at maximal ideals.*

We finish this section with the connection between Artinian and Noetherian rings. We start with a simple observation:

Remark 2.4.9. A vector space V over k is Artinian (as k-module) if and only if it is Noetherian. Indeed both conditions are equivalent to having finite dimension over k.

This can be readily generalized:

Lemma 2.4.10. *Let A be a ring, and assume that there are maximal ideals $\mathcal{M}_1, \ldots, \mathcal{M}_r$,* not necessarily distinct, *such that $\mathcal{M}_1 \cdots \mathcal{M}_r = 0$. Then A is Noetherian if and only if it is Artinian.*

Proof. If $r = 1$, A is a field. Else, for $i = 2, \ldots, r$ consider the A-modules

$$N_i = (\mathcal{M}_1 \cdots \mathcal{M}_{i-1})/(\mathcal{M}_1 \cdots \mathcal{M}_i).$$

Then A is Noetherian (resp., Artinian) if and only if all N_i are. But each N_i is annihilated by $\mathcal{M}_i$, so can be regarded as a module over $k_i = A/\mathcal{M}_i$.

Moreover each N_i is Noetherian (resp., Artinian) as A-module if and only if it is as k_i-module. So we conclude by the previous remark. □

We finally get our comparison theorem.

Theorem 2.4.11. *Let A be a ring. Then A is Artinian if and only if it is Noetherian and every prime of A is maximal.*

Proof. If A is Artinian, then every prime of A is maximal by Corollary 2.4.2. Moreover the nilradical is the intersection of finitely many maximal ideals, and it is nilpotent by Lemma 2.4.4. So Lemma 2.4.10 implies that A is Noetherian.

Vice versa, assume that A is Noetherian; then $\mathcal{N}(A)$ is nilpotent by Proposition 2.3.10, and it is the intersection of finitely many primes by Corollary 2.3.9. By our assumption these primes are all maximal, so the hypothesis of Lemma 2.4.10 are again satisfied and A is Artinian. □

Example 2.4.12.

(a) The typical picture of a local Artinian ring is $k[x]/(x^n)$.

(b) For more examples, one can take any Noetherian ring A and localize it at a prime $P \subset A$. The ring A_P is local with maximal ideal $\mathcal{M} = P \cdot A_P$. Then the ring $A_P/\mathcal{M}^n$ is local and Artinian.

Indeed it is Noetherian, and its prime ideals correspond to primes $Q \subset A$ such that $P^n \subset Q \subset P$. The only such prime ideal is P itself, so every prime of $A_P/\mathcal{M}^n$ is maximal.

(c) Let A be an infinite direct sum of fields. Then every prime of A is maximal, but A is not Artinian (or Noetherian).

2.5. Length

We begin with a

Definition 2.5.1. Let M be an A-module. A *chain* for M is a sequence

$$0 = M_0 \subsetneq M_1 \subsetneq \cdots \subsetneq M_n = M$$

of nested submodules. The integer n is called the *length* of the chain. If one cannot insert another submodule to increase the length, the chain is called a *series of composition* for M.

Example 2.5.2. It is not the case that a module always has a series of composition. The simplest example is the case where A is a field and M is a vector space of infinite dimension. For another example $\mathbb{Z}$ as a module over itself will do, since any nonzero ideal (n) of $\mathbb{Z}$ properly contains another ideal, for instance $(2n)$.

Remark 2.5.3. A chain (M_i) is a series of composition if and only if every quotient M_{i+1}/M_i is simple, that is, it does not have any nontrivial submodule.

Remark 2.5.4. If M is both Artinian and Noetherian, a series of composition can be obtained as follows. Let $M_0 = 0$ and M_1 be a minimal nonzero submodule; this exists because M is Artinian. Then let M_2 be minimal among those containing M_1 and so on. The chain (M_i) will stabilize because M is Noetherian, and it will do so at some $M_k = M$, so we have the desired series of composition. We shall see soon that this condition is also necessary.

Definition 2.5.5. If M admits a series of composition, we say that M has *finite length*, and we define the *length* of M, denoted $\ell(M)$, to be the minimum length of a series of composition for M. After the following result, it turns out that *every* series of composition is long exactly $\ell(M)$.

Theorem 2.5.6. *All series of compositions for M (if any) have the same length. If $\ell(M)$ is finite, any chain be enlarged to a series of composition.*

Proof. Let (M_i) be a series of composition for M having length $\ell(M)$. We first prove that a submodule $N \subsetneq M$ has *stricly smaller* length. Indeed let

$$N_i = M_i \cap N$$

and consider the chain (N_i). Since M_{i+1}/M_i is simple, we either have $N_{i+1}/N_i = M_{i+1}/M_i$ or 0. So, if we omit from (N_i) the repeated terms, we have a series of composition for N, long at most $\ell(M)$.

Actually the length must be smaller, otherwise by induction $N_i = M_i$ for each i, and so $N = M$.

Now let (M_i) be any chain for M, of length k. Then

$$\ell(M) > \ell(M_{k-1}) > \cdots \ell(M_1) > \ell(M_0) = 0,$$

so $\ell(M) \geq k$. In particular if this is a series of composition we must have $\ell(M) = k$ by minimality, so all series of composition have the same length.

If (M_i) is any chain, one can enlarge it only a finite number of times, because the length is bounded by $\ell(M)$. So after a finite number of steps we reach a series of composition. □

Corollary 2.5.7. *M has finite length if and only if it is both Artinian and Noetherian.*

Proof. We have already noted one implication in Remark 2.5.4. For the other, assume that M has finite length. Then every chain, either ascending or descending, is long at most $\ell(M)$, hence it stabilizes. □

Remark 2.5.8. A ring has finite length over itself if and only if it is Artinian, by Theorem 2.4.11.

We can be more precise than that.

Theorem 2.5.9 (Jordan–Hölder)**.** *Let M be an A-module of finite length and (M_i), (M'_i) two series of composition. There is a bijective correspondence between the sets of subquotients $\{M_{i+1}/M_i\}$ and $\{M'_{j+1}/M'_j\}$ such that corresponding couples are isomorphic.*

Proof. Let k be the minimal index such that $M_1 \subset M'_k$. Then $M_1 \cap M'_{k-1} = 0$ because M_1 is simple, hence we have an injective homomorphism

$$M_1 \to M'_k/M'_{k-1}.$$

This must be an isomorphism because M'_k/M'_{k-1} is simple.

Let $N := M/M_1$, $\pi\colon M \to N$ the projection. Then N has the filtration $\{M_i/M_1\}_{i \geq 2}$ and the filtration $\{\pi(M'_j)\}_{j \neq k-1}$, so we can conclude by induction on $\ell(M)$. □

The length of modules behaves additively on exact sequences:

Proposition 2.5.10. *Let*

$$0 \longrightarrow M_1 \longrightarrow M_2 \longrightarrow M_3 \longrightarrow 0$$

be a short exact sequence of A-modules. Then

$$\ell(M_2) = \ell(M_1) + \ell(M_3)$$

(including the statement that the left-hand side is finite if and only if the right-hand side is).

Proof. Assume that M_1 and M_3 have finite length, and take two series of composition for them. Using the bijective correspondence between submodules of M_3 and submodules of M_2 containing the image of M_1, the two fit together to give a chain for M_2. The associated subquotients are simple, so we get a series of composition for M_2, hence the thesis. If either M_1 or M_3 has infinite length, so does M_2 by the same argument. $\square$

In the case where A is Noetherian and M is finitely generated we have a clear description of when the length of M is finite and the shape of the factors that can appear. First we need a

Definition 2.5.11. Let M be an A-module. The set of *associated primes* $\mathrm{Ass}(M)$ is defined to be

$$\mathrm{Ass}(M) = \{P \subset A \text{ prime} \mid P = \mathrm{Ann}(m) \text{ for some nonzero } m \in M\}.$$

Remark 2.5.12. The simple fact that if $N \subset M$ then $\mathrm{Ass}(N) \subset \mathrm{Ass}(M)$ will be used often in the sequel.

Proposition 2.5.13. *Let P be a maximal element in the set*

$$\{\mathrm{Ann}(m) \mid m \in M, m \neq 0\}.$$

Then P is a prime ideal.

Proof. Let $P = \mathrm{Ann}(m)$ and take $a, b \in A$ such that $ab \in P$. Then $abm = 0$, hence $a \in \mathrm{Ann}(bm)$. Of course $P \subset \mathrm{Ann}(bm)$, so by maximality we must have $a \in P$ or $\mathrm{Ann}(bm) = A$. In the first case we are done; in the second we have $1 \in \mathrm{Ann}(bm)$, so $b \in \mathrm{Ann}(m) = P$. $\square$

Corollary 2.5.14. *If A is Noetherian and $M \neq 0$, then $\mathrm{Ass}(M) \neq \emptyset$.*

Lemma 2.5.15. *A prime P belongs to $\mathrm{Ass}(M)$ if and only if M has a submodule N isomorphic to A/P.*

Proof. Let $P = \mathrm{Ann}(m)$ be a prime, and consider the homomorphism

$$\phi\colon A \to M$$

given by $\phi(a) = am$. Then the image of ϕ is Am, and the kernel is P, so

$$A/P \cong Am \subset M.$$

For the reverse implication we can reduce to the case where $M = N = A/P$; then $P = \mathrm{Ann}(1)$. $\square$

We can now state the result for Noetherian rings.

Theorem 2.5.16. *Let A be Noetherian and let M be a finitely generated A-module. Then there exists a chain (M_i) for M such that*

$$M_{i+1}/M_i \cong A/P_i$$

for some primes $P_i \subset A$.

Proof. Since M is Noetherian, we can find a maximal submodule N which satisfies the conclusion of the theorem (the set of such submodules is not empty, for instance any simple submodule will do). Let us prove that $N = M$. If $N \subsetneq M$, then by Corollary 2.5.14, M/N has some associated prime P. By Lemma 2.5.15 we find some $N' \subset M$ containing N such that

$$N'/N \cong A/P.$$

But then also N' satisfies the conclusion of the Theorem, a contradiction. □

Corollary 2.5.17. *Under the hypothesis of the Theorem, M has finite length if and only if each P_i is maximal. In this case (M_i) is a series of composition.*

Proof. By Proposition 2.5.10 and induction, it is enough to prove that $\ell(A/P)$ is finite if and only if P is maximal. Each A-submodule of A/P is also an A/P-submodule; hence A/P has finite length if and only if it is Artinian as a ring. By Proposition 2.4.1 an Artinian domain is a field, so we have the first conclusion.

If all P_i are maximal, the factors M_{i+1}/M_i are fields, so they are simple; it follows that (M_i) is a series of composition. □

Corollary 2.5.18 (Generic freeness)**.** *Let A be a Noetherian integral domain, M a finitely generated A-module. Then there exists a nonzero $f \in A$ such that M_f is free over A_f.*

Proof. Let (M_i) be a series of composition as in Theorem 2.5.16, so that $M_{i+1}/M_i \cong A/P_i$. For each prime $P_i \neq 0$, choose a nonzero element $f_i \in P_i$. Then $f := \prod f_i \neq 0$ because A is an integral domain.

If $P_i \neq 0$, we have $f \in P_i$. This implies $(A/P_i)_f = 0$, so the locallization of the factor M_{i+1}/M_i at f becomes trivial. This means that M_f has a series of composition where each successive quotient is isomorphic to A_f. It is easy to check that a short exact sequence where the last element is free splits, hence M_f is isomorphic to a direct sum of some copies of A_f. □

2.6. Exercises

1. Assume that $A[x]$ is a principal ideal domain. Then A is a field.

2. Prove the analogue of Theorem 1.5.6 for principal ideal domains.

3. Give examples to show that the conclusion of Proposition 2.2.12 cannot be strengthened to say that $S^{-1}M$ is Noetherian (or Artinian) as an A-module.

4. If A is Noetherian, then $A[[x]]$ is.

5. Let M be a Noetherian A-module, $f\colon M \to M$ a surjective homomorphism. Then f is an isomorphism.

6. Let M be an Artinian A-module, $f\colon M \to M$ an injective homomorphism. Then f is an isomorphism.

7. Let M be an A-module of finite length, $f\colon M \to M$ a homomorphism. Show that for m big enough there is a decomposition $M = \ker f^m \oplus \operatorname{im} f^m$.

8. Let A be a ring and assume that every *prime* ideal of A is finitely generated. Then A is Noetherian. This result is known as *Cohen's theorem*. (If A is not Noetherian, prove that there are maximal ideals among those which are not finitely generated.)

9. Let A be a domain and assume that every *prime* ideal of A is principal. Then A is a PID.

10 ([**Pol**]). A subring of a Noetherian ring is not necessarily Noetherian. Namely, let $A \subset k[x, y]$ be the ring of those polynomials that do not involve monomials of the form x^k for some $k > 0$. Show that A is a ring and exhibit an ideal that is not finitely generated.

11. Let $B = A[b_1, \ldots, b_n]$ be a finitely generated A-algebra, and M a finitely generated B-module. If $b_1, \ldots, b_n \in \sqrt{\operatorname{Ann}_B(M)}$, show that M is finitely generated as A-module.

12. Let $A \subset \mathbb{Q}[x]$ be the ring of polynomials f such that $f(x) \in \mathbb{Z}$ for all $x \in \mathbb{Z}$. Show that $\mathbb{Z}[x] \subsetneq A \subsetneq \mathbb{Q}[x]$. Show that A is not Noetherian—namely, the ideal

$$I := \{f \in A \mid f(0) \in 2\mathbb{Z}\}$$

is not finitely generated.

For the next exercises, we need a little terminology. A *Boolean ring* A is a ring such that $a^2 = a$ for all $a \in A$.

13 ([**Cla15**, Prop 9.4]). Let A be a Boolean integral domain. Show that A is a field, and in fact $A \cong \mathbb{Z}/2\mathbb{Z}$. Deduce that a prime ideal in a Boolean ring is maximal.

14 ([**Cla15**, Prop 9.5]). Let A be a local Boolean ring. Show that $A \cong \mathbb{Z}/2\mathbb{Z}$.

15 ([**Cla15**, Prop. 9.6]). Let A be a Noetherian boolean ring. Show that A is finite.

16. Using the previous exercises, find an example of a non-Noetherian ring such that all its localizations are Noetherian. Hence, being Noetherian is not a local property.

17. Let A be a graded ring, M a graded A-module (over $\mathbb{Z}$). Prove that every prime $P \in \mathrm{Ass}(M)$ is graded, and moreover $P = \mathrm{Ann}(m)$ for some homogeneous element $m \in M$.

18. Let k be a field. The ring $A = k[x_1, x_2, \dots]$ with the relations $x_1^2 = 0$ and $x_{k+1}^2 = x_k$ for $k \geq 0$ on has a single prime (hence maximal) ideal. Still it is not Noetherian, hence it is not Artinian. Exhibit an explicit infinite descending chain of ideals. Show that its analogue with a finite number of variables *is* Artinian.

19. Let k be a field and A a local Artinian k-algebra, with maximal ideal $\mathcal{M}_A$ and residue field $k_A = A/\mathcal{M}_A$. Let d be the degree of the field extension k_A/k, and let M be any finitely generated A-module. Then

$$\dim_k M = d\ell_A(M).$$

20 (By Eric Wofsey on Math Stack Exchange, see [**Wof**]). Let A be a Noetherian ring. Prove that a localization $S^{-1}A$ is a quotient of A if, and only if, there is a decomposition as a direct product $A \cong B \times C$ such that the B-coordinate of every element of S is a unit in B and there is some element of S whose C-coordinate is 0. In this case the map $A \to S^{-1}A$ can be identified with the projection to B.

21 ([**Jen80**]). This exercise requires some model theory, see for instance [**CK90**]. Let $\mathcal{U}$ be any ultrafilter on $\mathbb{N}$ that contains all cofinite sets—in particular $\mathcal{U}$ is not principal. The ultraproduct $A := \mathbb{Z}^{\mathbb{N}}/\mathcal{U}$ satisfies all first-order propositions as $\mathbb{Z}$ (say over the language $\{0, 1, +, \times\}$) by Łoś's Theorem—in particular it is a ring and integral domain. Prove that A is not Noetherian, hence being Noetherian is not a property expressible in first order logic. (The element $(1, 2, 4, 8, \dots)$ is divisible by all powers of 2.)

22 ([**Leq85**]). Let A be a ring, $\mathcal{M} \subset A$ a maximal ideal, and assume that every ideal $I \subset \mathcal{M}$ is finitely generated. Then A is Noetherian. (Show that for an ideal $I \not\subset \mathcal{M}$ we have $I \oplus \mathcal{M} \cong A \oplus (I \cap \mathcal{M})$.)

23. Prove the following lemma of Artin and Tate ([**AT51**]). Let $A \subset B \subset C$ be rings, where A is Noetherian. Assume that C is finitely generated as an A-algebra and as a B-module. Prove that B is also finitely generated as an A-algebra.

The next exercises, up to Exercise 28, follow [**Cla15**] in laying out a beautiful thread of related results about conditions which guarantee Noetherianity.

24. Let M be an A-module, N a submodule of M maximal with respect to the condition of not being finitely generated. Let $P = \operatorname{Ann}(M/N)$. Prove that P is prime. (Let $a, b \in A$ such that $ab \in P$. If $a, b \notin P$, then $N + aM$ is finitely generated, say by k elements $\{n_i + am_i\}$. Define $L := \{m \in M \mid am \in N\}$) and prove that $N = (n_1, \ldots, n_k) + aL$. Derive a contradiction.)

25. Prove *Jothilingam's theorem* ([**Jot00**], [**Nag05**]): let A be a ring, M a finitely generated module such that $P \cdot M$ is finitely generated for all prime ideals P of A. Then M is Noetherian. This generalizes Cohen's theorem from Exercise 8. (Assuming it is not, find a submodule N maximal with respect to the condition of not being finitely generated, and let P be the prime ideal of the previous exercise. Prove that $P = \operatorname{Ann}(m)$ for some $m \in M$. Show that $N + Am$ is finitely generated, say by k elements $\{n_i + a_i m\}$, and prove that $N = (n_1, \ldots, n_k) + PM$. Deduce that $P \cdot M$ is not finitely generated.)

26. Let M be an A-module, E_M the family of submodules of the form $I \cdot M$ for ideals $I \subset A$. Prove that submodules in E_M are finitely generated if and only if E_M satisfies the ascending chain condition.

27. Prove *Formanek's theorem* ([**For73**]): let A be a ring, M a finitely generated, faithful A-module. Let E_M be the family of submodules of the form $I \cdot M$ for ideals $I \subset A$. If E_M satisfies the ascending chain condition, then M is Noetherian, and A is Noetherian as well. (See the previous two exercises.)

28. Prove the *Eakin–Nagata theorem* ([**Eak68**], [**Nag68**]): let $A \subset B$ be rings, with B Noetherian, and assume that B is finitely generated as A-module. Then A is Noetherian as well.

Chapter 3

Factorization

In this chapter we study how the analogue of the unique factorization theorem for the ring $\mathbb{Z}$ can be generalized to different rings. In the first section we study the rings that satisfy the most obvious generalization of the unique factorization theorem. In particular we show that principal ideal domains belong to this class.

The next step would be to study the class of Dedekind domains: those for which unique factorization holds for *ideals*. It turns out that this is in many respect a more interesting class of rings than unique factorization domains, and their introduction is really a cornerstone of algebraic number theory. Before doing so, however, we study the theory of primary decomposition. This gives a factorization theorem for ideals which is much less precise that the one achieved in Dedekind domains, but has the advantage that it holds with the only assumption that the ring is Noetherian.

The section about Dedekind rings benefits from the primary decomposition theorem. Using this theory, we can derive the factorization in a quick and efficient way. After having introduced Dedekind rings, we prove a structure theorem for modules over them, generalizing our previous result for modules over Euclidean domains.

3.1. Unique factorization domains

One of the basic tools in elementary number theory is the *unique factorization theorem*, which shows that prime numbers are really the building bricks of all integers. Let us study the rings that satisfy simplest generalization of this result.

Definition 3.1.1. Let A be a ring. We say that a noninvertible element $x \in A$ is *irreducible* if whenever $x = ab$ for some $a, b \in A$, at least one among a and b is invertible. We say that x is *prime* if the ideal (x) is.

Remark 3.1.2. We note that x is prime explicitly means that if x divides a product ab, then it must divide at least one of the factors. In an integral domain, a prime element is always irreducible. Indeed assume that $x = ab$; in particular x divides ab, so it must divide one of the factors, say a. So $a = \xi x$, and $x = \xi bx$. Factoring x out we see that $\xi b = 1$, so b is invertible.

Definition 3.1.3. Let A be an integral domain. We say that A is a *unique factorization domain* (in short UFD) if

(1) every irreducible element of A is prime and

(2) every infinite sequence of principal ideals

$$(x_1) \subset (x_2) \subset \cdots \subset (x_n) \subset \cdots$$

is stationary. This means that eventually we have $x_{i+1} = \xi_i x_i$ for some invertible element ξ_i.

We will refer to such a sequence as a *divisibility chain*, so we can rephrase 2 saying that every infinite divisibility chain stabilizes. Note that this is just a weaker form of the ascending chain condition; in particular it is satisfied by Noetherian rings.

We can now show that UFDs satisfy a unique factorization theorem. This is not surprising (and justifies the name), since our requests for a ring A to be a UFD are precisely the hypothesis used to prove the classical theorem.

Theorem 3.1.4 (unique factorization). *Let A be a unique factorization domain, and let $x \in A$ be a noninvertible element. Then we have a factorization*

$$x = p_1 \cdots p_n,$$

where $p_1, \ldots, p_n \in A$ are prime. Moreover the factors p_i are unique up to the order and up to multiplication by a unit of A.

Proof. Let us first show the existence of a factorization. If x is prime we have nothing to do. Otherwise it must be reducible, so $x = x_1 y_1$ where both x_1 and y_1 are not units. If both x_1 and y_1 are prime, then we are done, otherwise we repeat the process. This eventually must stop, otherwise we would have an infinite divisibility chain in A.

As to the uniqueness, assume that x has two decompositions

$$x = p_1 \cdots p_n = q_1 \cdots q_m.$$

Then p_1 divides x: since it is prime it must divide one of the factors q_i, and up to reordering we assume that p_1 divides q_1. So $q_1 = \xi p_1$, and since q_1 is irreducible, ξ must be a unit.

We can fit it into q_2, so it is safe to assume that $p_1 = q_1$. Since A is an integral domain, we can factor them out and obtain that

$$p_2 \cdots p_n = q_2 \cdots q_m.$$

Now we can conclude by induction on n. □

At this point it is natural to investigate which rings are UFDs. Our first result says that principal ideal domains are, so we have the unique factorization for a familiar class of rings.

Theorem 3.1.5. *Let A be a principal ideal domain. Then A is a unique factorization domain.*

Proof. We begin by showing that every irreducible element is prime, so let $x \in A$ be irreducible. Assume that x divides a product ab but x does not divide a. The ideal (a, x) is principal so $(a, x) = (c)$ for some $c \in A$. This means in particular that $x = cd$ for some $d \in A$.

Since x is irreducible and d is not a unit (otherwise x would divide a), c must be invertible. In other words (a, x) is the whole ring, so we find $m, n \in A$ such that

$$ma + nx = 1. \tag{3.1.1}$$

Multiplying (3.1.1) by b we get

$$mab + nxb = b;$$

since x divides ab, we see that it must divide b.

For the other part, assume that

$$x_1, x_2, \ldots, x_n, \ldots$$

is an infinite divisibility chain, and let $I = (x_1, x_2, \ldots)$. Since A is a principal ideal domain, we must have $I = (a)$ for some $a \in A$. So a admits a *finite* writing as

$$a = x_1 y_1 + \cdots + x_n y_n.$$

By the divisibility condition, every summand is a multiple of x_n, so a is. Since vice versa a divides x_n we see that $I = (x_n)$.

But then if $m > n$ both x_m divides x_n and conversely; this means that the chain stabilizes. □

By now we have a reasonable amount of examples of UFDs, like $\mathbb{Z}$, $\mathbb{Z}[i]$ or $k[x]$ for every field k. But these are all principal ideal domains and we would like to extend the unique factorization to more general rings. In order to do this we need to generalize the classical Gauss's lemma, which is of interest itself. First let us give some definitions.

Definition 3.1.6. Let A be a unique factorization domain, and take $x, y \in A$. The *greatest common divisor* of x and y, denoted by $\mathrm{GCD}(x, y)$ is the product of all common irreducible factors of x and y. As usual if a prime appears both in x and y, its multiplicity within $\mathrm{GCD}(x, y)$ is the minimum of the multiplicities that it has in x and y.

Of course the greatest common divisor is determined up to a unit.

Remark 3.1.7. It is easy to see that $\mathrm{GCD}(x, y)$ is determined by the following properties:

(1) $\mathrm{GCD}(x, y)$ divides both x and y;

(2) if a divides both x and y, then it divides $\mathrm{GCD}(x, y)$.

The generalization to the greatest common divisor of three or more elements is straightforward.

Definition 3.1.8. Let A be a unique factorization domain and take $f(x) \in A[x]$. The *content* of f, denoted by $\mathrm{cont}(f)$, is the greatest common divisor of its coefficients. We say that f is *primitive* if $\mathrm{cont}(f) = 1$.

In order to prove Gauss's lemma, we start with some lemmas about the behavior of the content in products.

Lemma 3.1.9. *Let A be a UFD and $f, g \in A[x]$ be primitive polynomials. Then the product fg is primitive as well.*

Proof. Assume that a prime p of A divides all the coefficients of fg; then $\overline{f}\overline{g} = 0$ in $A/(p)[x]$. Since p is prime, $A/(p)$ is an integral domain, so $A/(p)[x]$ is an integral domain too. This means that either $\overline{f} = 0$ or $\overline{g} = 0$, so f and g cannot both be primitive. □

Corollary 3.1.10. *Let A be a UFD and take $f, g \in A[x]$. Then*

$$\mathrm{cont}(fg) = \mathrm{cont}(f)\,\mathrm{cont}(g).$$

Proof. Write $f = \mathrm{cont}(f) f_1$ and $g = \mathrm{cont}(g) g_1$, where f_1 and g_1 are primitive. □

We are now ready to prove Gauss's lemma.

Theorem 3.1.11 (Gauss's lemma). *Let A be a unique factorization domain with field of fractions k. Take $f, g \in A[x]$ with f primitive, and assume that f divides g in $k[x]$. Then f divides g in $A[x]$.*

Proof. Assuming the hypothesis we have

$$g(x) = f(x)h(x)$$

where h has coefficients in k. Clearing out denominators we find

$$mg(x) = f(x) \cdot (mh(x))$$

for some $m \in A$, where now $mh(x) \in A[x]$.

By the corollary we find

$$m \operatorname{cont}(g) = \operatorname{cont}(f) \operatorname{cont}(mh) = \operatorname{cont}(mh),$$

since f is primitive. In particular m divides $\operatorname{cont}(mh)$, so that $h(x)$ actually has coefficients in A. □

Corollary 3.1.12. *Let A be a unique factorization domain with field of fractions k. Let $f \in A[x]$ be a primitive polynomial. If f is irreducible over A, it is irreducible also over k.*

The above corollary is most effective when coupled with the following result.

Proposition 3.1.13 (Eisenstein's criterion [**Eis50**]). *Let A be a ring, $P \subset A$ a prime ideal. Take a polynomial $f(x) \in A[x]$ such that*

$$f(x) = a_n x^n + \cdots + a_1 x + a_0,$$

where $a_n \notin P$, all other $a_i \in P$ and $a_0 \notin P^2$. Then f is irreducible over A.

Proof. Assume that $f(x) = g(x)h(x)$ and reduce this modulo P. The reduction of $f(x)$ is just $\overline{a_n} x^n$, so

$$\overline{a_n} x^n = \overline{g}(x) \cdot \overline{h}(x).$$

Since A/P is an integral domain we see that $\overline{g}(x) = \overline{b_k} x^k$ for some $k < n$, so

$$g(x) = b_k x^k + \cdots + b_1 x + b_0,$$

with $b_i \in P$ for $i < k$, in particular $b_0 \in P$. A similar remark holds for h, so we see that $a_0 \in P^2$. □

More generally reduction modulo P, with an *ad hoc* argument, may prove irreducibility of a given polynomial even if Eisenstein's criterion does not apply.

Our next theorem, together with Theorem 3.1.5, is the main tool to find examples of UFDs. Before proving it we need a lemma.

Lemma 3.1.14. *Let A be a unique factorization domain with field of fractions k, and take a nonconstant polynomial $f(x) \in A[x]$. Then f is irreducible in $A[x]$ if and only if it is primitive and irreducible in $k[x]$.*

Proof. Assume that f is irreducible in $A[x]$. Then of course it must be primitive, otherwise it has a constant factor. To show that it is irreducible in $k[x]$ assume that

$$f(x) = g(x)h(x)$$

for some $g, h \in k[x]$ of positive degree. Multiplying g by a constant and dividing h by the same constant, we can arrange that g is in $A[x]$ and primitive. By Gauss's lemma we find that $h \in A[x]$, so f is reducible in $A[x]$ too. The converse implication is obvious. □

Theorem 3.1.15. *Assume that A is a unique factorization domain. Then $A[x]$ is.*

Proof. Let $f(x) \in A[x]$ be a nonconstant irreducible polynomial. To show that it is prime assume that $f(x)$ divides a product $g(x)h(x)$, with $g, h \in A[x]$. Let us call k the field of fractions of A. By Lemma 3.1.14 f is irreducible in $k[x]$, so it is prime since we already know that $k[x]$ is a unique factorization domain.

It follows that f divides one of the factors in $k[x]$, say g. Applying Lemma 3.1.14 again we see that f is primitive, so Gauss's lemma tells us that f divides g in $A[x]$. The case where f is constant is obvious.

Next we consider an infinite divisibility chain

$$f_1, f_2, \ldots, f_n, \cdots \in A[x]. \tag{3.1.2}$$

Let $c_i = \mathrm{cont}(f_i)$; then we easily see that $c_1, c_2, \ldots$ is an infinite divisibility chain in A. Hence it must stabilize, so up to trimming a finite number of terms in the sequence (3.1.2) we can assume that the f_i have all the same content.

On the other hand we know that the chain in (3.1.2) will stabilize in $k[x]$, so there exist constants $\xi_i \in k$ such that eventually $f_{i+1} = \xi_i f_i$. Since the f_i are in $A[x]$ and all have the same content we easily see that the constants $\xi_i \in A^*$, so the chain stabilizes in $A[x]$. □

Corollary 3.1.16. *If A is a unique factorization domain, then $A[x_1, \ldots, x_n]$ is a unique factorization domain as well.*

Example 3.1.17. With this result at hand we see that rings like $k[x, y]$ or $\mathbb{Z}[x]$ are unique factorization domains, while they are not principal ideal domains.

Another nontrivial example is $A = k[x_1, x_2, \ldots, x_n, \ldots]$, that is, a polynomial ring in an infinite number of indeterminates. To see that A is a unique factorization domain we just have to remark that both the conditions defining unique factorization domains will involve only a finite number of variables.

Explicitly, take $f \in A$ irreducible, and assume that f divides a product gh. Then the polynomials f, g and h will live inside a smaller ring $k[x_1, \dots, x_n]$, and f will remain irreducible there, hence prime. It follows that f divides one among g and h in the smaller ring, and so a fortiori in A.

The argument for the divisibility chain is similar. Given a chain

$$f_1, f_2, \dots, f_n, \dots \in A$$

we see that, since all f_i divide f_1, they all live inside a ring $k[x_1, \dots, x_n]$, so we are done like before.

We end this section with other characterizations of UFDs. They are all consequences of the simple observation that an irreducible element of a ring *that admits a factorization into primes* is itself prime.

The following proposition justifies the name of unique factorization domains.

Proposition 3.1.18. *An integral domain A is a unique factorization domain if and only if every element $a \in A$ factorizes uniquely as a product of primes, up to the order and units.*

Proof. One implication is Theorem 3.1.4. For the other, note that any irreducible element is prime, by the above observation. Moreover, the unique factorization into primes easily implies that a divisibility chain stabilizes. Indeed in any divisibility chain the number of prime factors decreases. □

The following result gives a much more useful criterion to establish that a ring is a UFD.

Theorem 3.1.19 (Kaplansky)**.** *Let A be an integral domain. Then A is a unique factorization domain if and only if every nonzero prime ideal contains a nonzero prime element.*

Proof. Assume A is a unique factorization domain and take a prime $P \subset A$. For any $a \in P$, $a \neq 0$, factorize a as

$$a = p_1 \cdots p_n,$$

with the p_i primes. Then one of the p_i belongs to P.

Conversely, let S be the multiplicative set consisting of all elements which factorize uniquely as a product of nonzero primes. We want to show that $S = A \setminus \{0\}$.

Assume we can find a nonzero element $x \notin S$. Then first note that $S \cap (x) = \emptyset$. Indeed assume

$$xy = p_1 \cdots p_n;$$

each prime divides either x or y, and assuming y is not invertible, in which case $x \in S$, at least one prime divides y, say p_n. Then we can divide both sides by p_n and proceed by induction. After a finite number of steps we find $x \in S$, which we have excluded.

So the ideal (x) remains nontrivial in $S^{-1}A$, in particular it is contained in a maximal ideal. The inverse image of this maximal ideal is a maximal ideal $\mathcal{M}$ of A such that $\mathcal{M} \cap S = \emptyset$. This is a contradiction, since by hypothesis $\mathcal{M}$ contains a prime element. $\square$

As a consequence of Kaplansky theorem, we recover again the result that every principal ideal domain is a unique factorization domain.

3.2. Primary decomposition

In this section we will see a far reaching generalization of unique factorization, which is valid for arbitrary Noetherian rings. Our treatment follows [**AM69**], which is slightly more general than other sources, proving the uniqueness theorems for arbitrary rings.

Example 3.2.1. The simplest ring in which unique factorization fails is $A = \mathbb{Z}[\sqrt{-5}]$. Indeed in this ring we have

$$6 = 2 \cdot 3 = (1 + \sqrt{-5})(1 - \sqrt{-5}).$$

It is not difficult to check by hand that all elements involved in these two factorization are irreducible, and that the two factorizations are essentially distinct, meaning that 2 and $1 \pm \sqrt{-5}$ do not differ by an invertible element.

A clever way to check this without computations is the following. The ring A has an automorphism $\overline{\cdot}$ sending $u = a + b\sqrt{-5}$ to $\overline{u} := a - b\sqrt{-5}$. Therefore the map

$$\begin{aligned} N\colon\ A &\longrightarrow \mathbb{Z} \\ u &\longmapsto u\overline{u} \end{aligned}$$

is multiplicative. Assume that for instance 2 factors in a nontrivial way in A; since $N(2) = 4$ we see that a factor α must have $N(\alpha) = 2$, which is easily seen to be impossible in A. Similarly all other factors are irreducible.

To check that 2 and $1 \pm \sqrt{-5}$ do not differ by an invertible element observe first that if α is invertible, it must satisfy $N(\alpha) = \pm 1$. But then the fact that $N(2) = 4$ and $N(1 \pm \sqrt{-5}) = 6$ shows that they are indeed distinct irreducible factors.

By the above example, we see that unique factorization can fail even in nice rings. Note that when we speak about unique factorization we always consider equivalent elements which differ by an invertible; in other words generators of the same principal ideal. This might lead to the idea that the right generalization of unique factorization is unique factorization for ideals. That is, we might hope that under mild hypothesis every ideal could be written uniquely as the product of prime ideals.[1]

This is actually true in a class of rings called *Dedekind rings*, but for our purposes they are still too special.

Example 3.2.2.

(a) Consider the ring $A = k[x, y]$, and the ideals $P = (x, y)$ and $I = (x, y^2)$. Then P is prime and $P^2 \subset I \subset P$ so we cannot hope to factorize I as the product of prime ideals.

(b) In the ring $A = k[x, y, z]$ consider the prime ideals $P = (x, y)$ and $Q = (y, z)$, and let $I = P \cap Q$. Then $y \in I$, but $y \notin P \cdot Q$. Again, since

$$P \cdot Q \subset I \subset P, Q$$

there is no way to factorize I as a product.

The examples above show three phenomena. First, even if in a UFD, unique factorization for ideals can fail. Second, one of the obstacles in the factorization is the existence of ideals lying between two consecutive powers of a prime. Third, even for radical ideals the fact that the intersection and product differ can be an obstacle.

The solution is the following: we enlarge the class of "building blocks" from prime powers to something better behaved, and we take intersection instead of product as the main operation.

Definition 3.2.3. Let $Q \subset A$ be an ideal. We say that Q is *primary* if $xy \in Q$ implies that either $x \in Q$ or some power $y^n \in Q$.

Remark 3.2.4. If Q is primary its radical $P = \sqrt{Q}$ is a prime. In this case we say that Q is P-primary.

Remark 3.2.5. As for prime and maximal ideals, there is characterization in terms of the quotient ring. Namely a proper ideal $Q \subsetneq A$ is primary if and only if every zero divisor in A/Q is nilpotent.

Although this concept is our replacement for the prime powers, it is not true that a prime power is necessarily primary. Still

Proposition 3.2.6. *Let $I \subset A$ an ideal and assume that $\mathcal{M} = \sqrt{I}$ is maximal. Then I is primary.*

---

[1] By the way, this is the historical origin of the concept of an ideal. Dedekind proved that such a unique factorization holds in $\mathbb{Z}[\sqrt{-5}]$, and so thought that unique factorization could be restored provided one allowed the use of "ideal elements".

Proof. Every prime ideal containing I must contain its radical, so the only prime containing I is $\mathcal{M}$. In particular A/I is local with maximal ideal $m = \mathcal{M}/I$. So every noninvertible element of A/I lies in m, hence it is nilpotent. □

In particular this applies to powers of a maximal ideal.

Example 3.2.7.

(a) The example above of $I = (x, y^2)$ inside $A = k[x, y]$ shows that a primary ideal is not necessarily a prime power. Indeed I is primary since $A/I \cong k[y]/(y^2)$.

(b) The example of a prime power which is not primary is more subtle. First we have to take a prime ideal which is not maximal. Take $A = k[x, y]/(xy)$ and the prime $P = (\overline{x})$. Then P^2 is not primary. Indeed $\overline{xy} = 0 \in P^2$ and $\overline{x} \notin P^2$, so there should be a power of $\overline{y}$ in P^2, which is clearly not the case.

(c) In a principal ideal domain the primary ideals are exactly the prime powers, which is easily checked using factorization of elements.

Definition 3.2.8. We say that I admits a *primary decomposition* if there exist primary ideals $Q_1, \ldots, Q_n$ such that

$$I = \bigcap_{i=1}^{n} Q_i.$$

We shall see that primary decomposition always exists in Noetherian rings, and that it satisfies a sort of uniqueness (this is independent of the Noetherian hypothesis).

For Noetherian rings there is an easier concept to work with.

Definition 3.2.9. The ideal $I \subset A$ is said *irreducible* if whenever $I = J \cap K$ for two ideals J, K of A, we must have $I = J$ or $I = K$.

Proposition 3.2.10. *Let A be a Noetherian ring, I an irreducible ideal. Then I is primary.*

Proof. Working in A/I we reduce to the case where $I = 0$. Let x be a zero divisor, say $xy = 0$; we must show it is nilpotent. Consider the chain

$$\text{Ann}(x) \subset \text{Ann}(x^2) \subset \cdots;$$

this must stabilize, hence at some point $\text{Ann}(x^n) = \text{Ann}(x^{n+1})$.

We claim that $(x^n) \cap (y) = 0$. Indeed assume $ax^n \in (y)$; then $a \in \text{Ann}(x^{n+1}) = \text{Ann}(x^n)$. Since 0 is irreducible and $y \neq 0$ we must have $x^n = 0$. □

Remark 3.2.11. The converse is definitely not true. For a complete picture on the irreducibility of primary ideals, see [**Fer**, Section 9.4].

Now we have the following easy result.

Proposition 3.2.12. *Every ideal of a Noetherian ring is the intersection of finitely many irreducible ideals.*

Proof. Let X be the set of ideals which do not satisfy the conclusion, and let $I \in X$ be maximal. In particular I is not irreducible, so we can write

$$I = J \cap K$$

for bigger ideals J and K. But then J and K satisfy the conclusion, hence I does. □

Corollary 3.2.13. *Every ideal of a Noetherian ring admits a primary decomposition.*

As in the case of the unique factorization we see that the existence part is easy, and it is a simple finiteness argument. The uniqueness is more subtle, and at first it is not even clear what is the right statement.

First, it is clear that we should group together primary ideals with the same radical, as we do for prime powers is the case of unique factorization domains. This is achieved by the following

Lemma 3.2.14. *Let $Q_1, \ldots, Q_n$ be P-primary ideals. Then the intersection*

$$Q = Q_1 \cap \cdots \cap Q_n$$

is P-primary.

Proof. First note that

$$\sqrt{Q} = \sqrt{Q_1} \cap \cdots \cap \sqrt{Q_n} = P.$$

Now let $xy \in Q$ and assume $x \notin Q$. Then x does not belong to some Q_i, so $y \in \sqrt{Q_i} = \sqrt{Q}$. □

Definition 3.2.15. Let

$$I = \bigcap_{i=1}^{n} Q_i.$$

be a primary decomposition. We call it *minimal* if

(1) The radicals $P_i = \sqrt{Q_i}$ are all distinct, and

(2) no Q_i includes an intersection $Q_{i_1} \cap \cdots \cap Q_{i_k}$.

Clearly, starting from any primary decomposition we can get a minimal one. Namely, we group together all primary factors with the same radical, thanks to Lemma 3.2.14, and then simply discard the primary factors which do not satisfy 2. So we treat the uniqueness problem only for minimal decompositions.

Remark 3.2.16. With the usual factorization over $\mathbb{Z}$ the case 2 does not arise. But in more general rings it can happen that two different prime ideals are included one in the other, so redundant decompositions actually appear.

Recall that for a module M the set $\mathrm{Ass}(M)$ consists of those prime ideals of the form $\mathrm{Ann}(m)$ for some $m \in M$.

Definition 3.2.17. Let $I \subset A$ be an ideal. The *associated primes* of I are the associated primes of the module A/I.

The definition is standard, and hopefully will not cause confusion, even though I is of course a module itself.

Remark 3.2.18. For $\overline{x} \in A/I$ we have

$$\mathrm{Ann}(\overline{x}) = (I : x),$$

hence $\mathrm{Ass}(I)$ is equally described as the set of primes of the form $(I : x)$ for $x \in A$.

Theorem 3.2.19 (Uniqueness theorem, 1st form). *Let*

$$I = \bigcap_{i=1}^{n} Q_i$$

be a minimal primary decomposition, $P_i = \sqrt{Q_i}$. *Then the set* $\{P_i\}$ *is determined by* I. *More precisely the ideals* P_i *are exactly the primes of the form* $\sqrt{(I : x)}$ *for* $x \in A$.

Proof. For $x \in A$ we have

$$\sqrt{(I : x)} = \bigcap_{i=1}^{n} \sqrt{(Q_i : x)}.$$

Now if $x \in Q_i$ we have $\sqrt{(Q_i : x)} = A$, while for $x \notin Q_i$ we have $\sqrt{(Q_i : x)} = P_i$ by definition of a primary ideal. Hence this intersection can be rewritten as

$$\sqrt{(I : x)} = \bigcap_{x \notin Q_i} P_i. \tag{3.2.1}$$

If moreover $\sqrt{(I : x)}$ is prime we must have $\sqrt{(I : x)} = P_i$ for some i, by Proposition 1.1.20 ii).

In the other direction, by minimality of the primary decomposition, we can find and element x_i such that $x_i \in \bigcap_{j \neq i} Q_j$ but $x_i \notin Q_i$. Then

$$\sqrt{(I : x_i)} = P_i$$

by (3.2.1). □

Corollary 3.2.20. *Assume that A is Noetherian and let*

$$I = \bigcap_{i=1}^{n} Q_i.$$

be a minimal *primary decomposition, $P_i = \sqrt{Q_i}$. Then the set $\{P_i\} = \operatorname{Ass}(I)$.*

Proof. If $P \in \operatorname{Ass}(I)$ we have $P = (I : x)$ for some $x \in A$, and P is radical, so $P = \sqrt{(I : x)}$.

In the other direction take a prime P_i and write

$$R = \bigcap_{j \neq i} Q_i,$$

so that $I = Q_i \cap R$. Then

$$\frac{R}{I} = \frac{R}{Q_i \cap R} \cong \frac{R + Q_i}{Q_i} \subset \frac{A}{Q_i}.$$

Note that A/Q_i has the only associated prime P_i by the first part of the proof. Since $\operatorname{Ass}(R/I) \subset \operatorname{Ass}(A/Q_i)$ and is not empty, we see that P_i is an associated prime of R/I, and a fortiori of A/I. □

Note that the above results only give uniqueness of the *radicals* of the factors of a primary decomposition. We shall say that the primes P_i appearing in a minimal primary decomposition of I *belong* to I. The factors themselves are not determined by I only, as the following examples show, but there is a weak sort of uniqueness, which we shall investigate.

Example 3.2.21.

(a) It is very easy to find different minimal primary decompositions for ideals in a polynomial ring. A typical example is $I = (x^2, xy)$ in the ring $k[x, y]$, which can be written

$$I = (x) \cap (x^2, xy, y^2) = (x) \cap (x^2, y).$$

All ideals appearing in the decomposition are indeed primary: (x) is prime, while both (x^2, xy, y^2) and (x^2, y) have the radical (x, y), which is maximal.

(b) For another example in the same ring take

$$I = (y - x^2) \cap (x^2, xy, y^2) = (y - x^2) \cap (x, y^2).$$

(c) Consider[2] the non-Noetherian ring $A = k[x_1, x_2, \dots]$ and the ideal $I = (x_1^2, x_2^2, \dots)$. Then, as
$$\sqrt{I} = \mathcal{M} = (x_1, x_2, \dots),$$
which is maximal, I is $\mathcal{M}$-primary. In particular I is certainly decomposable, and $\mathcal{M}$ belongs to I. Still $\mathcal{M} \notin \operatorname{Ass}(I)$, showing that the Noetherian hypothesis in Corollary 3.2.20 is necessary.

To see this assume $\mathcal{M} = (I : f)$ for some $f \in A$, say $f = f(x_1, \dots, x_n)$. Clearly $f \notin I$; but then
$$x_{n+1} f(x_1, \dots, x_n) \notin I,$$
a contradiction.

(d) Take the ring $A = C(\mathbb{R})$ of continous functions on $\mathbb{R}$. For a closed interval D, say of length ℓ, we shall see that the ideal
$$I = \left\{ f \in A \mid f|_D = 0 \right\}$$
does not admit a primary decomposition.

First note that I is radical, hence each primary component is actually prime. Take a prime $P \supset I$. We can choose a function $f_1 \in I$ which vanishes exactly on D and is positive outside, say by taking it piecewise linear. Then write
$$f_1 = f_2 g_2,$$
where f_2 vanishes exactly on the left half of D and g_2 vanishes exactly on the left half of D. So either f_2 or g_2 belongs to P, say f_2. We can repeat the process writing
$$f_2 = f_3 g_3$$
and so on. The conclusion is that P contains functions which vanish on arbitrarily small subintervals of D.

Now assume
$$I = P_1 \cap \cdots \cap P_n.$$
For each i we can choose a function $h_i \in P_i$ which vanishes exactly on an interval of length less that ℓ/n. But then the product $h_1 \cdots h_n \in I$ and cannot be zero on the whole D, a contradiction.

Remark 3.2.22. Assume I admits a primary decomposition; then any prime containing I contains a prime belonging to I. Indeed assume
$$P \supset I = Q_1 \cap \cdots \cap Q_n;$$
then P contains some Q_i by Proposition 1.1.20 ii), and taking radicals we see that $P \supset P_i$.

[2]This example was suggested by Georges Elencwajg on MathOverflow, see [**Ele**].

In particular the minimal primes of I are the primes which are minimal among those belonging to I. (Recall that a minimal prime of I is a prime which is minimal among those containing I.)

With this language we can prove the following result, which looks somewhat unrelated to primary decomposition.

Corollary 3.2.23 (of Theorem 3.2.19). *Let D be the set of zero divisors of A. If 0 admits a primary decomposition, then D is the union of the primes belonging to 0.*

Proof. It is immediate to see that

$$D = \bigcup_{x \neq 0} \sqrt{(0 : x)}.$$

Let

$$0 = \bigcap_{i=1}^{n} Q_i$$

be a minimal decomposition of 0, and let $P_i = \sqrt{Q_i}$. Then by (3.2.1) we see that each term of the union $\sqrt{(0 : x)}$ is contained in some prime P_i.

On the other hand, any prime belonging to 0 has the form $\sqrt{(0 : x)}$, thereby proving the reverse inclusion. □

Definition 3.2.24. Let P be a prime belonging to I. If P is not minimal we say that P is an *embedded prime* of I.

It will turn out that only the primary factors corresponding to minimal components are uniquely determined. To see this it is convenient to understand how primary decomposition behaves with respect to localization.

Let S be a multiplicative set. We know that there is a bijective correspondence between prime ideals P such that $P \cap S = \emptyset$ and ideals of $S^{-1}A$, given by $P \to S^{-1}P$ (Corollary 1.6.9). We want to extend this to primary ideals.

Proposition 3.2.25. *There is a bijective correspondence between primary ideals $Q \subset A$ such that $Q \cap S = \emptyset$ and primary ideals of $S^{-1}A$.*

Proof. First we remark that for any ideal I the conditions $I \cap S \neq \emptyset$ and $\sqrt{I} \cap S \neq \emptyset$ are equivalent, since S is multiplicatively closed. So we need only consider primary ideals Q such that $\sqrt{Q} \cap S = \emptyset$.

The correspondence is the usual one. Denote by

$$\iota \colon A \to S^{-1}A$$

the canonical homomorphism. Then Q corresponds to the extended ideal $S^{-1}Q$ and in the other direction $Q' \subset S^{-1}A$ corresponds to $\iota^{-1}(Q')$.

By Proposition 1.6.6 we know that for an ideal $Q' \subset S^{-1}A$ we have $Q' = S^{-1}Q$, where $Q := \iota^{-1}(Q')$. In the other direction assume Q is primary. By the same Proposition

$$\iota^{-1}(Q \cdot S^{-1}A) = \bigcup_{s \in S} (Q : s),$$

and if $S \cap \sqrt{Q} = \emptyset$ we have $(Q : s) = Q$ by definition.

It remains to check that this correspondence sends primary ideals to primary ideals, which is easy and left to the reader. □

Remark 3.2.26. With quotients the matters are even simpler. Since the fact that Q is primary can be detected looking at the quotient A/Q, the bijective correspondence between ideals containing I and ideals of A/I restricts to a bijective correspondence between primary ideals.

For the purpose of the rest of the section, denote $\widehat{I} := \iota^{-1}(S^{-1}I)$, where

$$\iota\colon A \to S^{-1}A$$

is the canonical homomorphism.

Corollary 3.2.27. *Let I be an ideal admitting the minimal primary decomposition*

$$I = \bigcap_{i=1}^{n} Q_i,$$

and let $P_i = \sqrt{Q_i}$. Then a minimal primary decomposition for $\widehat{I}$ is

$$\widehat{I} = \bigcap_{P_i \cap S = \emptyset} Q_i.$$

Definition 3.2.28. Let X be a subset of the primes belonging to I. We say that X is *isolated* if whenever $P_1 \subset P_2$ are two prime ideals belonging to I and $P_2 \in X$, then also $P_1 \in X$.

We can now prove

Theorem 3.2.29 (Uniqueness theorem, 2nd form)**.** *Let I an ideal admitting the minimal primary decomposition*

$$I = \bigcap_{i=1}^{n} Q_i$$

and let $P_i = \sqrt{Q_i}$. For any isolated set X of primes belonging to I the intersection

$$\bigcap_{P_i \in X} Q_i$$

is independent of the decomposition. In particular if P_i is minimal, Q_i is determined by I.

Proof. Consider the multiplicative set

$$S := A \setminus \left(\bigcup_{P \in X} P \right).$$

The definition of isolated set is made so that $P_i \cap S = \emptyset$ if and only if $P_i \in X$. So by Corollary 3.2.27

$$\bigcap_{P_i \in X} Q_i = \widehat{I}$$

is determined by I. □

Having established the two uniqueness theorems, we treat the question nonuniqueness. For this purpose it is convenient to introduce a tool of independent interest.

Definition 3.2.30. Let P be a prime ideal of A and let $S := A \setminus P$. The ideal

$$P^{(n)} := \widehat{P^n} = \iota^{-1}(S^{-1}P^n)$$

is called the nth *symbolic power* of P.

These symbolic powers, among other things, give a measure of how far is P^n from being P-primary.

Proposition 3.2.31. *The symbolic power $P^{(n)}$ is P-primary. More precisely if P^n is decomposable, $P^{(n)}$ is the P-primary component of any primary decomposition of P^n.*

Proof. Since $S^{-1}P$ is maximal in A_P, $S^{-1}P^n$ is $S^{-1}P$-primary, hence its contraction $P^{(n)}$ is P-primary by Proposition 3.2.25.

Now assume

$$P^n = \bigcap_{i=1}^{k} Q_i;$$

by Corollary 3.2.27 we know that

$$P^{(n)} = \bigcap_{P_i \subset P} Q_i.$$

But the assumptions $P_i \subset P$ and $P^n \subset Q_i \subset P_i$ show, taking radicals, that $P_i = P$. Hence this intersection only has one factor Q_i for $P_i = P$, and we are done. □

The following result shows that nonuniqueness of embedded primary components is not pathological, but is really what one has to expect.

Theorem 3.2.32. *Assume A is Noetherian and let $I \subset A$ be an ideal with a minimal decomposition*

$$I = \bigcap_{i=1}^{n} Q_i;$$

as usual let $P_i = \sqrt{Q_i}$. If P_i is an embedded prime, there are infinitely many different choices for the P_i-primary component of I.

Proof. Passing to the quotient we can assume $I = 0$. We show that $P_i^{(k)} \subset Q_i$ for k big enough, and that all symbolic powers $P_i^{(k)}$ are distinct. The two claims together imply the thesis.

The first claim follows immediately by Proposition 2.3.10 applied to the extension of Q_i in the localization A_{P_i}.

For the second one, observe that A_{P_i} is not Artinian since it contains prime ideals which are not maximal (the extension of any prime strictly contained in P_i). Let $\mathcal{M}_i := P_i A_{P_i}$. Since A_{P_i} is Noetherian, no power $\mathcal{M}_i^k$ can be 0, otherwise we could apply Lemma 2.4.10. By Nakayama's lemma we deduce that

$$\mathcal{M}^{k+1} \subsetneq \mathcal{M}^k.$$

So their contractions are also distinct by Proposition 3.2.25. □

3.3. Primary decomposition for modules

We now generalize the results of the previous section to the case of submodules of a given module. Since the proofs are almost identical, we shall be a little sketchy. This generalization is fruitful: we have started studying the problem of factorization of elements and we end up with a sort of structure theorem for modules!

Our first remark is that the primes P containing an ideal I are exactly those for which the extension $I \cdot A_P$ is not trivial. In other words the localization $(A/I)_P$ should not be zero. This condition makes sense also for modules (we should remark that primary decomposition for the ideal I will correspond to primary decomposition for the module A/I). We isolate these primes with a definition.

Definition 3.3.1. Let M be an A-module. The *support* of M is the set

$$\operatorname{Supp}(M) := \{P \subset A \text{ prime} \mid M_P \neq 0\}.$$

Remark 3.3.2. Assume that M is finitely generated; then the condition $P \in \operatorname{Supp}(M)$ is equivalent to saying that $\operatorname{Ann}(M) \subset P$. In fact, $P \in \operatorname{Supp}(M)$ is equivalent to having some nonzero $m \in M_P$. This means that there is no $s \in A \setminus P$ such that $sm = 0$, that is, $\operatorname{Ann}(m) \subset P$.

If M is finitely generated, say $M = (m_1, \dots, m_r)$, then

$$\operatorname{Ann}(M) = \bigcap_{i=1}^{r} \operatorname{Ann}(m_i),$$

so some $\operatorname{Ann}(m_i) \subset P$ if and only if $\operatorname{Ann}(M) \subset P$ by Proposition 1.1.20 ii.

The support of a module behaves nicely in exact sequences. In fact, from Proposition 1.6.17 it follows immediately:

Proposition 3.3.3. *Let*

$$0 \longrightarrow M' \longrightarrow M \longrightarrow M'' \longrightarrow 0$$

be an exact sequence. Then $\operatorname{Supp}(M) = \operatorname{Supp}(M') \cup \operatorname{Supp}(M'')$.

The concept of support is strictly related to that of associated primes. Recall that in 2.5.11 we have defined associated primes for a module M as

$$\operatorname{Ass}(M) = \{P \subset A \text{ prime} \mid P = \operatorname{Ann}(m) \text{ for some } m \in M\}.$$

Moreover, for an ideal I in a Noetherian ring the minimal primes among those containing I are the minimal associated primes of A/I. We are going to work out the corresponding facts for modules. To start, the following fact is reassuring.

Lemma 3.3.4. *Any associated prime of M belongs to* $\operatorname{Supp}(M)$.

Proof. Let $P \in \operatorname{Ass}(M)$, so we have an injective homomorphism

$$A/P \hookrightarrow M.$$

Then M_P contains the residue field $k(P) = \mathcal{F}(A/P)$, so is not zero. □

Next we define primary modules. As we have remarked, the property of Q being primary can be detected by the quotient A/Q.

Definition 3.3.5. Let M be an A-module, $a \in A$. We say that a is a *zero divisor* in M if the multiplication by a

$$\begin{aligned} a\cdot\colon\ M &\longrightarrow M \\ m &\longmapsto am \end{aligned}$$

is not injective. We say that a is *nilpotent* in M if this map is nilpotent.

We say that a submodule $Q \subset M$ is *primary* if every zero divisor in M/Q is nilpotent in M/Q.

Lemma 3.3.6. *If Q is primary in M, then $(Q : M)$ is a primary ideal.*

Proof. Let $xy \in (Q : M)$ with $x \notin (Q : M)$. Then multiplication by y is a zero divisor on M/Q, so it is nilpotent. That is, a power of y is in $(Q : M)$. □

Definition 3.3.7. For a submodule $Q \subset M$ we define the *radical* of Q in M as the ideal

$$\sqrt{(Q : M)} = \{a \in A \mid a^n M \subset Q \text{ for some } n\}.$$

If Q is primary in M and $\sqrt{(Q : M)} = P$ we say that Q is P-primary. Throughout the rest of the section the ambient module M is fixed.

A *primary decomposition* for $N \subset M$ is a decomposition

$$N = Q_1 \cap \cdots \cap Q_n,$$

where the Q_i are primary submodules of M.

Definition 3.3.8. Let $N \subset M$ be a submodule. We say that N is *irreducible* if whenever $N = R \cap S$ is written as the intersection of two submodules, we must have $N = R$ or $N = S$.

Proposition 3.3.9. *Let M be a Noetherian module. Then every submodule of M is a finite intersection of irreducible submodules.*

Proof. Noetherian induction, as in the proof of Proposition 3.2.12. □

Proposition 3.3.10. *Let M be a Noetherian module. Then an irreducible submodule $N \subset M$ is primary.*

Proof. We can assume $N = 0$. Let a be a zero divisor on M, say $am = 0$. Denote ϕ_a the multiplication by a endomorphism of M. We look at the ascending chain

$$\ker \phi_a \subset \ker \phi_a^2 \subset \cdots .$$

This eventually stabilizes, so we must have $\ker \phi_a^n = \ker \phi_a^{n+1}$ for some n.

Then $a^n M \cap Am = 0$. Indeed assume $a^n m' \in Am$; then

$$m' \in \ker \phi_a^{n+1} = \ker \phi_a^n,$$

so $a^n m' = 0$. Since 0 is irreducible and $Am \neq 0$ we must have $a^n M = 0$, proving that ϕ_a is nilpotent. □

Corollary 3.3.11. *Let M be a Noetherian module. Then every submodule of M admits a primary decomposition.*

This settles the existence part. The uniqueness, again, is very similar to the case of ideals. First, it is easy to see that the intersection of two P-primary submodules is again P-primary, so we can group together all primary components with the same radical in M. Then we remove components which do not cut out anything and find a minimal primary decomposition.

Theorem 3.3.12 (Uniqueness theorem, 3rd form). *Let*

$$N = \bigcap_{i=1}^{n} Q_i.$$

be a minimal *primary decomposition,* $P_i = \sqrt{(Q_i : M)}$. *Then the set* $\{P_i\}$ *is determined by* N. *More precisely the ideals* P_i *are exactly the primes of the form* $\sqrt{(N : m)}$ *for* $m \in M$.

We omit the proof, which is identical to that of Theorem 3.3.12. There is also an equivalent to 3.2.23, again with the same proof.

Corollary 3.3.13. *Let* M *be an* A*-module, and let* D *be the set of zero divisors of* A *in* M—*that is, elements* $a \in A$ *such that there exists a nonzero* $m \in M$ *with* $am = 0$. *Assume that* $0 \in M$ *admits a minimal primary decomposition*

$$0 = \bigcap_{i=1}^{n} Q_i.$$

Then $D = \bigcup P_i$, *where* $P_i = \sqrt{(Q_i : M)}$.

Corollary 3.3.14. *Assume that* A *is Noetherian and* M *is finitely generated, and let*

$$N = \bigcap_{i=1}^{n} Q_i.$$

be a minimal *primary decomposition,* $P_i = \sqrt{(Q_i : M)}$. *Then the set* $\{P_i\} =$ $\mathrm{Ass}(M/N)$.

Proof. If $P \in \mathrm{Ass}(M/N)$ we have $P = (N : m)$ for some $m \in M$, and P is radical, so $P = \sqrt{(N : m)}$.

In the other direction, take a prime P_i and write

$$R = \bigcap_{j \neq i} Q_i,$$

so that $N = Q_i \cap R$. Then

$$\frac{R}{N} = \frac{R}{Q_i \cap R} \cong \frac{R + Q_i}{Q_i} \subset \frac{M}{Q_i}.$$

Note that M/Q_i has the only associated prime P_i. Since $\mathrm{Ass}(R/N)$ is not empty and contained in $\mathrm{Ass}(M/Q_i)$, we see that P_i is an associated prime of R/N, and a fortiori of M/N. □

Corollary 3.3.15. *If* A *is Noetherian and* M *is finitely generated, every prime* $P \in \mathrm{Supp}(M/N)$ *contains a prime associated to* M/N.

Proof. We can assume $N = 0$. Write

$$0 = \bigcap_{i=1}^{n} Q_i.$$

Since $M_P \neq 0$ there exists $m \in M$ such that $\mathrm{Ann}(m) \subset P$; in particular

$$\bigcap_{i=1}^{n} (Q_i : m) \subset P.$$

But then P contains some $(Q_i : m)$ by Proposition 1.1.20 ii), and a fortiori $P \supset (Q_i : M)$. Taking radicals we obtain $P \supset P_i$. □

Definition 3.3.16. This ensures that the primes that are minimal in the sets $\mathrm{Supp}(M/N)$ and $\mathrm{Ass}(M/N)$ are the same; we shall call them the *minimal primes* of N in M.

We keep following closely the previous section.

Proposition 3.3.17. *Let S be a multiplicative set. There is a bijective correspondence between primary submodules $Q \subset M$ such that $\sqrt{(Q : M)} \cap S = \emptyset$ and primary submodules of $S^{-1}M$.*

Proof. First note that if $\sqrt{(Q : M)} \cap S = \emptyset$, then $(Q : s) = Q$ for any $s \in S$. Now apply Proposition 1.6.16 to obtain the bijective correspondence. Again, one has to do the easy check that Q is primary if and only if $S^{-1}Q$ is. □

Let

$$\iota \colon M \to S^{-1}M$$

be the canonical homomorphism and for $N \subset M$ denote $\widehat{N} := \iota^{-1}(S^{-1}N)$.

Corollary 3.3.18. *Let $N \subset M$ be a submodule admitting the minimal primary decomposition*

$$N = \bigcap_{i=1}^{n} Q_i$$

and let $P_i = \sqrt{(Q_i : M)}$. Then a minimal primary decomposition for $\widehat{N}$ is

$$\widehat{N} = \bigcap_{P_i \cap S = \emptyset} Q_i.$$

We can now prove

Theorem 3.3.19 (Uniqueness theorem, 4th form)**.** *Let $N \subset M$ be a submodule admitting the minimal primary decomposition*

$$N = \bigcap_{i=1}^{n} Q_i$$

and let $P_i = \sqrt{(Q_i : M)}$. *For any isolated subset* X *of* $\{P_i\}$ *the intersection*

$$\bigcap_{P_i \in X} Q_i$$

is independent of the decomposition. In particular if P_i *is minimal,* Q_i *is determined by* N.

Proof. Consider the multiplicative set

$$S := A \setminus \left(\bigcup_{P \in X} P \right).$$

By definition, $P_i \cap S = \emptyset$ if and only if $P_i \in X$. So, by Corollary 3.3.18,

$$\bigcap_{P_i \in X} Q_i = \widehat{N}$$

is determined by N. □

3.4. Factorization in Dedekind rings

We now look at the class of rings where the theory of primary decomposition takes its simplest form.

Definition 3.4.1. Let A be a Noetherian integral domain. We say that A is a *Dedekind ring* (or *Dedekind domain*) if for every prime $P \subset A$ the only nonzero ideals of the ring A_P are the powers of the maximal ideal.

Remark 3.4.2. The local rings having as only nonzero ideals the powers of the maximal ideal are called discrete valuation rings (DVR). We will study them in detail in Section 7.3.

Remark 3.4.3. Since localization commutes with quotients, it is immediate to see that a nontrivial quotient of a Dedekind ring that is integral is in fact a field.

The definition above is in some sense temporary. We will give a more useful characterization of Dedekind rings in Theorem 5.1.19, after studying integral extensions. For now we are interested in Dedekind rings because the theory of factorization of ideals gets much simpler there.

Example 3.4.4.

(a) Every principal ideal domain is a Dedekind ring. Indeed, Noetherianity is clear, and the other condition follows from unique factorization.

(b) We are not able at the present point to give more examples, but we shall see in Chapter 5 that many rings of interest in arithmetics, like $\mathbb{Z}[\sqrt{19}]$ or $\mathbb{Z}[(1 + \sqrt{5})/2]$ are Dedekind rings.

We now apply the theory of primary decomposition to Dedekind rings. We start with two simple remarks.

Remark 3.4.5. Every nonzero prime ideal of a Dedekind ring is maximal. Indeed if $P \subsetneq Q$ is a proper inclusion of nonzero prime ideals, both PA_Q and QA_Q are prime, hence the first cannot be a power of the second.

Remark 3.4.6. Let $I \subset A$ be a nonzero ideal of a Dedekind ring. Since A is Noetherian, I admits a primary decomposition

$$I = Q_1 \cap \cdots \cap Q_n;$$

assume this is minimal. Since every nonzero prime of A is maximal, there cannot be embedded primes; so the primary components Q_i are uniquely determined by I.

This is good enough, but there is more.

Proposition 3.4.7. *Let P be a nonzero prime of a Dedekind ring. The P-primary ideals are exactly the powers of P.*

Proof. Since P is maximal, P^n is primary by Proposition 3.2.6.

Vice versa, assume Q is P-primary. Let

$$\iota\colon A \to A_P$$

be the localization morphism. By Proposition 3.2.25,

$$Q = \iota^{-1}(QA_P);$$

since QA_P is a power of PA_P, we are done. □

Corollary 3.4.8 (Factorization of ideals). *Let A be a Dedekind ring. Any nonzero ideal $I \in A$ can be uniquely written as*

$$I = P_1^{a_1} \cdots P_n^{a_n},$$

where the P_i are primes (up to reordering the factors).

Proof. By Proposition 3.4.7 and the previous remarks, it is clear that I can be uniquely written as an intersection

$$I = P_1^{a_1} \cap \cdots \cap P_n^{a_n}.$$

Recall from the proof of the Chinese remainder theorem that

$$J \cap K = J \cdot K$$

whenever J and K are coprime ideals, that is $J + K = A$. By induction, we only need to check that this is satisfied for $J = P_1^{a_1} \cdots P_k^{a_k}$ and $K = P_{k+1}^{a_{k+1}}$. It is an easy check that two ideals are coprime if and only if their radicals are. So we must prove that $J' := P_1 \cdots P_k$ and P_{k+1} are coprime.

Since P_{k+1} is maximal, this is true unless $J' \subset P_{k+1}$. By Proposition 1.1.20 ii) this implies that some $P_i \subset P_{k+1}$, which is a contradiction since the two are distinct and both maximal. □

In the rest of this section, we explore the consequences of unique factorization for ideals. In particular, we will obtain some results which compare Dedekind rings with unique factorization domains.

Let A be a Dedekind ring; if A is a principal ideal domain, unique factorization for ideals just becomes unique factorization for elements. This is unsurprising, since we know that every principal ideal domain is a unique factorization domain. A little less trivial is the converse.

Proposition 3.4.9. *Let A be a Dedekind ring. Then A is a principal ideal domain if and only if it is a unique factorization domain.*

Proof. One direction is clear. So assume A is both a Dedekind ring and a unique factorization domain. Let P be a nonzero prime ideal of A; then P contains a prime element p by Theorem 3.1.19. Since every prime of A is maximal, the inclusion $(p) \subset P$ must be an equality, showing that P is principal. It follows that *every* ideal of A is principal, since it can be factorized as a product of primes. □

A way to restate this is that the intersection of the classes of unique factorization domains and Dedekind rings consist exactly of the principal ideal domains. We will give a slightly stronger form in Corollary 5.1.20.

We now introduce a measure of how far a Dedekind ring is from being a principal ideal domain.

Definition 3.4.10. Let A be an integral domain, $k = \mathcal{F}(A)$ its field of fractions. Any A-submodule of k of the form $I/a := \frac{1}{a} \cdot I$, for some ideal $I \subset A$ and $a \in A$, is called a *fractional ideal*. The fractional ideal I/a is called *principal* if I is.

Remark 3.4.11. Assume $M \subset k$ is a finitely generated A-submodule. Then there is a common denominator for elements in M, so for some $a \in A$ we have $aM \subset A$. Then $I := aM$ is an ideal of A, so $M = \frac{1}{a}I$ is a fractional ideal.

The converse holds assuming A is Noetherian. So for Noetherian rings we can think of fractional ideals as finitely generated submodules of $\mathcal{F}(A)$.

Fractional ideals of A can be multiplied by defining

$$(I/a) \cdot (J/b) := IJ/ab.$$

With this operation, fractional ideals form a commutative monoid, whose unit is the ring A itself. Principal ideals form a submonoid.

In the case of Dedekind rings, things are better behaved.

Proposition 3.4.12. *Let A be a Dedekind ring, $I \subset J$ ideals. Then there exists an ideal K such that $I = JK$.*

Proof. Using factorization for I and J it is enough to prove the following claim. Factorize I as

$$I = P_1^{a_1} \cdots P_k^{a_k}$$

and let P be a prime, say $P = P_1$ (here we allow a_1 to be 0). If $I \subset P^b$, then $b \leq a_1$.

To see this, remark that every prime P_j with $j \neq 1$ is not contained in P by maximality. So all such primes become trivial in the localization A_P, and

$$IA_P = P^{a_1} A_P.$$

Since this is contained in $P^b A_P$, we have $b \leq a_1$. If this was not the case, we would have

$$P^{a_1+1} A_P = P^{a_1} A_P,$$

which by Nakayama's lemma gives $P^{a_1} A_P = 0$; this is a contradiction since A_P is an integral domain. □

Corollary 3.4.13. *Let A be a Dedekind ring. For every ideal $I \subset A$ there exists another ideal J such that IJ is principal.*

Proof. Take any nonzero $a \in I$. Then $(a) \subset I$ and we can apply the proposition. □

The important consequence of this corollary is that if A is a Dedekind ring, the quotient of the monoid of fractional ideals by the submonoid of principal ideals is actually a commutative *group*. This group is an important invariant of A, and is called the *ideal class group* of A.

We shall denote the ideal class group of A by $G(A)$. If it is finite, its order is denoted $h(A)$.

By Proposition 3.4.9, $G(A)$ is trivial if and only if A is a principal ideal domain, or equivalently A is a unique factorization domain.

Remark 3.4.14. As trivial as it can seem, the following observation is very important. If A is Dedekind ring, and I is an ideal of order k in $G(A)$, I^k is a principal ideal. In particular, if we know that $G(A)$ is finite, many questions about ideals in A can be reduced to the case of principal ideals.

Another result which gives a hint how far Dedekind rings are from principal ideal domains is the following.

Proposition 3.4.15. *Let A be Dedekind ring, $I \subset A$ a nonzero ideal. Then A/I is a principal ideal domain.*

Proof. Let J/I be an ideal of A/I; we must prove that it is principal. Using unique factorization for ideals we can assume J is prime. Since $I \subset J$, we know that J divides I by Proposition 3.4.12.

Factorize I as

$$I = P_1^{a_1} \cdots P_k^{a_k},$$

say $J = P_1$. By the Chinese remainder theorem we can find $x \in A$ such that

$$\begin{aligned} x &\equiv 0 \pmod{P_1} \\ x &\not\equiv 0 \pmod{P_1^2} \\ x &\equiv 1 \pmod{P_2} \\ &\vdots \\ x &\equiv 1 \pmod{P_k}. \end{aligned}$$

We claim $P_1 = I + (x)$, from which the thesis follows. The inclusion $I + (x) \subset P_1$ is clear. On the other hand, by Proposition 3.4.12, $I + (x)$ divides both I and (x); the factorization of I and the choice of x show that the only common factor of these two ideals is indeed P_1. □

Corollary 3.4.16. *Every ideal in a Dedekind ring is generated by at most two elements.*

Proof. Let J be a nonzero ideal and take any nonzero $x \in J$. Applying the proposition to (x) we see that the ideal $J/(x)$ in $A/(x)$ is generated by a single element $\overline{y}$, which we can lift to $y \in J$. Then $J = (x, y)$. □

Remark 3.4.17. More precisely, we see from the proof that given any nonzero $x \in I$, we can find $y \in I$ such that $I = (x, y)$.

3.5. The structure of modules over Dedekind rings

We now see, following [**Sch03**], how to generalize the classification of finitely generated modules over a principal ideal domain to the case of Dedekind rings. We start by understanding what ideals look like as modules.

Remark 3.5.1. Let I, J be two ideals of the Dedekind ring A. If $\overline{I} = \overline{J}$ in $G(A)$, we have $J = xI$ for some $x \in \mathcal{F}(A)$. Hence I and J are isomorphic as A-modules.

Lemma 3.5.2. *Let A be a Dedekind ring, I and J two nonzero ideals. Then*

$$I \oplus J \cong A \oplus IJ$$

as A-modules.

Proof. First assume $I + J = A$. Then the natural map

$$t\colon\ I \oplus J \longrightarrow A$$
$$(i,j) \longmapsto i + j$$

is surjective. Moreover $\ker t \cong I \cap J$. But

$$I \cap J \cong IJ$$

since I and J are coprime, hence we have the short exact sequence

$$0 \longrightarrow IJ \longrightarrow I \oplus J \longrightarrow A \longrightarrow 0\,.$$

We only need to check that this exact sequence splits, that is, there exist an A-linear map $s\colon A \to I \oplus J$ such that $t \circ s = \mathrm{id}_A$. But this is clear since A is free as an A-module.

For the general case, using the above remark, it is enough to show that given any two ideals I and J there exists an ideal J' with same class of J in $G(A)$, and such that I and J' are coprime.

Start from any $x \in J$; by Proposition 3.4.12 there exists an ideal K such that $JK = (x)$. Moreover by Proposition 3.4.15 there exists $y \in A$ such that

$$IK + (y) = K.$$

Multiplying by J we get

$$IJK + (y)J = JK = (x). \tag{3.5.1}$$

Again by Proposition 3.4.12 we have

$$(y)J = (x)J'$$

for some ideal J'. This last relation tells us that J and J' are in the same class in $G(A)$.

Plugging this last relation into (3.5.1) yields

$$(x)I + (x)J' = (x),$$

whence $I + J' = A$. □

Recall from Exercise 13 of Chapter 1 that an A-module is called *projective* if it is a direct summand of a free module. As shown in that exercise, projective modules are equivalently characterized by a universal property. In particular this implies that if

$$0 \longrightarrow M \longrightarrow N \longrightarrow P \longrightarrow 0$$

is a short exact sequence and P is projective, the sequence splits, so $N \cong M \oplus P$.

Proposition 3.5.3. *Let A be Dedekind ring. Every ideal of A is a projective A-module.*

Proof. Let $I \subset A$ be an ideal; we can assume $I \neq 0$. Take any $x \in I$; by Proposition 3.4.12 there exists an ideal J such that $IJ = (x)$. Then by Lemma 3.5.2 we have

$$I \oplus J \cong A \oplus (x) \cong A \oplus A. \qquad \square$$

Before going into the proof of the main result, we need to generalize our notion of rank to modules which are not free.

Definition 3.5.4. Let A be an integral domain, $k := \mathcal{F}(A)$, and let M be an A-module. We define the *rank* of M as the dimension of the k-vector space

$$M \otimes_A k.$$

Remark 3.5.5. This notion of course agrees with our previous definition of rank, but note that it is only defined for integral domains.

Remark 3.5.6. If M is finitely generated we have

$$\operatorname{rk} M = \dim_k M \otimes_A k = \dim_k \operatorname{Hom}_A(M, k),$$

since the latter vector space is dual of the former.

Lemma 3.5.7. *Let A be a Dedekind ring, M a finitely generated torsion-free A-module. Then M is isomorphic to nonzero ideal if, and only if,* $\operatorname{rk} M = 1$.

Proof. Let I be a nonzero ideal; for $x \in I$ we have

$$(x) \subset I \subset A.$$

Since $(x) \cong A$ as A-modules, we have

$$1 = \operatorname{rk}(x) \leq \operatorname{rk} I \leq \operatorname{rk} A = 1.$$

In the other direction assume $\operatorname{rk} M = 1$. In particular there exists a nontrivial homomorphism

$$\phi\colon M \to k.$$

Since M is finitely generated, the image of ϕ is a fractional ideal; multiplying ϕ by a constant we can assume $\phi(M) \subset A$ is an ideal, call it I.

The exact sequence

$$0 \longrightarrow \ker(\phi) \longrightarrow M \longrightarrow I \longrightarrow 0$$

splits, since I is projective, so

$$\operatorname{rk}\ker(\phi) + \operatorname{rk} I = \operatorname{rk} M;$$

by the first part of the proof we conclude that $\operatorname{rk}\ker(\phi) = 0$. Since $\ker(\phi)$ is torsion-free, it must be zero, hence $M \cong I$. $\square$

Example 3.5.8. The lemma is false if we don't assume that M is finitely generated. For instance, $\mathbb{Q}$ has rank 1 as a $\mathbb{Z}$-module, but of course it is not isomorphic to any ideal. To compute the rank of $\mathbb{Q}$, just note that the map

$$m\colon \mathbb{Q} \otimes_{\mathbb{Z}} \mathbb{Q} \to \mathbb{Q}$$

given by $m(a \otimes b) = ab$ is an isomorphism.

We are now ready to generalize the classification of finitely generated modules over principal ideal domains, from Exercise 2 of Chapter 2, and its consequence, Theorem 2.1.7.

Theorem 3.5.9. *Let A be a Dedekind ring, M a finitely generated module which is isomorphic to a direct sum of ideals. For every submodule $N \subset M$ we can find $m_1, \dots, m_n \in M$ and ideals $J_k \subset I_k$ for $k = 1, \dots, n$ such that*

$$M \cong I_1 m_1 \oplus \cdots \oplus I_n m_n$$
$$N \cong J_1 m_1 \oplus \cdots \oplus J_n m_n.$$

Proof. As in the proof of Lemma 3.5.7, we can find a homomorphism

$$\phi\colon M \to A$$

with image an ideal I; let $K := \ker \phi$, $J := \phi(N)$.

We have a commutative diagram of exact sequences

$$\begin{array}{ccccccccc} 0 & \longrightarrow & K \cap N & \longrightarrow & N & \longrightarrow & J & \longrightarrow & 0 \\ & & \downarrow & & \downarrow & & \downarrow & & \\ 0 & \longrightarrow & K & \longrightarrow & M & \longrightarrow & I & \longrightarrow & 0. \end{array}$$

Moreover, the rows are split exact sequences by Proposition 3.5.3.

First we argue that N is itself isomorphic to a direct sum of ideals. This is done by induction on $\operatorname{rk} N$, the induction basis being trivial. Since the first row splits,

$$\operatorname{rk}(K \cap N) = \operatorname{rk} N - 1$$

by Lemma 3.5.7. By induction, $K \cap N$ is isomorphic to a direct sum of ideals, so N is.

We then prove the thesis by induction on $\operatorname{rk} M$. As above we have

$$\operatorname{rk} K = \operatorname{rk} M - 1,$$

and by the first part of the proof K is isomorphic to a direct sum of ideals. By induction, $K \cap N \subset K$ satisfies the thesis. Since the rows are split, the thesis for N follows. □

Corollary 3.5.10. *Let A be a Dedekind ring, M a finitely generated A-module. Then $M \cong T \oplus P$, where T is torsion and P is projective. More precisely*

$$T \cong A/I_1 \oplus \cdots \oplus A/I_k$$

for some ideals $I_j \subset A$, and

$$P \cong A^r \oplus L$$

for some ideal L and some $r \geq 0$.

Of course here L can be A itself.

Proof. By Theorem 3.5.9 we know that M is a direct sum of modules of the form I/J for some ideals $J \subset I \subset A$.

If $J \neq 0$ we have that

$$I = J + (x)$$

for some $x \in A$ by Proposition 3.4.15. Hence

$$\frac{I}{J} \simeq \frac{J+(x)}{J} \cong \frac{(x)}{(x) \cap J};$$

the latter is a quotient of (x), hence isomorphic to a quotient of A. All this summands go into T.

The remaining summands are ideals of A, and we can group them together by Lemma 3.5.2, to obtain a module of the form $A^r \oplus L$. □

Corollary 3.5.11. *Let A be a Dedekind ring, M a finitely generated A-module. Then M is projective if, and only if, it is torsion-free.*

3.6. Exercises

1. Let k be a field of characteristic 0 and let $A = k[x^2, x^3]$. Prove that every element of A is a product of irreducible elements; still A is not a unique factorization domain because there exists $f \in A$ irreducible but not prime.

2. Let A be a unique factorization domain, $S \subset A$ a multiplicative set. Prove that $S^{-1}A$ is itself a unique factorization domain.

3. Find a subring of $k[x_1, \ldots, x_n]$ that is not a unique factorization domain.

4. Prove *Nagata's lemma*: let A be an integral domain such that every element of A is a finite product of irreducible elements, $S \subset A$ a multiplicative set generated by prime elements. Then A is a unique factorization domain if and only if $S^{-1}A$ is a unique factorization domain. Derive from this another proof of Theorem 3.1.15.

5. Let k be a field. The only homomorphisms

$$\phi\colon k[x,y]/(y^2 - x^3 - x) \to k[t]$$

are those given by $\phi(\overline{f}) = f(\alpha, \beta)$, where $\alpha, \beta \in k$ are such that

$$\beta^2 = \alpha^3 + \alpha.$$

6. Let A be a Noetherian ring and assume $a \in A$ is not contained in any minimal prime of A. If $ab = 0$, then b is nilpotent.

7. Let k be a field and $r \in k(x)$ a rational function, say $r(x) = a(x)/b(x)$, where a and b are coprime polynomials. Show that the element $a(y) - rb(y)$ is irreducible over the ring $A := k[r]$, and that x is a root for it.

As a weaker condition than unique factorization, one can consider rings such that every (nonzero, noninvertible) element can be written (not uniquely) as a product of irreducible ones. A ring that satisfies this condition is called an *atomic ring*. One may expect that this condition is equivalent to the ascending chain condition for *principal* ideals (ACCP), that is, condition 2 in Definition 3.1.3. In fact, being atomic is a strictly weaker condition. In the next exercises, up to Exercise 13, we will give a counterexample from [**Gra74**]. For more on atomic rings, including other examples, see also [**Zak82**].

8. Let A be a ring that satisfies ACCP. Show that A is atomic.

9. Let M be the additive monoid generated by all rationals of the form $\frac{1}{2^k p_k}$, where $p_0 = 3$, $p_1 = 5$, and so on, is an enumeration of the odd primes. Show that each element $m \in M$ can be written uniquely as

$$m = \frac{a}{2^u} + \frac{n_0}{3} + \cdots + \frac{n_s}{2^s p_s}, \tag{3.6.1}$$

where $a, u, n_i \in \mathbb{N}$, $n_i < p_i$ and either $s = 0$ or $n_s \neq 0$. Deduce that for each $m \in M$ there is a maximum dyadic rational (that is, a rational of the form $a/2^k$) $\psi(m)$ such that $\psi(m) \leq m$.

10. Let k be a field of characteristic 0, and consider the set T of elements having the form

$$f_0 + \sum_{i=1}^{n} f_i x^{m_i}, \tag{3.6.2}$$

where $m_i \in M$ and $f_i \in k$. Show that T is a ring and the set $S \subset T$ of elements having $f_0 \neq 0$ is a multiplicative set.

11. Let T, S be as in the previous exercise and consider the ring $A = S^{-1}T$. Show that each element of the form $x^{\frac{1}{2^k p_k}}$ is an irreducible element of A.

12. Let A be the ring of the previous exercise. Show that A *does not* satisfy ACCP.

13. Let A be the ring of Exercise 11. Show that A is atomic. (Take any element $a/b \in A$, with a in form (3.6.2)—we want to factorize it as a product of irreducibles. By factoring out a suitable element, show that it is not restrictive to assume that $\psi(m_i) = 0$ for some m_i, where ψ is the function of Exercise 9. Writing m_i as in (3.6.1), show that a factorization for a contains at most $n_0 + \cdots + n_s$ nontrivial factors.)

14. Let A be a Dedekind ring. Prove that the principal fractional ideals form a free abelian group. Deduce that $\mathbb{Q}^*/\pm 1$ and $k(x)^*/k^*$ are free abelian groups.

15. Let A be a Dedekind ring, I and J two ideals. Prove that if A/I is finite

$$\#(A/I) = \#(J/IJ).$$

16. Let $I_1, \ldots, I_n \subset A$ be ideals such that $I_i + I_j = A$ for all distinct pairs of i, j. Show that $\bigcap_{k=1}^n I_k = \prod_{k=1}^n I_k$.

17. Let A be an integral domain. The following conditions are equivalent:

(a) A is a Dedekind ring;

(b) every nonzero ideal of A factorizes uniquely as a product of prime ideals;

(c) the monoid $G(A)$ of fractional ideals modulo principal fractional ideals of A is a group.

The two following exercises are taken from the beautiful presentation [**GT02**].

18. Assume you have a solution to the Fermat's equation

$$a^n + b^n = c^n,$$

where $a, b, c \in k[x]$ for some field k of characteristic 0. Prove that either all polynomials are multiple of each other, or else $n \leq 2$. (Derive the equation and rearrange to find that a^{n-1} divides $cb' - bc'$, then make this symmetric.)

19. By the same strategy as the previous exercise, prove the following theorem of Mason. Let k be a field of characteristic 0. For a polynomial $f \in k[t]$, let $r(f)$ the number of distinct roots of f in an algebraic closure of k.

Let $a, b, c \in k[x]$ be polynomials without common roots such that $a+b = c$; then $\max\{\deg a, \deg b, \deg c\} \leq r(abc)$. The generalization of this fact to integer numbers is the famous *abc conjecture*; see [**GT02**].

20. Let A be a Noetherian ring, $I \subset A$ an ideal. Prove that if $IA_P = 0$ for all primes $P \in \operatorname{Ass}(A)$, then $I = 0$.

21. Let A be a Noetherian ring, M a finitely generated A-module, and S a multiplicative subset of A. Prove that there is a natural bijection between $\operatorname{Ass}_{S^{-1}A}(S^{-1}M)$ and $\operatorname{Ass}_A(S^{-1}M)$, and that in fact the latter is the set of primes in $\operatorname{Ass}_A(M)$ that do not meet S.

22. Prove some analogue of Proposition 3.3.3 for associated primes. Namely, let

$$0 \longrightarrow M' \longrightarrow M \longrightarrow M'' \longrightarrow 0$$

be an exact sequence of A-modules. Then $\operatorname{Ass}(M) \subset \operatorname{Ass}(M') \cup \operatorname{Ass}(M'')$. Find a counterexample to show that we do not always have equality.

23. Let A be a Noetherian ring, $I \subset A$ an ideal, not necessarily prime. Define the nth *symbolic power* of I as the intersection

$$I^{(n)} := \bigcap_{P \in \operatorname{Ass}(I)} \iota_P^{-1}(I^n A_P),$$

where P runs through the associated primes of I, and $\iota_P\colon A \to A_P$ is the localization map. Prove that this definition agrees with our previous Definition 3.2.30 when I is a prime ideal.

24. Let A be a Noetherian ring, $I \subset A$ an ideal. Prove that for any $m \geq 1$, $I^m \subset I^{(m)}$. Moreover, for $m, n \geq 1$, $I^m \subset I^{(n)}$ if and only if $m \geq n$ ([**BDRH$^+$09**, Lemma 8.1.4]).

25. Let M, N be two finitely generated A-modules. Prove that

$$\operatorname{Supp}(M \otimes N) = \operatorname{Supp}(M) \cap \operatorname{Supp}(N).$$

Chapter 4

Computational Methods

In this chapter we introduce some computational techniques that allow us to work out explicitly some interesting examples. These techniques are most well-suited for polynomial rings and their quotients. This chapter is indeed composed of two mostly independent parts.

The first one introduces the resultant. This is a tool that allows one to find the common factors of two polynomials in $A[x]$, where A is a unique factorization domain. When we take $A = k[x_1, \ldots, x_n]$ a polynomial ring itself, the resultant is helpful to find the common zeroes of a set of polynomials in an inductive way. The resultant is the core technique of classical elimination theory, which is concerned with solving systems of polynomial equations by eliminating one variable at a time. In Section 4.2, we give a link between the resultant and some invariants of field extensions.

In the second part, we introduce Gröbner bases. These are special sets of generators for an ideal I of a polynomial ring, which make computable the operation of deciding whether a given element $f \in I$. Most of the foundations on Gröbner bases were laid in Buchberger's thesis [**Buc65**], and they have become the most used tool in computational algebra since then. In Section 4.3 we give the definition and study the simplest properties; in particular we solve the *ideal membership* problem. Then in Section 4.4 we show that Gröbner bases can be used to solve some other common problems in commutative algebra, in particular getting a different approach to elimination theory. Our presentation of this part is inspired by [**CLO96**] and [**KR00**].

A word of caution about the implementation of the algorithms. For brevity, we shall be a bit sloppy when we describe algorithms. In particular, we will only give a textual description, without a pseudocode implementation. We leave to the reader the task of formalizing and implementing the algorithms.

Moreover, we don't strive for effectiveness. The algorithms that we give are guaranteed to terminate, but we do not make any study of the expected (or worst case) time of execution. We do not even try to give the quickest and most efficient known algorithms, and we shall be content of presenting the basics.

4.1. The resultant

Let A be a unique factorization domain. We know by Theorem 3.1.15 that $A[x]$ is again a unique factorization domain, so it makes sense to ask, given $f, g \in A[x]$, whether they have a common divisor of positive degree in x.

This is essentially a problem in linear algebra. Let $m := \deg f$, $n := \deg g$ and write

$$\begin{aligned} f(x) &= a_m x^m + \cdots + a_1 x + a_0 \\ g(x) &= b_n x^n + \cdots + b_1 x + b_0. \end{aligned}$$

The whole idea of the resultant stems from the simple remark that f and g have a common factor of positive degree if and only if their lowest common multiple has smaller degree than their product. Or, equivalently, if and only if they have *any* common multiple of degree smaller then their product.

So we ask if we can find nonzero polynomials $r, s \in A[x]$, with $\deg r < n$ and $\deg s < m$, such that

$$f(x)r(x) + g(x)s(x) = 0. \tag{4.1.1}$$

Let $k = \mathcal{F}(A)$ be the field of fractions of A; then equation (4.1.1) has solutions $r, s \in A[x]$ if and only if it has solutions $r, s \in k[x]$, because we can always clear denominators. But finding solutions in $k[x]$ is a problem in linear algebra.

Namely, we can assign indeterminate coefficients to r and s and observe that (4.1.1) becomes a linear system in these indeterminates. Explicitly, the matrix of this system is just

$$\mathrm{Syl}(f,g) := \begin{pmatrix} a_m & \cdots & a_1 & a_0 & 0 & & \cdots & & 0 \\ 0 & a_m & \cdots & a_1 & a_0 & 0 & \cdots & & 0 \\ & & & \vdots & & & & & \\ 0 & & \cdots & & 0 & a_m & \cdots & a_1 & a_0 \\ b_n & & \cdots & b_1 & b_0 & 0 & & \cdots & 0 \\ 0 & b_n & & \cdots & b_1 & b_0 & 0 & \cdots & 0 \\ & & & \vdots & & & & & \\ 0 & & \cdots & 0 & b_n & & \cdots & b_1 & b_0 \end{pmatrix},$$

where the a_i appear in n lines and the b_i in m lines.

Definition 4.1.1. The matrix $\mathrm{Syl}(f,g)$ is called the *Sylvester matrix* of f and g. Its determinant is the *resultant* of f and g, and is denoted $R(f,g)$.

Since a homogeneous square linear system over a field has a nontrivial solution if and only if the associated matrix has determinant 0, we deduce from our earlier remarks:

Theorem 4.1.2. *The polynomials f and g have a common factor in $A[x]$ of positive degree if and only if $R(f,g)=0$.*

The theorem has some interesting corollaries when one takes A to be a polynomial ring itself. The first observation is that the resultant of f and g is just a polynomial in A, whose coefficients can be explicitly written in terms of those of f and g. These expressions do not depend on the ring where we are working, hence we have the following specialization result.

Proposition 4.1.3. *Let A be a unique factorization domain, take $f,g \in A[x,y]$ and let $R = R(f,g) \in A[x]$ denote the resultant with respect to the y variable. Write*

$$\begin{aligned} f(x,y) &= a_m(x)y^m + \cdots + a_1(x)y + a_0(x) \\ g(x,y) &= b_n(x)y^n + \cdots + b_1(x)y + b_0(x). \end{aligned}$$

Finally let $\alpha \in A$. Assume $a_m(\alpha) \neq 0$ and $b_n(\alpha) \neq 0$; then the resultant of $f(\alpha,y)$ and $g(\alpha,y)$ is $R(\alpha)$.

Remark 4.1.4. The request that $a_m(\alpha) \neq 0$ and $b_n(\alpha) \neq 0$ is necessary to be sure that the degrees of $f(\alpha,y)$ and $g(\alpha,y)$ do not drop; if the degrees change, the expression for the coefficients of the resultant also changes.

Corollary 4.1.5. *With the same notation, $f(\alpha,y)$ and $g(\alpha,y)$ have a common factor of positive degree in y if and only if $R(f,g)(\alpha)=0$.*

By induction, this result can be generalized to specialization of more than one variable. We leave this to the reader, and only note the following particular case.

Corollary 4.1.6. *Let k be an algebraically closed field, $f, g \in k[x_1, \dots, x_s, y]$, and consider the resultant $R(f,g) \in k[x_1, \dots, x_s]$ with respect to the y variable. Let $\alpha_1, \dots, \alpha_s \in k^s$ such that*

$$R(f,g)(\alpha_1, \dots, \alpha_s) = 0.$$

Then either

(i) *the highest coefficient of f or g vanishes at $\alpha_1, \dots, \alpha_s$, or*

(ii) *$f(\alpha_1, \dots, \alpha_s, y)$ and $g(\alpha_1, \dots, \alpha_s, y)$ have a common root in k.*

Proof. If the degree of f and g does not drop, the resultant of $f(\alpha_1, \dots, \alpha_s, y)$ and $g(\alpha_1, \dots, \alpha_s, y)$ is $R(f,g)(\alpha_1, \dots, \alpha_s)$. This vanishes if and only if $f(\alpha_1, \dots, \alpha_s, y)$ and $g(\alpha_1, \dots, \alpha_s, y)$ have a common factor of positive degree. Since k is algebraically closed, this is equivalent to having a common root. □

The above corollary can be used to solve the *main problem of elimination theory*: given the system of polynomial equations

$$\begin{cases} f_1(x_1, \dots, x_n) = 0 \\ \vdots \\ f_m(x_1, \dots, x_n) = 0 \end{cases}$$

find the solutions in k^n, by eliminating one variable at a time. If we let $I = (f_1, \dots, f_m)$, then the problem reduces to finding generators for $I \cap k[x_1, \dots, x_{n-1}]$. If we are able to do this, we eventually reduce to a problem in $k[x_1]$, solve for one variable, and finally backtrack to find the general solution.

We will only consider the case $m = n = 2$, and leave the general case to Exercise 2. In this case, the system becomes just

$$\begin{cases} f(x,y) = 0 \\ g(x,y) = 0, \end{cases}$$

and we can apply Corollary 4.1.6. If (x_0, y_0) is a solution of the system, then $f(x_0, y)$ and $g(x_0, y)$ have a common linear factor, hence x_0 is a root of $R(f,g)$.

Example 4.1.7. Consider the system given by

$$\begin{cases} f(x,y) = x^2y^2 - 25x^2 + 9 = 0 \\ g(x,y) = 4x + y = 0. \end{cases}$$

Then $R(f,g) = 16x^4 - 25x^2 + 9$, which is biquadratic and has roots $x = \pm 3, \pm 4$. Setting these values for x in g and solving for y, we find the four solutions $(1,-4)$, $(-1,4)$, $(\frac{3}{4},-3)$ and $(-\frac{3}{4},3)$.

The next results give some alternative expressions for the resultant. They are all based on the slogan that to prove an identity between polynomials, it is enough to prove it in a universal case.

Proposition 4.1.8. *Let $f, g \in A[x]$ be polynomials over the UFD A, and assume that we can factorize*

$$f(x) = a(x-\alpha_1)\cdots(x-\alpha_m)$$
$$g(x) = b(x-\beta_1)\cdots(x-\beta_n)$$

in $A[x]$ (the roots can be repeated). Then

$$R(f,g) = a^n b^m \cdot \prod_{i,j}(\alpha_i - \beta_j).$$

Proof. It is enough to prove the identity in the ring

$$A' = A[x_1,\ldots,x_m,y_1,\ldots,y_n],$$

where $\alpha_i = x_i$ and $\beta_j = y_j$. Indeed, once we know this, the identity remains true when we specialize x_i to α_i and y_j to β_j, thanks to Proposition 4.1.3.

To prove this, let $k = \mathcal{F}(A)$ and isolate one of the variables, say y_1. We regard the resultant $R(f,g)$ as an element of $k(x_1,\ldots,x_m,y_2,\ldots,y_n)[y_1]$. As a polynomial in y_1 it becomes 0 for $y_1 = x_i$, hence it must be divisible by $y_1 - x_i$ in that ring. By Gauss's lemma it remains divisible by $y_1 - x_i$ in A'.

Since the same is true for $y_2,\ldots,y_n$ we conclude that $R(f,g)$ is divisible by $\prod_{i,j}(x_i - y_j)$. Both this product and $R(f,g)$ are polynomials of total degree n in the set of indeterminates $\{x_i\}$, and m in the $\{y_j\}$, so

$$R(f,g) = c\prod_{i,j}(x_i - y_j)$$

for some $c \in A$. Finally comparing the coefficient of any given monomial we find $c = a^n b^m$. □

Corollary 4.1.9. *With notation as above we have*

$$R(f,g) = a^n \prod_{i=1}^{m} g(\alpha_i).$$

Corollary 4.1.10. *For any $f, g, h \in A[x]$ we have*

$$R(f,gh) = R(f,g)\cdot R(f,h).$$

Proof. It is enough to prove the identity in a bigger ring. So we can substitute A by the algebraic closure of $\mathcal{F}(A)$. Hence we can assume that f, g and h factor completely.

In this case the assertion follows by Proposition 4.1.8. $\square$

The above corollary is very useful, because computations with determinants of high rank easily become very cumbersome.

Assuming that f is monic, we can give yet another characterization of the resultant. In this case ring $B = A[x]/(f)$ is a free A-module with basis $\mathcal{B} = \{1, x, \ldots, x^{m-1}\}$: on the one hand, one can use f to express any monomial of higher degree in terms of this $\mathcal{B}$; on the other hand, elements of $\mathcal{B}$ are independent over A, since any dependence relation would yield a polynomial of degree smaller than f that becomes 0 in B.

Any A-linear endomorphism of B is then represented by a matrix, hence has a well defined determinant, which a posteriori is independent of the choice of a basis. Let

$$\phi_g \colon B \to B$$

be the homomorphism given by multiplication by g.

Proposition 4.1.11. *With the above notation, we have*

$$R(f, g) = \det \phi_g.$$

Proof. As in the previous proof, we can substitute A by the algebraic closure of $\mathcal{F}(A)$, hence we can assume that g factors completely. Note that both members of the identity are multiplicative in g: the left-hand side by Corollary 4.1.10, and the right-hand side thanks to the properties of the determinant and the fact that $\phi_{gh} = \phi_g \circ \phi_h$.

So it is enough to prove the assertion when g has degree 1, say $g(x) = bx + c$. As usual set

$$f(x) = x^m + a_{m-1}x^{m-1} + \cdots + a_1 x + a_0.$$

Since

$$\phi_g(x^i) = bx^{i+1} + cx^i$$

for $i \leq m-2$ and

$$\begin{aligned}\phi_g(x^{m-1}) &= bx^m + cx^{m-1} \\ &= (c - ba_{m-1})x^{m-1} - ba_{m-2}x^{m-2} - \cdots - ba_0,\end{aligned}$$

a matrix for ϕ_g is the transpose of

$$M = \begin{pmatrix} c & b & 0 & & \cdots & & 0 \\ 0 & c & b & & \cdots & & 0 \\ \vdots & & & & & & \\ 0 & & \cdots & 0 & c & & b \\ -ba_0 & & \cdots & & -ba_{m-2} & & c - ba_{m-1} \end{pmatrix}.$$

We have to prove that $\det M$ is the same as the determinant of

$$\mathrm{Syl}(f,g) = \begin{pmatrix} 1 & a_{m-1} & & & \cdots & a_1 & a_0 \\ b & c & 0 & & \cdots & & 0 \\ 0 & b & c & 0 & \cdots & & 0 \\ \vdots & & & & & & \\ 0 & & \cdots & & 0 & b & c \end{pmatrix}.$$

Write the last row of M as

$$-b(a_0, \ldots, a_{m-1}) + (0, \ldots, 0, c).$$

Using the multilinearity of the determinant we obtain

$$\det M = c^m - b\det(N),$$

where

$$N = \begin{pmatrix} c & b & 0 & & \cdots & & 0 \\ 0 & c & b & & \cdots & & 0 \\ \vdots & & & & & & \\ 0 & & \cdots & 0 & c & & b \\ a_0 & & \cdots & & a_{m-2} & & a_{m-1} \end{pmatrix}.$$

By Laplace expansion on the first column of the Sylvester matrix, we get

$$\det \mathrm{Syl}(f,g) = c^m - b\det(N),$$

hence the thesis. □

Remark 4.1.12. It is not a priori clear that $\det \phi_g$ is multiplicative in f as well; this would simplify the proof. Anyway, this follows as a byproduct.

4.2. Discriminants

In this section we look at a particular case of resultant, called the discriminant. Then we review some constructions from field theory, and we see how the discriminant of a basis of a field extension gives a generalization of this concept.

Recall that for a polynomial over any ring we can define the formal derivative by requiring that

$$\frac{d}{dx}x^k = kx^{k-1}$$

for monomials, and then extending by linearity.

Definition 4.2.1. Let A be an integral domain, $f \in A[x]$, say

$$f(x) = a_n x^n + a_{n-1}x^{n-1} + \cdots + a_0,$$

with $a_n \neq 0$. Then the *discriminant* of f is defined as

$$\operatorname{disc}(f) := (-1)^{\binom{n}{2}} \frac{1}{a_n} R(f, f').$$

Remark 4.2.2. The change of sign and the fraction may appear arbitrary, but this is the established definition.

As an example, let us compute the discriminant of a quadratic polynomial $f(x) = ax^2 + bx + c$ over a field of characteristic $\neq 2$. The derivative is $f'(x) = 2ax + b$, hence the Sylvester matrix is

$$\operatorname{Syl}(f, g) := \begin{pmatrix} a & b & c \\ 2a & b & 0 \\ 0 & 2a & b \end{pmatrix},$$

so

$$R(f, f') = ab^2 - 2a(b^2 - 2ac) = -a(b^2 - 4ac),$$

and finally $\operatorname{disc}(f) = b^2 - 4ac$, which is the familiar formula.

In general, note that if $n \neq 0$ in A (for instance if A has characteristic 0), the discriminant is a homogeneous polynomial of degree $2n - 2$ in the coefficients of f.

Proposition 4.2.3. *Let $f \in A[x]$, say*

$$f(x) = a_n x^n + a_{n-1}x^{n-1} \cdots + a_0,$$

with $a_n \neq 0$, and assume that f' has degree $n - 1$. Let $\alpha_1, \ldots, \alpha_n$ be the roots of f in a splitting field of $\mathcal{F}(A)$, listed with multiplicities. Then

$$\operatorname{disc}(f) = a_n^{n-2} \prod_{i<j} (\alpha_i - \alpha_j)^2.$$

Proof. We can enlarge A and assume that the roots of f belong to A. So we can factor f as

$$f(x) = a_n(x - \alpha_1) \cdots (x - \alpha_n).$$

The usual rules for differentiation yield

$$f'(x) = a_n \sum_{i=1}^{n} (x - \alpha_1) \cdots \widehat{(x - \alpha_i)} \cdots (x - \alpha_n),$$

where the term $\widehat{(x - \alpha_i)}$ is omitted. Corollary 4.1.9 then gives the thesis. $\square$

Note that the sign $(-1)^{\binom{n}{2}}$ in the definition of the discriminant is exactly the sign needed to reorder the terms in the last part of the proof.

Corollary 4.2.4. *Under the same hypothesis, f has repeated roots if, and only if,* $\operatorname{disc}(f) = 0$.

We next discuss a generalization of the discriminant in the context of field theory. For the rest of the section, we fix a finite separable extension of fields L/K, and denote by $\overline{K}$ a fixed algebraic closure of K, which is determined up to isomorphism. Recall from Section A.7 that in this situation we have the trace homomorphism $\operatorname{Tr}_{L/K}\colon L \to K$, and the norm $\operatorname{N}_{L/K}\colon L^* \to K^*$.

Looking at L as a vector space over K, we can consider the dual space $L^\vee$. For any $\alpha \in L$ we get a linear functional

$$T_\alpha\colon L \to K$$

defined by $T_\alpha(\beta) = \operatorname{Tr}_{L/K}(\alpha\beta)$. This gives a K-linear map

$$T\colon L \to L^\vee$$

sending α to T_α. By Proposition A.7.6, T is an isomorphism. Indeed,

$$\ker T = \{\alpha \in L \mid \operatorname{Tr}_{L/K}(\alpha\beta) = 0 \text{ for all } \beta \in L\} = 0.$$

In other words, the K-bilinear pairing on L defined by $(\alpha, \beta) \to \operatorname{Tr}_{L/K}(\alpha\beta)$ is nondegenerate.

In particular, given a basis $u_1, \dots, u_n$ of L as a K-vector space, we can find a *dual* basis $v_1, \dots, v_n$, such that

$$\operatorname{Tr}_{L/K}(u_i v_j) = \delta_{ij}.$$

Definition 4.2.5. We shall say that the two bases $u_1, \dots, u_n$ and $v_1, \dots, v_n$ are *dual* with respect to the trace.

We are now ready to study the promised generalization of the discriminant.

Definition 4.2.6. Let L/K be a finite separable field extension of degree n, and let $\sigma_1, \dots, \sigma_n$ be the embeddings of L into $\overline{K}$. Given $a_1, \dots, a_n \in L$, consider the matrix $A = (\sigma_i(a_j))_{ij}$. The *discriminant* of $a_1, \dots, a_n$ is defined by

$$\operatorname{disc}(\{a_1, \dots, a_n\}) := (\det(A))^2.$$

Remark 4.2.7. Because of the square, the discriminant does not depend on the ordering of the a_i. Still, note that it depends on the fields L and K, that are now fixed, even though the notation omits this dependence.

There is another convenient formula for the discriminant. Just note that $\det(A)^2 = \det(A^T \cdot A)$, and the matrix $A^T \cdot A$ has entries

$$(A^T \cdot A)_{ij} = \sum_{k=1}^{n} \sigma_k(a_i)\sigma_k(a_j) = \mathrm{Tr}_{L/K}(a_i a_j).$$

So we have the following expression:

$$\mathrm{disc}(\{a_1, \ldots a_n\}) = \det(\mathrm{Tr}_{L/K}(a_i a_j))_{ij}. \tag{4.2.1}$$

Remark 4.2.8. The last formula shows in particular that $\mathrm{disc}(\{a_1, \ldots a_n\}) \in K$.

Proposition 4.2.9. *The discriminant* $\mathrm{disc}(\{a_1, \ldots a_n\})$ *is* 0 *if, and only if,* $a_1, \ldots, a_n$ *are linearly dependent over* K.

Proof. Assume that

$$c_1 a_1 + \cdots + c_n a_n = 0$$

for some nontrivial choice of $c_1, \ldots, c_n \in K$. Then we have

$$c_1 \sigma_i(a_1) + \cdots + c_n \sigma_i(a_n) = 0$$

for every i, so the vector $(c_1, \ldots, c_n)$ lies in the kernel of the matrix $A = (\sigma_i(a_j))_{ij}$. It follows that

$$\mathrm{disc}(\{a_1, \ldots, a_n\}) = (\det(A))^2 = 0.$$

Conversely, assume that $\mathrm{disc}(\{a_1, \ldots, a_n\}) = 0$. Using the expression (4.2.1) we see that the rows of the matrix $(\mathrm{Tr}_{L/K}(a_i a_j))_{ij}$ are linearly dependent, hence for some choice of $c_1, \ldots, c_n \in K$ we have

$$T_{a_i}(c_1 a_1 + \cdots + c_n a_n) = 0$$

for all i. If $a_1, \ldots, a_n$ are linearly independent, the functionals T_{a_i} span the whole $L^\vee$, hence

$$c_1 a_1 + \cdots + c_n a_n = 0,$$

a contradiction. □

Proposition 4.2.10. *Let* $a_1, \ldots, a_n \in L$ *and let* $b_1, \ldots, b_n \in L$ *be linear combinations with integer coefficients of the* a_i, *say* $\vec{b} = M\vec{a}$, *where* M *is a matrix with integer coefficients. Then*

$$\mathrm{disc}(\{b_1, \ldots, b_n\}) = (\det M)^2 \,\mathrm{disc}(\{a_1, \ldots, a_n\}).$$

Proof. For any embedding σ of L in the algebraic closure of K we have $\sigma(\vec{b}) = M\sigma(\vec{a})$, hence

$$\begin{aligned}\mathrm{disc}(\{b_1, \ldots, b_n\}) &= \det(\sigma_i b_j)^2 = (\det M)^2 (\sigma_i a_j)^2 \\ &= (\det M)^2 \,\mathrm{disc}(\{a_1, \ldots, a_n\}).\end{aligned}$$

□

We can now make the link with our previous notion of discriminant. Take an element $\alpha \in \overline{K}$ of degree n, and let $L = K(\alpha)$. So L is spanned as a K-vector space by the powers $1, \alpha, \dots, \alpha^{n-1}$.

Proposition 4.2.11. *Let α be algebraic over K with minimal polynomial f_α of degree n, and take $L = K(\alpha)$. Then*

$$\operatorname{disc}(f_\alpha) = \operatorname{disc}(\{1, \alpha, \dots, \alpha^{n-1}\}).$$

Proof. Let $\alpha_i = \sigma_i(\alpha)$ be the conjugates of α, so that

$$f_\alpha(x) = (x - \alpha_1) \cdots (x - \alpha_n).$$

Then the matrix

$$(\sigma_i(\alpha^j))_{ij} = (\alpha_i^j)_{ij}$$

is in Vandermonde form, and its determinant is

$$\prod_{i<j} (\alpha_i - \alpha_j).$$

The square of this is by definition $\operatorname{disc}(\{1, \alpha, \dots, \alpha^{n-1}\})$, so the thesis follows from Proposition 4.2.3. □

Corollary 4.2.12. *With notations as above, we have*

$$\operatorname{disc}(\{1, \alpha, \dots, \alpha^{n-1}\}) = (-1)^{\binom{n}{2}} \mathrm{N}_{L/K}(f'_\alpha(\alpha)).$$

Proof. Indeed this is the same of

$$\operatorname{disc}(f_\alpha) = (-1)^{\binom{n}{2}} R(f_\alpha, f'_\alpha),$$

and by Corollary 4.1.9 we have

$$R(f_\alpha, f'_\alpha) = \prod_{i=1}^{n} f'_\alpha(\alpha_i) = \mathrm{N}_{L/K}(f'_\alpha(\alpha)).$$ □

We will come back to discriminants in the context of number fields in Section 6.1.

4.3. Gröbner bases

In this section we work in a polynomial ring $A = k[x_1, \dots, x_n]$, where k is a field. We want to solve some algorithmic problems in A. The most basic problem is *ideal membership*: given $a \in A$ and an ideal $I = (f_1, \dots, f_k)$, how do we test if $a \in I$?

The answer is easy if $n = 1$. In this case A is a principal ideal domain, and we can find a generator f of I using the Euclidean algorithm. Having done that, we just perform the division of a by f and check if the remainder is 0. Alternatively, if we do not want to rely on the fact that A is a principal ideal domain, we can perform the division of a by f_1, obtaining

$$a = q_1 f_1 + r_1,$$

then divide r_1 by f_2 and so on. If the last remainder r_k is 0, $a \in I$, and conversely.

This idea seems good to be generalized to a higher number of variables, but we run into two problems. The first one, which is easy to solve, is that there is no obvious way to perform the division algorithm between polynomials in more than one variable. The first step of the division algorithm is to perform the division between the highest-order monomials, and when there is more than one variable there is no single choice of what *highest-order monomial* means.

This is easily repaired with the following

Definition 4.3.1. A *monomial order* is a total order $\preccurlyeq$ on the set of monomials in the indeterminates $x_1, \ldots, x_n$ satisfying:

(1) if x^α, x^β and x^γ are monomials and $x^\alpha \preccurlyeq x^\beta$, then $x^{\alpha+\gamma} \preccurlyeq x^{\beta+\gamma}$;

(2) there exists no infinite descending chain

$$x^{\alpha_1} \succ x^{\alpha_2} \succ \cdots \succ x^{\alpha_r} \succ \cdots .$$

Property 2 just states that $\preccurlyeq$ is a *well-ordering*, so every nonempty set of monomials admits a minimum. Property 1 is called *multiplicativity.*

Remark 4.3.2. There is a natural bijective correspondence between the set of monomials and $\mathbb{N}^n$, sending x^α to the multiindex α, and correspodingly we can give a definition of a monomial order on $\mathbb{N}^n$. Multiplication of monomials becomes addition of vectors in $\mathbb{N}^n$, and we shall switch between the two points of view according to convenience.

Remark 4.3.3. A simple consequence of the axioms is that for every $\alpha \in \mathbb{N}^n$ we have $x^\alpha \succcurlyeq 1$. Indeed if $1 \succ x^\alpha$, we get by multiplicativity the infinite descending sequence

$$1 \succ x^\alpha \succ x^{2\alpha} \succ \cdots .$$

Applying multiplicativity again, it follows that if x^α divides x^β, then $x^\alpha \preccurlyeq x^\beta$.

In other words, every monomial order is a total order which refines the partial order given by divisibility of monomials. In particular, the standard order is the only monomial order in one variable.

We now give some examples of monomial orders. These are better described as orders on $\mathbb{N}^n$.

Example 4.3.4.

(a) The first example is the *lexicographic order* LEX. In order to compare two multiindices $\alpha = (\alpha_1, \ldots, \alpha_n)$ and $\beta = (\beta_1, \ldots, \beta_n)$ one first compares α_1 with β_1. If $\alpha_1 < \beta_1$ one puts $\alpha \prec \beta$, and conversely. If $\alpha_1 = \beta_1$, we compare α_2 with β_2, and proceed until we find a different entry (unless of course all entries coincide and so $\alpha = \beta$).

In other words, $\alpha \succ \beta$ if and only if the first nonzero entry of $\alpha - \beta$ is positive. From this characterization, property 1 follows immediately. Indeed summing γ to both α and β does not change their difference. Property 2 follows easily by induction on the number of variables.

(b) The *graded lexicographic order* DEGLEX is similar to LEX, but to compare α and β we first compare their total degree. Here the degree of $\alpha = (\alpha_1, \ldots, \alpha_n)$ is

$$\deg(\alpha) := \alpha_1 + \cdots + \alpha_n.$$

If $\deg(\alpha) > \deg(\beta)$ we declare $\alpha \succ \beta$. For a monomial of equal total degree, we use lexicographic ordering.

(c) Finally, we shall consider the *graded reverse lexicographic order*, denoted DEGREVLEX. As with DEGLEX, we first compare the total degree of α and β. If the degree is the same, we use the reverse lexicographic order: in this order $\alpha \succ \beta$ if and only if the *last* nonzero entry of $\alpha - \beta$ is *negative.*

The only thing which is not obvious is property 2, which is proved as follows. In any descending chain, the total degree is decreasing. Since this is a natural number, it must eventually stabilize. But there are only finitely many monomials of a given total degree, hence we cannot have infinite descending chains.

Remark 4.3.5. All algorithms that we consider require the choice of a monomial order, and for most of them any one will do. Still, the performance of the algorithm may depend on the order chosen; this is why we choose not to pick any particular one from the very beginning.

For the rest of the section, let us fix a monomial order $\preccurlyeq$.

Definition 4.3.6. Let $f \in k[x_1, \ldots, x_n]$, and write

$$f = \sum_{\alpha \in S} a_\alpha x^\alpha,$$

where all $a_\alpha \in k$ are nonzero. The set S is called the monomial support of S, and if $\alpha \in S$ we say that x^α is a monomial of f.

The monomial of f which is maximum with respect to $\preccurlyeq$ is called the *leading monomial* and it is denoted by $\mathrm{LM}(f)$. Its coefficient in f is called the *leading coefficient* of f and it is denoted by $\mathrm{LC}(f)$. The term $\mathrm{LC}(f)\mathrm{LM}(f)$ is called the *leading term* of f and it is denoted by $\mathrm{LT}(f)$.

If $\mathrm{LM}(f) = x^\alpha$ we define the *multidegree* of f as

$$\mathrm{mdeg}(f) := \alpha.$$

We now describe the division algorithm between two polynomials f and g. Our aim is to write

$$f = q \cdot g + r, \tag{4.3.1}$$

where no monomial of r is divisible by $\mathrm{LM}(g)$. We will construct q inductively, one monomial at a time.

Start from the leading monomial of f. If $\mathrm{LM}(f)$ is divisible by $\mathrm{LM}(g)$, say

$$\mathrm{LM}(f) = x^\alpha \mathrm{LM}(g),$$

put the term

$$h = \frac{\mathrm{LC}(f)}{\mathrm{LC}(g)} x^\alpha$$

in q, and replace f by $f - h \cdot g$.

If instead $\mathrm{LM}(f)$ is not divisible by $\mathrm{LM}(g)$, we just move the leading term of f to r. Either way, the multidegree of f has lowered. We repeat the process until we reach the point where $f = 0$, which will happen because $\preccurlyeq$ is a monomial order and the sequence of multidegrees is decreasing. At the end of the process, we have produced polynomials q and r such that (4.3.1) holds, and r only contains monomials that are not divisible by $\mathrm{LM}(g)$, by construction.

We can extend this process to define the division of f by a finite set of polynomials $g_1, \ldots, g_n$. This is done by dividing f by g_1, then the remainder of this by g_2 and so on.

Remark 4.3.7. The result of the division actually depends on the monomial order we have chosen.

This process is best explained by way of an example.

Example 4.3.8. Take the lexicographic order on $k[x, y]$ and consider the division of $f = x^2y + xy^2 + y^2$ by $f_1 = xy - 1$ and $f_2 = y^2 - 1$. Dividing f by f_1 we get $f = (x + y)f_1 + x + y^2 + y$. Next, we put the x term into the remainder and divide $y^2 + y$ by f_2 to obtain

$$f = (x + y)f_1 + f_2 + (x + y + 1).$$

We cannot do any more steps since no monomial in $x+y+1$ is divisible by the leading term of either f_1 or f_2.

Having discussed the issue of the monomial order, we can now see a second, more serious, problem. Let's take an ideal $I = (f_1, \dots, f_k)$ and assume we want to check whether $g \in I$. The naive idea would be to compute the successive divisions of g by $f_1, \dots, f_k$. Namely, we perform the first division to obtain

$$g = q_1 f_1 + r_1,$$

then divide r_1 by f_2 obtaining

$$r_1 = q_2 f_2 + r_2$$

and so on. At the last step we are left with a remainder r_k, which we can consider as the remainder of the division of g by the system of polynomials $f_1, \dots, f_k$.

The problem is that this remainder is not determined by the polynomials alone, but depends on the order we put on the f_i. It may even happen that for some ordering the remainder is 0, thereby showing that $g \in I$, but for another ordering it is not (see Exercise 6).

Example 4.3.9. We perform the same computation of Example 4.3.8 but this time we invert the order and we divide by f_2 first. Since $\mathrm{LC}(f)$ is not divisible $\mathrm{LC}(f_2)$, x^2y goes in the remainder. By dividing what remains, we get $f = (x+1)f_2 + x^2y + x + 1$. But then x^2y is multiple of $\mathrm{LC}(f_1)$ and the other terms have too small degree, so the final result is

$$f = xf_1 + (x+1)f_2 + 2x + 1.$$

Notice that the remainder, $2x+1$, is different from the one obtained in Example 4.3.8.

To start solving this issue, we need some information about monomial ideals. We recall from Exercise 11 in Chapter 1 the following

Definition 4.3.10. The ideal $I \subset k[x_1, \dots, x_n]$ is called *monomial* if it is generated by monomials.

In the same exercise, you proved that if f belongs to a monomial ideal, then so does each of its monomials. If $I \subset k[x_1, \dots, x_n]$ is any ideal, we denote by $\mathrm{LT}(I)$ the monomial ideal generated by $\mathrm{LT}(f)$ for $f \in I$.

Remark 4.3.11. If I is an ideal, $\mathrm{LT}(I)$ is finitely generated by Hilbert's basis theorem 2.3.3. Hence it is generated by finitely many monomials, and each of these monomials is multiple of $\mathrm{LT}(f)$ for some $f \in I$. We deduce that there exist $f_1, \dots, f_m \in I$ such that $\mathrm{LT}(f_1), \dots, \mathrm{LT}(f_m)$ generate $\mathrm{LT}(I)$.

Now, it is not in general the case that any set of generators $f_1, \ldots, f_m$ of I will do.

Definition 4.3.12. A set of elements $g_1, \ldots, g_m$ of $I \subset k[x_1, \ldots, x_n]$ is called a *Gröbner basis* of I (with respect to a given momonial order) if $\mathrm{LT}(g_1), \ldots, \mathrm{LT}(g_m)$ generate $\mathrm{LT}(I)$.

The relevance of Gröbner bases for our task is given by the following result.

Proposition 4.3.13. *Let $g_1, \ldots, g_m$ be a Gröbner basis for I and take any $f \in k[x_1, \ldots, x_n]$. Then there exists a unique $r \in k[x_1, \ldots, x_n]$ such that*

(i) *No monomial of r is divisible by any of $\mathrm{LT}(g_1), \ldots, \mathrm{LT}(g_m)$.*

(ii) *We can write $f = g + r$ with $g \in (g_1, \ldots, g_m)$.*

Proof. Existence is clear by the division algorithm. For the uniqueness, assume $f = g + r = g' + r'$, with $g, g' \in I$ and r, r' as stated. Then $r - r' \in I$, so in particular $\mathrm{LT}(r - r') \in \mathrm{LT}(I)$. Since $g_1, \ldots, g_m$ is a Gröbner basis, this is excluded by i) unless $r = r'$. □

In particular, we see that the remainder of the division of f by $g_1, \ldots, g_m$ does not the depend on the order in which we perform the division if $g_1, \ldots, g_m$ is a Gröbner basis—although the other coefficients in the division might change.

Corollary 4.3.14. *Let $g_1, \ldots, g_m$ be a Gröbner basis for I and take any $f \in k[x_1, \ldots, x_n]$. Then $f \in I$ if, and only if, the remainder of f by $g_1, \ldots, g_m$ is 0.*

Remark 4.3.15. This in particular implies that $g_1, \ldots, g_m$ generate the ideal I, a fact that we did not assume in the definition of a Gröbner basis.

This solves the problem of determining if $f \in I$ in an algorithmic way, provided we can find a Gröbner basis for I. Now, such a basis certainly exists by Remark 4.3.11. Before asking how to find one such basis algorithmically, we look at a concrete example.

Example 4.3.16.

(a) The set $\{xy - 1, y^2 - 1\}$ in $k[x, y]$ is not a Gröbner basis for the LEX order, since we showed in the previous examples that the remainder is not unique.

(b) We take $k[x, y, z]$, again with the LEX order. The set

$$G = \{g_1, g_2\} = \{x + y^2 - 2yz + 3z - z^5, y + 3z^2 - z^3\}$$

is a Gröbner basis. In fact, let $I = (G)$, take $f \in \mathrm{LT}(I)$ and assume that $\mathrm{LT}(f)$ does not belong to the ideal generated by $\mathrm{LT}(g_1) = x$ and

$\mathrm{LT}(g_2) = y$. Then $\mathrm{LT}(f)$ is a monomial in z and by the property of lexicographic ordering f is a polynomial in z alone.

On the other hand, $f = f_1 g_1 + f_2 g_2$, hence it vanishes if we specialize $y = -3z^2 + z^3$ and $x = -y^2 + 2yz - 3z + z^5$. Since f does not involve either x or y, there is nothing to specialize, hence $f = 0$ identically. In fact, all we used was that we could solve for y in g_2 and then solve for x in g_1.

We are now left with the task of producing a Gröbner basis for an ideal $I \subset k[x_1, \ldots, x_n]$ given a set of generators $F = \{f_1, \ldots, f_m\}$. If F is not already a Gröbner basis, there exists $f \in I$ such that $\mathrm{LT}(f)$ does not belong to $(\mathrm{LT}(f_1), \ldots, \mathrm{LT}(f_m))$.

If we have a way of producing such f algorithmically, we can add f to F and ask again if we have a Gröbner basis. If not, we can enlarge F again, and so on. The process must end since $k[x_1, \ldots, x_n]$ is Noetherian, leaving us with a Gröbner basis.

Hence, we are reduced to the following problem: *given a set $f_1, \ldots, f_m \in k[x_1, \ldots, x_n]$, either show that it is a Gröbner basis, or produce a polynomial that shows that it is not.*

Let f, g be two polynomials, and consider a linear combination $h = af + bg$ with $a, b \in k[x_1, \ldots, x_n]$. If the highest terms of af and bg do not cancel out, $\mathrm{LT}(h) \in (\mathrm{LT}(f), \mathrm{LT}(g))$. So we start by looking for cases where a cancellation occurs.

Definition 4.3.17. The S-polynomial of f and g, denoted $S(f, g)$, is given by

$$S(f, g) = \frac{m}{\mathrm{LT}(f)} f - \frac{m}{\mathrm{LT}(g)} g,$$

where m is the lowest common multiple of $\mathrm{LM}(f)$ and $\mathrm{LM}(g)$ (which makes sense since both are monomials).

By construction, $S(f, g)$ has degree less than m, because of cancellation.

Lemma 4.3.18. *Let $f_1, \ldots, f_m$ be polynomials of the same multidegree (meaning that $\mathrm{LM}(f_1) = \cdots = \mathrm{LM}(f_m)$). If a linear combination*

$$h = \sum_{i=1}^{m} c_i f_i$$

with $c_i \in k$ has a smaller multidegree, then h is a linear combination of the $S(f_i, f_j)$ with coefficients in k.

Proof. Let $a_i = \mathrm{LC}(f_i)$ and $p_i = f_i/a_i$. The S-polynomials are simply $S(f_i, f_j) = f_i/a_i - f_j/a_j = p_i - p_j$, so that the expression for h becomes

$$\begin{aligned} h = {} & c_1a_1(p_1 - p_2) + (c_1a_1 + c_2a_2)(p_2 - p_3) + \cdots + \\ + {} & (c_1a_1 + \cdots + c_{m-1}a_{m-1})(p_{m-1} - p_m) + \\ + {} & (c_1a_1 + \cdots + c_ma_m)p_m. \end{aligned}$$

Our hypothesis amounts to $\sum c_ia_i = 0$, hence we have the desired expression for h. □

We can now state the main technical result of this section.

Theorem 4.3.19 (Buchberger's criterion). *Let $I \subset k[x_1, \ldots, x_n]$ be an ideal with generators $F = \{f_1, \ldots, f_m\}$. Then F is a Gröbner basis if and only if for each i, j, the remainder of $S(f_i, f_j)$ divided by F is 0.*

Note that in the theorem the order in which we use the f_i to perform the division is arbitrary, and in fact any ordering will do.

Proof. Let $h \in I$ so that

$$h = \sum_{i=1}^{m} f_ig_i. \tag{4.3.2}$$

Let α be the (multi)degree of the highest monomial appearing in any of the f_ig_i. If h has itself degree α, its leading monomials lie in $(\mathrm{LT}(f_1),\ldots,\mathrm{LT}(f_m))$. By contradiction, assume that we find h such that this is not the case, so cancellation happens in the highest degree.

Let S be the set of indices i for which f_ig_i has degree α. Then the previous lemma ensures that

$$\sum_{i\in S} f_ig_i = \sum_{i,j\in S} c_{ij}S(f_ig_i, f_jg_j)$$

for some constants c_{ij}. Note that $S(f_ig_i, f_jg_j)$ is itself a linear combination of f_i and f_j with coefficients in $k[x_1, \ldots, x_n]$, but has degree lower than α. Hence we can rewrite (4.3.2) as

$$h = \sum_{i=1}^{m} f_ig_i',$$

but this time the maximum degree of any combination f_ig_i' is less than α.

We can repeat the process at will, constructing an infinite descending sequence of degrees, thereby contradicting the definition of monomial order. □

As a consequence of Buchberger's criterion, we finally have Buchberger's algorithm to produce a Gröbner basis for an ideal I starting from a set of generators $F = \{f_1, \dots, f_n\}$.

(1) For each pair of polynomials f_i, f_j, we consider the S-polynomial $S(f_i, f_j)$ and perform the division with respect to F.

(2) If for some pair i, j the remainder is not 0, we add $S(f_i, f_j)$ to the set and start again from 1.

(3) Otherwise F is a Gröbner basis by Buchberger's criterion.

If we want to check if a given polynomial f lies in I we add a final step:

(4) Once we have a Gröbner basis G, perform the division of f by G. If the remainder is 0, then $f \in I$, and conversely.

Remark 4.3.20. As a corollary, we also obtain an algorithm for ideal inclusion or equality. In fact, say $I = (f_1, \dots, f_n)$ and $J = (g_1, \dots, g_m)$. Then to check if $I \subset J$ it is enough to check whether each $f_i \in J$.

4.4. More algorithmic operations

In the previous section, we have learned to solve a few problems algorithmically. We work inside $k[x_1, \dots, x_n]$ and we assume that ideals are given by a finite set of generators. By computing Gröbner bases with respect to any monomial order, we are able to

(1) determine if a polynomial f belongs to an ideal I, or

(2) for two ideal I and J, determine whether $I \subset J$, and hence also detect equality.

It is also trivial to compute generators for $I \cup J$ and $I \cdot J$—hence our next problem is to compute generators of $I \cap J$.

To do this, we first give a small introduction to *elimination theory*. As we discussed in Section 4.1, the core idea of elimination theory is to solve a system of polynomial equations by trying to eliminate one variable at a time. The main result that links it to Gröbner bases is the following:

Theorem 4.4.1 (Elimination theorem). *Let $I \subset k[x_1, \dots, x_n]$ and G a Gröbner basis for I with respect to* LEX. *Then, for every $i \leq n$, $G \cap k[x_i, \dots, x_n]$ is a Gröbner basis for $I \cap k[x_i, \dots, x_n]$.*

In particular, we can find generators for $I \cap k[x_i, \dots, x_n]$ by taking the elements of G that do not involve the first $i - 1$ variables.

Proof. Let $f \in I$ be a polynomial that does not involve the first $i - 1$ variables. The leading monomial $\mathrm{LM}(f)$ is divisible by $\mathrm{LM}(g)$ for some

$g \in G$, and clearly $\mathrm{LM}(g)$ does not involve the first $i-1$ variables. Since this is the leading monomial for LEX, all other monomials have the same property, hence $g \in G \cap k[x_i, \ldots, x_n]$. □

Example 4.4.2. [**Mor08**] Consider the system of equations

$$\begin{cases} x^2 + y + z = 1 \\ x + y^2 + z = 1 \\ x + y + z^2 = 1. \end{cases}$$

To solve this system, we take the ideal $I = (f_1, f_2, f_3)$, where

$$f_1(x,y,z) = x^2 + y + z - 1$$
$$f_2(x,y,z) = x + y^2 + z - 1$$
$$f_3(x,y,z) = x + y + z^2 - 1.$$

The three elements are not a Gröbner basis, but we can apply Buchberger's algorithm to find one. A Gröbner basis for I is given by

$$\begin{aligned} g_1(x,y,z) &= x + y + z^2 - 1, \\ g_2(x,y,z) &= y^2 - y - z^2 + z, \\ g_3(x,y,z) &= 2yz^2 + z^4 - z^2, \\ g_4(x,y,z) &= z^2(z-1)^2(z^2 + 2z - 1). \end{aligned}$$

The last equation can be solved for z giving the four solutions $z = 0$, $z = 1$ and $z = -1 \pm \sqrt{2}$. We can then substitute back z in g_2 or g_3 to solve for y and finally back into g_1 to solve for x. For instance, when $z = 0$, we recover $y = 0$ or $y = 1$ from g_2, and accordingly $x = 1$ or $x = 0$ from g_1. Work out the other solutions!

Theorem 4.4.1 guarantees that we can always find a way to eliminate variables in a system of equations, and then we can backtrack to solve for one variable at a time.

As we promised, another application for this result is the computation of generators for $I \cap J$. We use the following

Proposition 4.4.3. *Let I and J be ideals of $k[x_1, \ldots, x_n]$. Inside the ring $k[t, x_1, \ldots, x_n]$ consider the ideal $L = t \cdot I + (1-t) \cdot J$. Then*

$$L \cap k[x_1, \ldots, x_n] = I \cap J.$$

Proof. If $f \in I \cap J$ we can write $f = tf + (1-t)f$, which shows one inclusion. For the other one, let $g \in k[x_1, \ldots, x_n]$ and assume that

$$g = ta + (1-t)b$$

for some $a \in I$, $b \in J$. Then writing $g = t(a-b)+b$, we see that $a = b$, so they both lie in $I \cap J$. □

Using this result, we now have an algorithm to compute generators for $I \cap J$, since it is easy to compute generators for $t \cdot I + (1-t) \cdot J$ and then we can apply Theorem 4.4.1 to eliminate the t variable.

A similar reduction exists for the radical of an ideal.

Proposition 4.4.4. *Let k be a field, $I \subset k[x_1, \ldots, x_n]$ an ideal. Then $f \in \sqrt{I}$ if, and only if, 1 belongs to the ideal generated by I and $1 - tf$ in the ring $k[x_1, \ldots, x_n, t]$.*

Proof. Let $A = k[x_1, \ldots, x_n]$, and consider the multiplicative set $S \subset A$ generated by f. Then $f \in \sqrt{I}$ if and only if I becomes trivial in $S^{-1}A$. But there is an isomorphism

$$\phi\colon\ A[t]/(1-tf) \longrightarrow S^{-1}A,$$

$$t \longmapsto 1/f,$$

hence this is equivalent to the condition that I and $1-tf$ generate the whole $A[t]$. □

Using this result, we get another algorithmic application of Gröbner bases: to check whether $f \in \sqrt{I}$, we only need to test whether $1 \in (I, 1-tf)$, which is something we can easily do by computing a Gröbner basis for the latter ideal.

We end with another example that shows the far-reaching applicability of Gröbner bases.

Example 4.4.5. [**Mir11**] Many problems in Euclidean geometry can be reduced to systems of polynomial equations by choosing coordinates. These, in turn, can be solved using elimination theory as we outlined above. This is the observation at the heart of automatic theorem proving in Eucliden geometry.

For a concrete example, consider the circle theorem of Apollonius: *let ABC be a triangle right in A. The midpoints of the three sides and the foot of the altitude drawn from A to BC all lie on one circle.*

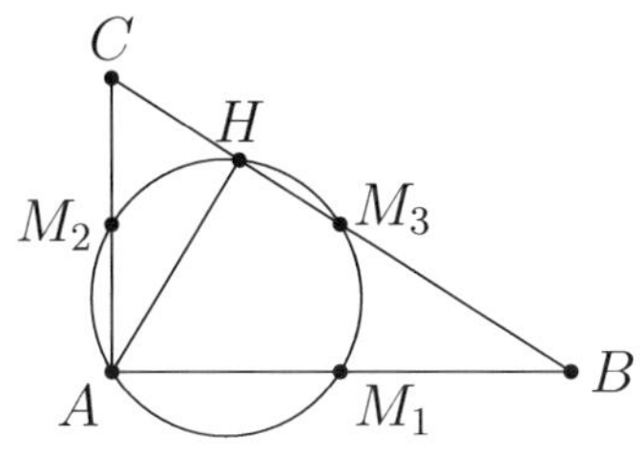

To prove this, we place A in the origin and choose coordinates $(u_1, 0)$ for B and $(0, u_2)$ for C. The midpoints are $M_1 = (x_1, 0)$, $M_2 = (0, x_2)$, and $M_3 = (x_3, x_4)$. We also let $H = (x_5, x_6)$ be the foot of the altitude and $O = (x_7, x_8)$ be the center of the circle through the midpoints.

The construction gives a system of equations. First we have $2x_1 = u_1$ and three other equations for the midpoints. Then, the condition $AH \perp BC$ becomes $x_5u_1 = x_6u_2$, and the fact that $H \in BC$ becomes $x_5u_2 + x_6u_1 = u_1u_2$. Finally, we get two quadratic equations for $\overline{M_1O} = \overline{M_2O}$ and $\overline{M_1O} = \overline{M_3O}$.

The whole system is

$$\begin{cases} 2x_1 - u_1 = 0 \\ 2x_2 - u_2 = 0 \\ 2x_3 - u_1 = 0 \\ 2x_4 - u_2 = 0 \\ x_5u_1 - x_6u_2 = 0 \\ x_5u_2 + x_6u_1 - u_1u_2 = 0 \\ (x_1 - x_7)^2 + x_8^2 - x_7^2 - (x_8 - x_2)^2 = 0 \\ (x_1 - x_7)^2 + x_8^2 - (x_7 - x_3)^2 - (x_8 - x_4)^2 = 0. \end{cases}$$

The thesis is given by the equation

$$(x_5 - x_7)^2 + (x_6 - x_8)^2 - (x_1 - x_7)^2 - x_8^2 = 0.$$

Using Gröbner bases, one can show that the latter polynomial lies in the ideal generated by the polynomials appearing in the system. This can be automated using the algorithms of this chapter, and this technique allows one to transform statements in Euclidean geometry into algorithmic problems.

4.5. Exercises

1. Let k be an algebraically closed field, and $I \subset k[x_1, \ldots, x_n]$ a radical ideal, say $I = (f_1, \ldots, f_k)$. For each pair $i \neq j$, where $1 \leq i \leq k$, consider the resultant $r_{i,j} := R(f_i, f_j)$, where the resultant is taken with respect to the variable x_1, so that $r_{i,j} \in k[x_2, \ldots, x_n]$. Assume that each f_i has positive degree in x_1. Show that $I \cap k[x_2, \ldots, x_n] = \sqrt{(r_{i,j})}$.

2. Try to apply resultants to find solutions to a generic system of polynomial equations. What difficulties do you find? Work out a concrete case by using resultants to solve the system in Example 4.4.2.

3. A Gröbner basis $G = \{g_1, \ldots, g_r\}$ is called *reduced* if each g_i is monic (that is, $\mathrm{LC}(g_i) = 1$) and for each i, $\mathrm{LM}(g_i)$ does not divide any monomial of any g_j for $j \neq i$.

Show that given a Gröbner basis for an ideal I, one can construct explicitly a reduced one. Moreover, show that a reduced Gröbner basis for I is unique (up to order). Derive a simple algorithm to test ideal equality in polynomial rings.

4. Let $I \subset k[x, y]$ be the ideal generated by $xy - x$ and $x^2 - y$. Compute a Gröbner basis for I with respect to LEX.

5. Let $I \subset k[x, y]$ be the ideal generated by x^2 and $xy + y^2$. Compute a Gröbner basis for I with respect to LEX.

6. Inside $k[x, y]$, consider the polynomial $f = x^2y^3 - 2xy^2$ and the ideal $I = (f_1, f_2)$, where $f_1 = x^2y - 2x$ and $f_2 = y^3 + 4$. Using the LEX order, perform the division of f by f_1, then f_2. Then, do the same computation using f_2, then f_1. Show that in one case, the remainder is 0, while in another it is nonzero.

7 (**[RZ10]**)**.** Consider the polynomials $f(x, y) = (y^2 - y)x^2 + x$ and $g(x, y) = (y^2 - y)x + y$ in $k[x, y]$. Compute the resultant with respect to the x variable, $R(f, g) = y^2(y - 1)^2$. Letting $I = (f, g)$, use Gröbner bases to compute that $I \cap k[y] = (y)$. What happens when you use resultants to solve the system given by f and g? Explain the discrepancy between the two techniques that we have introduced for the elimination problem.

8. Let $w \in \mathbb{R}^n$ be a vector such that the components w_i are linearly independent over $\mathbb{Q}$. Given $\alpha, \beta \in \mathbb{N}^n$, define $\alpha \prec_w \beta$ if $w \cdot \alpha < w \cdot \beta$, where $\cdot$ denotes scalar product. Show that $\prec_w$ is a monomial order, called a *weight order.*

9. Let $w \in \mathbb{R}^j$, and consider the order defined as follows. Given $\alpha, \beta \in \mathbb{N}^n$, define $\alpha \prec_w \beta$ if $w \cdot (\alpha_1, \ldots, \alpha_j) < w \cdot (\beta_1, \cdots, \beta_j)$, where $\cdot$ denotes scalar product. If the two products are equal, fallback to comparing α and β by DEGREVLEX. This is called an *elimination order* for $\{1, \ldots, j\}$. Show that it has the property that if a variable $x_1, \ldots, x_j$ does not appear in $\mathrm{LM}(f)$, then it does not appear in f at all.

10. Let $\prec_w$ be the elimination order defined in the previous exercise. Show that it can be used to compute the elimination ideal for the first j variables. Namely, let $I \subset k[x_1, \ldots, x_n]$ and G a Gröbner basis for I with respect to $\prec_w$. Then $G \cap k[x_{j+1}, \ldots, x_n]$ is a Gröbner basis for $I \cap k[x_{j+1}, \ldots, x_n]$.

11. Consider a finite undirected graph G where each pair of vertices is connected by at most one edge. For each vertex v_i introduce a variable

x_i and consider the system of equations given by $x_i^3 = 1$ (for each i) and $x_i^2 + x_i x_j + x_j^2 = 0$ whenever v_i and v_j are adjacent.

Show that G can be colored with 3 colors (that is, each vertex can be assigned one of 3 colors such that adjacent vertices get different colors) if, and only if, the system has a solution. Derive an algorithm to test whether a graph is 3-colorable.

12. Verify that the proof of correctness and termination of Buchberger's algorithm does not require the full strength of Hilbert's basis theorem, but only the special case that states that *monomial* ideals in $k[x_1, \ldots, x_n]$ are finitely generated.

13. The following result is known as *Dickson's lemma*. Consider the partial order on $\mathbb{N}^n$ defined by $a \preccurlyeq b$ if and only if $a_i \leq b_i$ for all $i = 1, \ldots, n$. Then every nonempty subset $S \subset \mathbb{N}^n$ has finitely many (at least one) minimal elements. Prove Dickson's lemma directly, and then show that it is equivalent to the form of Hilbert's basis theorem for monomial ideals required in the previous exercise.

In order to implement the algorithms described in this chapter, one has to be able to implement operations on the base field k. A common choice is to take the finite field $\mathbb{F}_q$ over $q = p^d$ elements. This can be described as the quotient $\mathbb{F}_p[x]/(f)$, where f is an irreducible polynomial of degree d over $\mathbb{F}_p$, so one can perform computations in $k = \mathbb{F}_q$ by finding one such polynomial f, and then using the division algorithm in $\mathbb{F}_p[x]$. This inspires one to find algorithms that generate an irreducible polynomial of a given degree d over $\mathbb{F}_p$. The next exercises, up to Exercise 15 discuss this problem.

14. Let p be a prime, $\mathbb{F}_p = \mathbb{Z}/p\mathbb{Z}$, and take $d \geq 2$. Compute the proportion of polynomials of degree d in $\mathbb{F}_p$ that are irreducible. Namely, the number of monic polynomials of degree d is p^d, and the number of irreducible ones is asymptotic to p^d/d as $d \to \infty$.

15. By the previous exercise, a viable—even if a little costly—strategy to generate an irreducible polynomial of degree d is to generate random polynomials and test whether they are irreducible. In this exercise, we are going to provide such a test of irreducibility.

Namely, let $f \in \mathbb{F}_p[x]$ be a polynomial of degree d. Show that if f is reducible, then $\gcd(f, x^{p^r} - x)$ is nontrivial for some $r \leq \deg f/2$. Hence a test for irreducibility is just to compute all such gcd and check if any of them is not a constant.

16. For a polynomial $f \in k[x_1, \ldots, x_n]$ define the *Netwon's polytope* of f as the convex hull (in $\mathbb{R}^n$) of points $\alpha \in \mathbb{N}^n$ such that x^α is a nonzero monomial of f.

Show that if $m = x^\alpha$ is the leading monomial of f for any monomial order, then α is extremal in the Netwon polytope of f (that is, if we omit α in the above construction, we get a strictly smaller convex hull).

17. Find a nontrivial solution in $\mathbb{C}^3$ to the system

$$\begin{cases} xy^2 - z - z^2 = 0 \\ x^2y - y = 0 \\ y^2 - z^2 = 0. \end{cases}$$

18. Gröbner bases can be used to prove Heron's formula for the area of a triangle

$$s^2 = p(p-a)(p-b)(p-c), \tag{4.5.1}$$

where s is the area, a, b, c are the side lengths and $p = \frac{a+b+c}{2}$ is the semiperimeter.

To prove it, choose a reference system such that the vertices are in $(0,0)$, $(0,a)$ and (x,y). By the Pythagorean theorem, we have the relations

$$b^2 = x^2 + y^2 \text{ and}$$
$$c^2 = (a-x)^2 + y^2,$$

and the area satisfies $2s = a \cdot y$. Let $I \subset \mathbb{R}[s,a,b,c,x,y]$ be the ideal generated by these relations. Use elimination theory to find the intersection $I \cap \mathbb{R}[s,a,b,c]$, and show that it is generated by the single equation (4.5.1).

19. Let $A = k[x_1, \ldots, x_n]$. Follow the various connections in this chapter and write an explicit algorithm to test whether a polynomial $f \in A$ belongs to the radical $\sqrt{I}$ of an ideal $I = (g_1, \ldots, g_k) \subset A$.

20. Let $A = k[x_1, \ldots, x_n]$ and $B = k[y_1, \ldots, y_m]$, and let $f \colon A \to B$ be a homomorphism. Let $f_i := f(x_i) \in B$, and define the ideal

$$I_f := (x_1 - f_1, \ldots, x_n - f_n) \in k[x_1, \ldots, x_n, y_1, \ldots, y_m].$$

Show that $\ker f = I_f \cap A$. Use this result and the elimination theorem to deduce an algorithm to compute $\ker f$.

21. A polynomial $f \in k[x_1, \ldots, x_n]$ is called *symmetric* if it is invariant with respect to the action of the symmetric group $\mathfrak{S}_n$ that permutes the indeterminates $x_1, \ldots, x_n$. The *elementary symmetric polynomials* are the polynomials $s_1, \ldots, s_n$ defined by

$$s_k = \sum_{i_1 < i_2 < \cdots < i_k} x_{i_1} \cdots x_{i_k}.$$

Of course, these polynomials are in fact symmetric.

Prove that every symmetric polynomial is a polynomial in the elementary symmetric polynomials $s_1, \dots, s_n$, in other words,

$$k[x_1, \dots, x_n]_{\mathfrak{S}_n} = k[s_1, \dots, s_n].$$

In general, it rarely happens that the invariant ring for the action of a group has this simple structure, see [**Dol03**]. (Use the order LEX and do induction on the multidegree. Prove that the multidegree of a symmetric polynomial has the form $\alpha = (\alpha_1, \dots, \alpha_n)$, where $\alpha_1 \geq \alpha_2 \geq \dots \geq \alpha_n$.)

22. Let $I \subset k[x_1, \dots, x_n]$ be an ideal generated by binomials. Show that there exists a Gröbner basis for I which consists of binomials (follow the steps of Buchberger's algorithm). In fact, even a *reduced* Gröbner basis for I (see Exercise 3) is composed by binomials. For much more about binomial ideals, see [**HHO18**].

23. The result of the preceeding exercise does not hold for polynomials composed of three terms. In fact, take the ideal $I = (x_1+x_2+x_3, x_1+x_4+x_5)$ in the ring $k[x_1, \dots, x_5]$. Show that a reduced Gröbner basis for I with respect to DEGREVLEX contains the polynomial $x_2 + x_3 - x_4 - x_5$.

Chapter 5

Integral Dependence

In this chapter we study the so called integral extensions: given an inclusion of rings $A \subset B$ we want to understand what happens for elements of B which satisfy a monic polynomial equation with coefficients in A. This is the natural generalization of the concept of algebraic elements from field theory (see Section A.1). Many results from field theory can be transferred to this more general situation more or less literally, and we do so in the first section. For example we prove that the set of $x \in B$ that satisfy a monic polynomial equation with coefficients in A is a subring of B, called the *integral closure* of A. We also prove the *going up* and *going down* theorems, which describe the behavior of prime ideals in integral extensions, and then prove the famous *Noether normalization lemma*, which gives a structure theorem for integral extensions for k-algebras. Finally, we see how this theory allows us to construct many important examples of Dedekind domains.

5.1. Integral extensions

We want to mimic the theory of algebraic elements for fields in the more general context of rings.

Definition 5.1.1. Let $A \subset B$ be an extension of rings. We say that an element $b \in B$ is *integral* over A if it satisfies an equation

$$b^n + a_{n-1}b^{n-1} + \cdots + a_0 = 0, \tag{5.1.1}$$

where $a_i \in A$.

Definition 5.1.2. The subset of B of integral elements over A is called the *integral closure* of A inside B. If the integral closure amounts just to A, we say that A is *integrally closed* in B. If it is the whole of B, we say that B is *integral* over A.

Remark 5.1.3. We note that the equation (5.1.1) that we ask b to satisfy is *monic* (recall that a polynomial is called monic if its first coefficient is 1); this is irrelevant when working with fields since we can always divide everything by the leading coefficient.

We shall see soon that the integral closure of A is a subring of B as expected. Explicitly we have to see that if $a, b \in B$ are integral over A, then $a+b$ and ab are integral too. In order to do this we take a roundabout way: we do not exhibit a monic polynomial satisfied by $a+b$ or ab; rather we find an equivalent condition for an element to be integral.

To state the result in full generality we need

Definition 5.1.4. Let M be an A-module. We say that M is *faithful* if the induced homomorphism $A \to \operatorname{End}(M)$ is injective. Equivalently the ideal

$$\operatorname{Ann}(M) := \bigcap_{m \in M} \operatorname{Ann}(m)$$

is 0.

Example 5.1.5.

(a) If $A \subset B$ are rings, B is a faithful A-module. Indeed assume that $a \in \operatorname{Ann}(B)$, so $a \cdot b = 0$ for all $b \in B$; taking $b = 1$ we find $a = 0$.

(b) If I is a nonzero ideal of A, the inclusion $I \subset \operatorname{Ann}(A/I)$ shows that A/I is not a faithful A-module.

(c) The infinite direct sum

$$M := \bigoplus\nolimits_{p \text{ prime}} \mathbb{Z}/(p)$$

is a faithful Z-module, but any finitely generated submodule of M is not faithful.

Theorem 5.1.6. *Let $A \subset B$ be rings, $b \in B$. The following conditions are equivalent:*

(i) *b is integral over A;*

(ii) *the ring $A[b]$ is finitely generated as an A-module;*

(iii) *there exists a ring C containing b, with $A \subset C \subset B$, such that C is finitely generated as an A-module;*

(iv) *there exists a faithful $A[b]$-module M which is finitely generated as an A-module.*

Note that (part of) this criterion is completely analogous to the usual statement for fields: an element b is algebraic over the field k if and only if $k[b]/k$ is a finite extension (Remark A.1.3).

Proof. If b is integral over A, b satisfies an equation of the form (5.1.1). We can use the equation to express every power b^k for $k \geq n$ as a linear combination of $1, b, \ldots, b^{n-1}$ with coefficients in A, so i) implies ii).

That ii) implies iii) is obvious. To show that iii) implies iv) it is enough to take $M = C$.

Now assume iv). Consider the map

$$\mu_b\colon M \to M$$

given by multiplication by b. If we choose a finite set $m_1, \ldots, m_k$ of generators of M, we can write

$$\mu_b m_i = b \cdot m_i = \sum_{j=1}^{k} a_{ij} m_j,$$

for some $a_{ij} \in A$. It follows that the matrix

$$D := b \cdot I_{k\times k} - (a_{ij})$$

satisfies $D \cdot m = 0$ for every element

$$m = \sum_{i=1}^{k} \lambda_i m_i \in M.$$

Let D' be the matrix of cofactors of D; that is, the (i, j) coefficient of D' is $(-1)^{i+j}$ times the determinant of the minor of D obtained by removing the i-th column and the j-th row. Then it is well known from linear algebra that

$$D' \cdot D = \det D \cdot I. \tag{5.1.2}$$

Hence, from $D'D \cdot m = 0$ we deduce

$$\det D \cdot m = 0$$

for every $m \in M$. Note that the latter is a scalar equation. Since M is faithful, we see that $\det D = 0$; now just expand this determinant to find the desired monic equation for b, thereby proving i). □

Remark 5.1.7. Some readers may only be familiar with equation (5.1.2) as it relates to matrices with coefficients in a field. As in Proposition 4.1.8, this particular case implies that the same equation holds over any ring.

First, note that if A is an integral domain with quotient field k, the result for k implies that for A trivially. In particular equation (5.1.2) holds in the ring $\mathbb{Z}[x_{ij}]$, where x_{ij} are indeterminates, for $i, j = 1, \ldots, k$. Now for any ring A and for any matrix $D = (d_{ij})$ with coefficients in A we can find a homomorphism

$$\mathbb{Z}[x_{ij}] \to A$$

sending x_{ij} to d_{ij}, and the result follows by specialization.

The method of proof that we have used is known as the *determinant trick*. Notice that it also yields a more explicit form of Nakayama's lemma. More explicitly, let M be a finitely generated module over A—say $m_1, \dots, m_n$ are the generators. If I is an ideal such that $I \cdot M = M$, we can use the above identity to find an explicit element $x \in I$ such that $1 + x \in \operatorname{Ann} M$.

Theorem 5.1.8 (Nakayama's lemma, explicit form). *Let M be a module over the ring A, generated by $m_1, \dots, m_n$. Let I be an ideal such that $I \cdot M = M$, and write*

$$m_i = \sum_{j=1}^{n} a_{ij} m_j.$$

Let A be the matrix with coefficients (a_{ij}), $a_{ij} \in I$, and $y = \det(\mathrm{id} - A)$. Then $y \in 1 + I$ and $y \cdot M = 0$.

The standard form of Nakayama's lemma follows immediately: if $I \subset \mathcal{J}(A)$, $1 + x$ is a unit for every $x \in I$, hence M must be 0.

Proof. Let $B = \mathrm{id} - A$, so that B annihilates the vector $(m_1, \dots, m_n)$. By (5.1.2), $\det B \cdot \mathrm{id}$ also annihilates this vector, hence $\det B \cdot m = 0$ for all $m \in M$. All summands in the expansion of the determinant belong to I, except for the product of n factors 1. □

Theorem 5.1.6 has some important corollaries.

Corollary 5.1.9. [1] *Let $A \subset B$ be rings. The integral closure of A in B is a subring of B.*

Proof. Let $a, b \in B$ be integral over A. Then the ring $C = A[a, b]$ is finitely generated as an A-module; since $a + b$ and $a \cdot b$ belong to C we see that both are integral over A. □

Corollary 5.1.10. *Let $A \subset B$ be rings. If B is finitely generated as an A-algebra and it is integral over A, it is finitely generated as an A-module.*

[1] Indeed this is the origin of the word *ring*. Dedekind observed that he could obtain interesting sets of numbers, closed under addition and multiplication, of the form $\mathbb{Z}[\alpha]$, where α is integral over $\mathbb{Z}$. Once he had the first powers $1, \alpha, \dots, \alpha^{n-1}$, higher powers could be expressed recursively using an equation for α. In the important case where $\alpha^n = 1$, he just had to cycle through the powers; hence he decided to use the expression *ring of numbers* for such a set, and the name "ring" has then stuck.

Proof. Write

$$B = A[b_1, \ldots, b_n].$$

Each of the b_i is integral over A, so the claim follows by Theorem 5.1.6 and induction. □

Corollary 5.1.11. *Let $A \subset B \subset C$ be rings. If B is integral over A and C is integral over B, C is integral over A.*

Proof. Assume $c \in C$ satisfies an equation

$$c^n + b_{n-1}c^{n-1} + \cdots + b_0 = 0,$$

for some $b_i \in B$. The ring

$$B' := A[b_1, \ldots, b_{n-1}]$$

is finitely generated as an algebra and integral over A, so it is finitely generated as an A-module. Moreover $B'[c]$ is finitely generated as a B'-module. It follows that $B'[c]$ is finitely generated as an A-module, so c is integral over A. □

This easily implies that the integral closure is idempotent.

Corollary 5.1.12. *Let $A \subset B$ be rings. If C is the integral closure of A in B, C is integrally closed in B.*

Definition 5.1.13. We say that A is *integrally closed* (or *normal*) if it is integrally closed in the total ring of fractions $\mathcal{F}(A)$.

The simplest example is the following.

Proposition 5.1.14. *A unique factorization domain is integrally closed.*

Proof. Let A be a UFD, $s, t \in A$, and assume that $x = s/t$ satisfies (5.1.1). We can choose $s, t \in A$ without common factors; then by clearing denominators we get

$$s^n + a_{n-1}s^{n-1}t + \cdots + a_0t^n = 0.$$

Thus t divides s^n, which can only hold if t is invertible in A, hence $x \in A$. □

Example 5.1.15.

(a) Let A be an integral domain with field of fractions k, and let k'/k be an extension of fields. Then the integral closure of A inside k' is integrally closed. This is often applied with $A = \mathbb{Z}$, to produce a number of rings of interest in arithmetic.

Often it is difficult to determine exactly the integral closure, but we will see that this can be done in some cases. The case of quadratic extensions of $\mathbb{Q}$ is Example 6.1.2, while the case of cyclotomic extensions is treated in Section 6.8.

(b) By way of example, we can check that $\mathbb{Z}[i]$ is integral over $\mathbb{Z}$. Indeed the element $z = a + ib$ satisfies

$$z^2 - 2az + (a^2 + b^2) = 0.$$

Since $\mathbb{Z}[i]$ is a unique factorization domain, it is integrally closed, so it must be the integral closure of $\mathbb{Z}$ inside $\mathbb{Q}(i)$.

(c) A quotient of an integrally closed ring need not be integrally closed. For instance take $A = k[x, y]$, which is a unique factorization domain, and hence integrally closed. Consider the ideal $I = (y^2 - x^3)$; then A/I is not integrally closed.

Indeed, the element

$$f = \frac{y}{x}$$

satisfies $f^2 - x = 0$, but $f \notin A/I$.

Remark 5.1.16. In spite of the previous example, in some sense integral closure behaves well with respect to quotients.

More precisely, let $A \subset B$ be rings, $I \subset B$. If B is integral over A, B/I is integral over $A/A \cap I$. This is immediately checked, by reducing a monic equation for $b \in B$ modulo I.

For the localization we have:

Proposition 5.1.17. *Let $A \subset B$ be rings, $S \subset A$ a multiplicative set, and let $\overline{A}$ be the integral closure of A in B. Then $S^{-1}\overline{A}$ is the integral closure of $S^{-1}A$ in $S^{-1}B$.*

In particular, if B is integral over A, $S^{-1}B$ is integral over $S^{-1}A$.

Proof. Let $b \in \overline{A}$, $s \in S$. Since b is integral over A, b satisfies an equation

$$b^n + a_{n-1}b^{n-1} + \cdots + a_0 = 0,$$

whence

$$\left(\frac{b}{s}\right)^n + \frac{a_{n-1}}{s}\left(\frac{b}{s}\right)^{n-1} + \cdots + \frac{a_0}{s^n} = 0,$$

which shows that b/s is integral over $S^{-1}A$.

Conversely, assume b/s is integral over $S^{-1}A$. A monic equation for b/s can be written in the form

$$\left(\frac{b}{s}\right)^n + \frac{a_{n-1}}{t}\left(\frac{b}{s}\right)^{n-1} + \cdots + \frac{a_0}{t} = 0$$

for some $t \in S$ (after multiplying denominators). Then

$$(bt)^n + a_{n-1}st \cdot (bt)^{n-1} + \cdots + a_0 s^n t^{n-1} = 0,$$

so $bt \in \overline{A}$. Hence

$$\frac{b}{s} = \frac{bt}{st} \in S^{-1}A. \qquad \square$$

Corollary 5.1.18. *Let A be an integral domain. A is integrally closed if, and only if, every localization A_P at a prime ideal P is integrally closed.*

Proof. Let k be the field of fractions of A, $\overline{A}$ the integral closure of A inside k. Let $\iota\colon A \hookrightarrow \overline{A}$ be the inclusion; for each prime $P \subset A$ let $S = A \setminus P$ and consider the map

$$\iota_P\colon A_P \hookrightarrow S^{-1}\overline{A}$$

induced in the localization. By the proposition $S^{-1}\overline{A}$ is the integral closure of A_P in k.

By Corollary 1.6.20, ι is surjective if, and only if, each ι_P is, hence the thesis. $\square$

As another application of this concept, we can give an alternative characterization of Dedekind rings.

Theorem 5.1.19. *Let A be an integral domain. A is a Dedekind ring if, and only if, it satisfies the following conditions:*

(i) *A is Noetherian;*

(ii) *A is integrally closed;*

(iii) *every nonzero prime ideal of A is maximal.*

Proof. First assume that A is a Dedekind ring. A is Noetherian by definition, and satisfies iii) by Remark 3.4.5. We only need to show that A is integrally closed, but that is identical to the proof of Proposition 5.1.14, using unique factorization for ideals.

For the converse, assume A satisfies the conditions of the Theorem, and let P be a prime of A. Then A_P is again Noetherian and integrally closed, and moreover is local with maximal ideal $\mathcal{M} = PA_P$. By primary decomposition, every nonzero ideal $I \subset A_P$ is the intersection of primary ideals; since $\mathcal{M}$ is the only nonzero prime of A_P, each component must be $\mathcal{M}$-primary. Hence I is $\mathcal{M}$-primary itself, and in particular $\sqrt{I} = \mathcal{M}$.

First we show that $\mathcal{M}$ is principal. Start from any nonzero $a \in \mathcal{M}$; by Proposition 2.3.10 there is some power $\mathcal{M}^n \subset (a)$. Take n minimal, so $\mathcal{M}^{n-1} \not\subset (a)$, and take some element $b \in \mathcal{M}^{n-1} \setminus (a)$.

Let $x := a/b$; we claim that $\mathcal{M} = (x)$. Note that by construction

$$\frac{1}{x}\mathcal{M} \subset \frac{1}{a}\mathcal{M}^n \subset A,$$

since every element of $\mathcal{M}^n$ is multiple of a. On the other hand, assume

$$\frac{1}{x}\mathcal{M} \subset \mathcal{M};$$

since $\mathcal{M}$ is a faithful $A[1/x]$-module, it follows that x is integral over A by Theorem 5.1.6. Since A is integrally closed, $1/x = b/a \in A$, which we have excluded.

Hence

$$\frac{1}{x}\mathcal{M} = A,$$

which means $\mathcal{M} = (x)$. Now we exploit that the choice of b above was arbitrary in $\mathcal{M}^{n-1} \setminus (a)$.

Taking $b = x^{n-1}$ shows that a/x^{n-1} is another generator x' of $\mathcal{M}$, say $x' = xu$ for some $u \in A^*$. In particular $(a) = (x^n)$, so every principal ideal is a power of $\mathcal{M}$.

Since every element of A is a power of x up to a unit, every ideal is generated by powers of x, so it must be a power of $\mathcal{M}$. □

Corollary 5.1.20. *A unique factorization domain A is a principal ideal domain if, and only if, every prime $P \subset A$ is maximal.*

Proof. One implication is Proposition 2.1.4.

Conversely, let A be a unique factorization domain; then A is integrally closed by Proposition 5.1.14, and we are assuming that every prime is maximal. If we show that A is Noetherian, it is a Dedekind ring, and so we are done by Proposition 3.4.9.

The first part of the proof of Proposition 3.4.9 carries over, and tells us that prime ideals of A are principal. By Exercise 8 of Chapter 2, this is enough to show that A is Noetherian. □

5.2. Going up and down

In this section we consider the following situation. Let $A \subset B$ be an integral extension; what is the relation between prime ideals in A and B? It turns out that there is a beautiful description which relates chains of prime ideals in the two rings. This is due to Cohen and Seidenberg ([**CS46**]).

Let $P \subset A$ and $Q \subset B$ be primes. If $P = Q \cap A$, we shall say that Q lies *over* P, or vice versa P is *under* Q. The first case that we study is that of a single prime.

So assume $P \subset A$ is a given prime; we want to find whether there is any prime of B over it. Localizing at P we can assume that P is actually maximal, and by taking a quotient we can even assume that $P = 0$ and A is a field. In this case we have:

Proposition 5.2.1. *Let $A \subset B$ be an integral extension of integral domains. Then A is a field if, and only if, B is.*

Proof. Assume A is a field, and take $b \in B$, $b \neq 0$. Then b satisfies the equation

$$b^n + a_{n-1}b^{n-1} + \cdots + a_0 = 0,$$

and we can assume $a_0 \neq 0$. Then

$$\frac{1}{b} = -\frac{1}{a_0}(b^{n-1} + a_{n-1}b^{n-2} + \cdots + a_1) \in B.$$

Conversely, assume B is a field, and take a nonzero $a \in A$. Then $1/a \in B$, so it satisfies the equation

$$\left(\frac{1}{a}\right)^n + a_{n-1}\left(\frac{1}{a}\right)^{n-1} + \cdots + a_0 = 0,$$

whence

$$\frac{1}{a} = -(a_{n-1} + a_{n-2} \cdot a + \cdots + a_0 \cdot a^{n-1}) \in A. \qquad \square$$

We now have a series of formal consequences.

Corollary 5.2.2. *Let $A \subset B$ be an integral extension, Q a prime of B over P, that is, $Q \cap A = P$. Then P is maximal if, and only if, Q is.*

Corollary 5.2.3. *Let $A \subset B$ be an integral extension, P a prime of A. Then there exists a prime Q of B over P.*

Proof. Let $S = A \setminus P$; by Proposition 5.1.17 we know that

$$A_P = S^{-1}A \subset S^{-1}B$$

is integral. Let $\mathcal{M} \subset S^{-1}B$ be any maximal ideal. Then $\mathcal{M} \cap A_P$ is maximal, so it must be PA_P. We conclude by the bijective correspondence for prime ideals in the localization (Corollary 1.6.9). $\square$

Corollary 5.2.4. *Let $A \subset B$ be an integral extension, P a prime of A and assume $Q \subset Q'$ are primes of B over P. Then $Q = Q'$.*

Proof. By localization we can assume that P is maximal. Then Q and Q' must be both maximal, so they are equal. $\square$

Remark 5.2.5. The above results tell us that we can always lift a prime P of A, and that there cannot be two lifts one inside the other. It is by no means true that the lift is unique, though. For instance the primes $Q_1 = (2 + i)$ and $Q_2 = (2 - i)$ of $\mathbb{Z}[i]$ both satisfy $Q_i \cap \mathbb{Z} = (5)$.

We can do a little better and lift chains of prime ideals.

Theorem 5.2.6 (Going up). *Let $A \subset B$ be an integral extension. Let $P_1 \subsetneq \cdots \subsetneq P_n$ be a chain of primes of A and $Q_1 \subsetneq \cdots \subsetneq Q_m$ $(m \leq n)$ a chain of primes of B such that Q_i is over P_i. Then we can extend this to a chain $Q_1 \subsetneq \cdots \subsetneq Q_n$ of primes of B with the same property.*

Proof. We can forget the first $m-1$ primes, so assume $m=1$. Moreover by induction we only need to do the step $n=2$. Finally we can quotient by P_1 and Q_1, and assume that both are 0. By Corollary 5.2.3 we can lift P_2, and we are done. □

Remark 5.2.7. Producing a chain in A from a chain in B is easy. Indeed assume $Q_1 \subsetneq \cdots \subsetneq Q_n$ is a chain of prime ideals of B, and let $P_i = Q_i \cap A$. Then the P_i are again prime, and they are distinct by Corollary 5.2.4, so we have the desired chain $P_1 \subsetneq \cdots \subsetneq P_n$.

Example 5.2.8.

(a) Consider the inclusion $k \subset k[x]/(x^2)$; this is an integral extension since x satisfies $x^2 = 0$. Note that both 0 and (x) restrict to 0 on k, so the hypothesis that B is an integral domain is necessary in Proposition 5.2.1.

(b) The same example shows that over the prime 0 we have both 0 and (x), so Corollary 5.2.4 does not extend to ideals which are not prime; not even primary is enough. Indeed in the ring $B = k[x]/(x^2)$ the ideal (x) is prime and 0 is primary, as every divisor of 0 in B is nilpotent.

(c) The going up property fails for nonprime ideals. To see this, consider the integral extension $k[x^2, x^3] \subset k[x]$, and take the ideal $I = (x^2) \subset k[x^2, x^3]$. Every ideal $J \subset k[x]$ that contains I must contain x^3 as well, but $x^3 \notin I$, and hence there is no ideal of $k[x]$ that lies over I.

(d) For extensions which are not integral, there is no hope to have any result as the above. This will be more apparent after we discuss dimension.

For now just consider the inclusion $k[x] \subset k[x,y]$. For every polynomial $f = f(y)$ we have $(f) \cap k[x] = (0)$; in particular both 0 and (y) are primes above 0.

(e) Similarly consider the extension $k[x] \subset k[x, 1/x]$, which is not integral. Since x becomes invertible in $k[x, 1/x]$, there is no ideal of $k[x, 1/x]$, prime or not, over the prime ideal (x).

There is a similar result which allows us to extend *descending* chains of prime ideals, but that requires more work. First we have to generalize our definition of integral dependence.

Definition 5.2.9. Let $A \subset B$ be an inclusion of rings, $I \subset A$ an ideal. We say that an element $b \in B$ is *integral over* I if it satisfies a monic equation

$$b^n + a_{n-1}b^{n-1} + \cdots + a_0 = 0, \tag{5.2.1}$$

where all $a_r \in I$. The set of all such elements b is called the *integral closure* of I in B.

As one may expect, this integral closure turns out to be an ideal, and it is even expressible in terms of concepts we already know.

Proposition 5.2.10. *Let $A \subset B$ be rings, $\overline{A}$ the integral closure of A in B, and let $I \subset A$ be an ideal. Then the integral closure of I in B is $\sqrt{I \cdot \overline{A}}$.*

Proof. Every element which is integral over I is certainly in $\overline{A}$, so to simplify notation we can assume $\overline{A} = B$. We also denote $J := I \cdot B$.

If $b \in B$ satisfies (5.2.1), $b^n \in J$, hence $b \in \sqrt{J}$.

Vice versa, let $b \in \sqrt{J}$, so

$$b^n = a_i b_i + \cdots + a_k b_k$$

for some $n > 0$ and elements $a_i \in I$ and $b_i \in B$. The ring $B' = A[b_1, \ldots, b_k]$ is a finitely generated A-module, since all b_i are integral over A.

We can now apply the determinant trick, using the fact that $b^n B' \subset I \cdot B'$, to obtain a monic equation for b whose nonleading coefficients are all in I. □

Corollary 5.2.11. *Let $A \subset B$ be integral domains and assume A is integrally closed. Denote $k = \mathcal{F}(A)$, and take an element $b \in B$ integral over an ideal $I \subset A$. If we write the minimal polynomial of b over k as*

$$f(x) = x^n + a_{n-1}x^{n-1} + \cdots + a_0,$$

then all $a_i \in \sqrt{I}$.

Proof. Let k' be the field of fractions of B; we can assume that k' contains all the conjugates b_i of b. The coefficients of f are polynomials in the b_i (the elementary symmetric functions if the extension k'/k is separable).

Each b_i is integral over I, since it satisfies the same equation as b, so the coefficients of f are integral over I. By the previous proposition and the assumption that A is integrally closed, these coefficients are in $\sqrt{I}$. □

We need a last result:

Lemma 5.2.12. *Let $f\colon A \to B$ be a ring homomorphism, and let $P \subset A$ be a prime. There there exists a prime $Q \subset B$ such that $f^{-1}(Q) = P$ if, and only if, $f^{-1}(f(P) \cdot B) = P$.*

Proof. The condition is clearly necessary.

So assume $f^{-1}(f(P) \cdot B) = P$. Let $S = A \setminus P$; by localizing at S we can assume that A is local with maximal ideal P.

In this case let Q be any maximal ideal containing $f(P) \cdot B$; then $f^{-1}(Q) \supset P$, so we must have equality by maximality. $\square$

Theorem 5.2.13 (Going down). *Let $A \subset B$ be an integral extension of integral domains, and assume that A is integrally closed.*

Let $P_1 \supsetneq \cdots \supsetneq P_n$ be a chain of primes of A and $Q_1 \supsetneq \cdots \supsetneq Q_m$ $(m \leq n)$ a chain of primes of B such that Q_i is over P_i. Then we can extend this to a chain $Q_1 \supsetneq \cdots \supsetneq Q_n$ of primes of B with the same property.

Proof. As in the proof of going up, we can assume $m = 1$, $n = 2$. By Lemma 5.2.12, it is enough to show that

$$P_2 B_{Q_1} \cap A = P_2.$$

Take an element

$$b = \frac{c}{s} \in P_2 B_{Q_1},$$

with $c \in P_2 B$ and $s \in B \setminus Q_1$. By Proposition 5.2.10, c is integral over P_2, so by Corollary 5.2.11 the minimal polynomial of c over $k := \mathcal{F}(A)$ is

$$f(x) = x^n + a_{n-1}x^{n-1} + \cdots + a_0,$$

with $a_i \in P_2$.

If moreover $b \in A$ we find that a minimal equation for $s = c/b$ over k is

$$s^n + \frac{a_{n-1}}{b} s^{n-1} + \cdots + \frac{a_0}{b^n} = 0.$$

Again by Corollary 5.2.11 we see that the coefficients of this equation are in A. If $b \notin P_2$, all coefficients must lie in P_2, so $s^n \in P_2 \cdot B$.

But then

$$s^n \in P_1 \cdot B \subset Q_1,$$

which contradicts our choice of s. So $b \in P_2$ and we are done. $\square$

5.3. Noether normalization

In this section we prove the celebrated Noether normalization lemma, which is a structural result for integral extensions of algebras over a field.

Recall that an extension of fields can always be factored into a completely transcendental part, followed by an algebraic extension. Namely, if $K \subset L$ is an inclusion of fields, and L is finitely generated over K, there exists an intermediate field $T \subset L$ which is isomorphic to $K(x_1, \ldots, x_r)$ as K-algebras, and such that L is an algebraic extension of T. Something similar holds for algebras over a field.

Theorem 5.3.1 (Noether normalization lemma). *Let k be a field and A a finitely generated k-algebra. Then there exist $n \geq 0$ algebraically independent elements $x_1, \ldots, x_n \in A$ such that A is finitely generated as a module over $k[x_1, \ldots, x_n]$.*

Proof. Let $y_1, \ldots, y_m$ be a minimal set of generators of A. We argue by induction over m, the base case $m = 0$ being trivial.

If the y_i are already algebraically independent, there is nothing to prove, so assume that there exists a polynomial f over k such that $f(y_1, \ldots, y_m) = 0$. We can try to alter the set of generators and use the relation given by f to write one of the generators in terms of the other ones.

We perform the change of variable

$$z_i = y_i - y_1^{r^{i-1}}$$

for $i \geq 2$ (r to be chosen later). This is an invertible change of variable which has the effect of increasing the degree of y_1 inside f.

More precisely, in the equation

$$f(y_1, z_2 + y_1^r, \ldots, z_m + y_1^{r^{m-1}})$$

we can look at the exponent of y_1 in the various monomials. For a monomial of multidegree $(\alpha_1, \ldots, \alpha_m)$ (in the y_i), the exponent of y_1 after the change of variables is

$$\alpha_1 + \alpha_2 r + \cdots + \alpha_m r^{m-1}.$$

For r big enough, we see that there is a single monomial which will be of maximum degree for y_1 such that the z_i do not appear in this monomial. This means that the transformed equation shows that y_1 is integral over $B := k[z_2, \ldots, z_m]$.

By induction, B already has the desired form of an integral extension of a polynomial ring, and the thesis follows. □

Remark 5.3.2. When k is infinite, one can use a *linear* change of variables in the above proof. The only thing that is needed is that in the resulting polynomial relation, y_1 appears with maximum degree in a monomial that contains no other indeterminates.

Consider a change of variable of the form

$$z_i = y_i - \lambda_i y_1.$$

This change of variable preserves the total degree of monomials, so consider the homogeneous part of maximum degree of f, call it F. We want to ensure that in F there appears a monomial containing only y_1. Let t be the degree of F.

The coefficient of y_1^t in

$$F(y_1, z_2 + \lambda_2 y_1, \dots, z_m + \lambda_m y_1)$$

is a nonzero polynomial in $\lambda_2, \dots, \lambda_m$. Since k is infinite, there exists a choice of $\lambda_2, \dots, \lambda_m$ that makes this coefficient not vanish, which is enough to show that y_1 is integral over $k[z_2, \dots, z_m]$.

5.4. Integral extensions of Dedekind rings

The theory developed so far allows us to give some important examples of Dedekind rings. The main tool which we shall use is the following

Theorem 5.4.1. *Let A be a Dedekind ring with field of fractions K, and let L be a finite separable extension of K. Let B be the integral closure of A in L. Then B is itself a Dedekind ring.*

A few remarks are in order before proving the theorem. Recall from Section A.7 that in this situation we have an additive homomorphism

$$\mathrm{Tr} := \mathrm{Tr}_{L/K} \colon L \to K$$

called the trace. Now assume $b \in B$, and let $\sigma \colon L \to \overline{K}$ be an embedding in an algebraic closure of K; then the conjugate $\sigma(b)$ is again integral over A, having the same minimal polynomial. It follows that the trace $\mathrm{Tr}(b)$ is integral over A, and since A is integrally closed we find $\mathrm{Tr}(b) \in A$. So we get by restriction an additive homomorphism

$$\mathrm{Tr} \colon B \to A.$$

Similarly we can see that $\mathrm{N}(B) \subset A$.

We isolate the main step, which is of independent interest.

Proposition 5.4.2. *Let A be a Noetherian integral domain with field of fractions K, and let L be a finite separable extension of K. Let B be the integral closure of A in L. Then B is a finitely generated A-module; in particular it is itself Noetherian.*

Proof. Let $n = [L : K]$ be the degree of the extension. First, observe that B contains a free rank n A-module. Namely, L is generated over K by a single element α, and after multipliying by a suitable element of K we can assume that $\alpha \in B$. So one can take the subring $A(\alpha) \subset B$, which as an A-module is just

$$A(\alpha) = \left\langle 1, \alpha, \dots, \alpha^{n-1} \right\rangle_A.$$

For any such submodule $M \subset B$ with basis $u_1, \dots, u_n$, consider the dual basis $v_1, \dots, v_n$ with respect to the trace (Definition 4.2.5), and let

$$N := \langle v_1, \dots, v_n \rangle_A.$$

We claim that $B \subset N$; since N is a Noetherian A-module, it will follow that B is finitely generated.

To see the claim, first note that since $\{v_1, \ldots, v_n\}$ is a basis of L over K, any element $b \in B$ can be written as

$$b = a_1 v_1 + \cdots + a_n v_n$$

for some $a_1, \ldots, a_n \in K$. Multiplying by u_i and taking the trace we find

$$a_i = \operatorname{Tr}(bu_i) \in A,$$

since $bu_i \in B$. □

Proof of Theorem 5.4.1. We use the characterization of Dedekind rings from Theorem 5.1.19. It is clear that B is integrally closed. Moreover, there cannot be a strict inclusion $P \subsetneq Q$ of nonzero primes of B, by Corollary 5.2.4 and the fact that every nonzero prime of A is maximal. Finally, B is Noetherian by Proposition 5.4.2. □

The result applies in particular when A is a principal ideal domain, so we can form a lot of new Dedekind rings. A special case is obtained when $A = \mathbb{Z}$ and L is a number field, that is, a finite extension of $\mathbb{Q}$.

Definition 5.4.3. Let K be a finite extension of $\mathbb{Q}$. Then K is called a *number field*. The integral closure of $\mathbb{Z}$ inside K is called its *ring of integers*, and it is sometimes denoted $\mathcal{O}_K$. For brevity, we will also call it a *number ring*, although this terminology is not entirely standard.

This provides a lot of examples of Dedekind rings which are of great interest in arithmetic—one of the early successes of commutative algebra was developing the unique factorization for ideals in integral extensions of $\mathbb{Z}$. We will study these rings in greater detail in Chapter 6.

These are not the only examples, though: in Exercise 17 of Chapter 6 you will see an example of a Dedekind ring which is not the integral closure of a principal ideal domain in its fraction field.

In fact, Theorem 5.4.1 is true even in the inseparable case. In this situation, though, Proposition 5.4.2 fails: the integral closure of a Dedekind ring in a finite extension of its fraction field can fail to be finite (Exercise 19). We will thus pursue a different strategy that avoids the construction of the dual basis.

Theorem 5.4.4 (Krull–Akizuki, [**Alo99**]). *Let A be an integral domain with fraction field K, B a ring such that $A \subset B \subsetneq K$. Assume that*

(i) *A is Noetherian*

(ii) *every nonzero prime of A is maximal.*

Then i) *and* ii) *hold for B as well.*

In the terminology that we will introduce in Chapter 9, conditions i) and ii) say that A is a 1-dimensional Noetherian ring.

Proof. Let $a \in A$ be a nonzero element. Then A/aA is Noetherian, and every prime of this ring is maximal. By Theorem 2.4.11, A/aA is Artinian. We claim that B/aB *is Artinian* as well.

Given this claim, we can easily conclude the proof. In fact, let $I \subset B$ be any nonzero ideal. Then $I \cap A \neq 0$ (why?), and we can choose a nonzero $a \in I \cap A$. Then the image of I in B/aB is finitely generated, so I is finitely generated as well, which proves that B is Noetherian. If moreover I is prime, its image in B/aB is maximal, hence I is maximal as well.

To prove the claim, consider the ideals

$$I_n := a^n B \cap A + aA$$

of A. Let $\overline{I_n}$ be the image of I_n in A/aA. This is a decreasing sequence of ideals, hence it stabilizes. By the correspondence of ideals, the same must hold for $\{I_n\}$, so there is a a minimum m such that $I_n = I_m$ for all $n \geq m$. We want to derive from this that

$$a^m B \subset a^{m+1} B + A. \tag{5.4.1}$$

Both members of this relation are A-submodules of K, hence the inclusion holds if and only if it holds between their localizations at any maximal ideal of A. This allows us to prove (5.4.1) under the additional hypothesis that A is local. In this case, let $\mathcal{M}$ be its maximal ideal. We can also assume that $a \in \mathcal{M}$, otherwise it is a unit and the inclusion is trivial.

Take any nonzero element $x = b/c \in B$, where $b, c \in A$. Each ideal in a primary decomposition for (c) is $\mathcal{M}$-primary, hence $\mathcal{M}^n \subset cA$ for some $n \geq m$. (Why?) But then $a^{n+1}x \in A$ (in fact, already $a^n x \in A$), so

$$a^{n+1}x \in a^{n+1}B \cap A + aA = I_{n+1} = I_{n+2} = a^{n+2}B \cap A + aA.$$

We can simplify this to

$$a^n x \in a^{n+1} B + A. \tag{5.4.2}$$

In fact, this holds for $n = m$. To see this, take n minimal such that (5.4.2) holds. If $n > m$, then

$$\begin{aligned} a^n x \quad &\in (a^{n+1}B + A) \cap a^n B = a^{n+1}B + (a^n B \cap A) \\ &\subset a^{n+1}B + (a^{n+1}B \cap A) + aA = a^{n+1}B + aA, \end{aligned}$$

so $a^{n-1}x \in a^n B + A$, which contradicts the fact that n is minimal. Hence, (5.4.2) holds with $n = m$, and finally since $x \in B$ is arbitrary, this implies (5.4.1).

Knowing this, we can finally prove our claim that B/aB is Artinian. In fact,

$$\frac{B}{aB} \cong \frac{a^m B}{a^{m+1}B} \subset \frac{a^{m+1}B + A}{a^{m+1}B} \cong \frac{A}{a^{m+1}B \cap A},$$

which proves that B/aB is an A/aA-module of finite length, hence it is Artinian as an A/aA-module, and a fortiori as a B/aB-module. □

Corollary 5.4.5. *Let A be an integral domain with fraction field K, L/K a finite field extension and B the integral closure of A in L. Assume that*

(i) *A is Noetherian*

(ii) *every nonzero prime of A is maximal.*

Then i) *and* ii) *hold for B as well. In particular, if A is a Dedekind domain, B is also a Dedekind domain.*

Proof. Take $b_1, \ldots, b_n \in B$ that generate L over K. Then the ring $C := A[b_1, \ldots, b_n]$ is finitely generated over A, hence Noetherian. Also, by Theorem 5.2.6, every nonzero prime of C is maximal. Since $C \subset B$, the thesis follows from Krull–Akizuki theorem. □

5.5. Exercises

1. Let A be a ring, $\overline{A}$ its integral closure inside the total ring of fractions $\mathcal{F}(A)$. Prove that

$$\overline{A} = \bigcup_{\substack{I \text{ fractional} \\ \text{ideal of } A}} (I :_{\mathcal{F}(A)} I).$$

2. Prove the following strengthening of Gauss's lemma. Let $A \subset B$ be integral domains, $f \in A[x]$ a monic polynomial, and assume we have a factorization $f = gh$, where $g, h \in B[x]$ are monic. Then the coefficients of g and h are integral over A. Can you prove the same result without assuming that A and B are integral domains?

3. Let A be an integrally closed integral domain, $f \in A[x]$ an irreducible monic polynomial. Then f is prime.

4. Let k be a field, $B = k[x, y]$, and define

$$A := \{f \in B \mid f(0,0) = f(1,1)\}.$$

Consider the ideal $Q_2 = (x-1, y-1) \subset B$, $P_2 = Q_2 \cap A$ and $P_1 = (x) \cap A$. Show that $P_1 \subset P_2$, but there is no prime $Q_1 \subset Q_2$ that lies over P_1, hence going-down fails for the inclusion $A \subset B$.

5. Compute the integral closure of $A := k[x, y]/(y^2 - x^3 - x^2)$, where k is a field. (Notice that y/x is integral over A.)

6. Let B be the integral closure of $A := k[x, y, z]/(y^2 - x^3 - x^2)$, where k is a field. Show that the inclusion $A \subset B$ does not satisfy the going-down property.

7. Let A be a ring with an action of a finite group G by ring homomorphisms. Denoting

$$A^G := \{a \in A \mid g \cdot a = a \text{ for all } g \in G\}$$

the invariant ring, prove that A is integral over A^G. In particular, if A is a finitely generated k-algebra, A^G is finitely generated as a k-algebra as well. In this case, we conclude that A^G is Noetherian.

The following exercises, up to Exercise 13, give some properties of the *conductor* ideal, a tool which is often used to study the properties of rings that are not integrally closed. In particular, we will use it to extend the property of ideal factorization from Dedekind rings to a slightly more general class of rings, with some restrictions.

8. Let $A \subset B$ be rings. The *conductor* ideal of A in B is defined as

$$\mathfrak{f}(B/A) := \mathrm{Ann}_A(B/A) = \{a \in A \mid aB \subset A\}.$$

Show that $\mathfrak{f}(B/A)$ is an ideal both of A and B. Moreover, if I is any ideal of B such that $I \subset A$, then $I \subset \mathfrak{f}(B/A)$. Deduce that $\mathfrak{f}(B/A)$ is the largest common ideal of A and B.

9. Let $A \subset B$ be rings, $S \subset A$ a multiplicative set. Show that

$$S^{-1}\mathfrak{f}(B/A) \subset \mathfrak{f}(S^{-1}B/S^{-1}A).$$

If moreover B is finitely generated as an A-algebra and integral over A, then equality holds.

10. Let A be an integral domain with field of fractions k, and take a ring B such that $A \subset B \subset k$. Then, as A-modules,

$$\mathfrak{f}(B/A) \cong \mathrm{Hom}_A(B, A).$$

11. Let A be a Noetherian domain and assume that every nonzero prime ideal of A is maximal. Let B be the integral closure of A—by Krull–Akizuki, this is a Dedekind ring. Let $I \subset B$ be an ideal that is relatively prime to the conductor $\mathfrak{f}(B/A)$. Show that $I \cap A$ is also relatively prime to $\mathfrak{f}(B/A)$ as ideals of A. Moreover, the homomorphism $A/I \cap A \to B/I$ is an isomorphism. Deduce that I is prime if and only if $I \cap A$ is prime.

12. Let $A \subset B$ be as in Exercise 11. Let $I \subset A$ be an ideal that is relatively prime to the conductor $\mathfrak{f}(B/A)$. Show that $I \cdot B$ is also relatively prime to $\mathfrak{f}(B/A)$ as ideals of B. Moreover, the homomorphism $A/I \to B/IB$ is an isomorphism. Deduce that I is prime if and only if $I \cdot B$ is prime.

13. Let $A \subset B$ be as in Exercise 11. Show that there is a bijection between ideals of A relatively prime to $\mathfrak{f}(B/A)$ and ideals of B relatively prime to $\mathfrak{f}(B/A)$, and that this correspondence is multiplicative and preserves the property of being prime. Deduce from the fact that B is a Dedekind domain a factorization theorem for A. Namely, ideals of A that are relatively prime to the conductor have the property of unique factorization into primes.

14. Use the Noether normalization lemma to prove this variant of *generic freeness* (see Corollary 2.5.18): let A be a Noetherian integral domain, B a finitely generated A-algebra; then there exists a nonzero $f \in A$ such that B_f is a free A_f-module.

15 ([**Huc76**, Prop 3.1])**.** Let A be a Noetherian local ring, $P \subset A$ a prime and I a P-primary ideal different from P itself. Let $B := A/I$ and C the integral closure of B inside its total ring of fractions. Prove that B is Noetherian, but C is not. (Let $b \in B$ be any element that is not a unit or a zero divisor. The nilradical $\mathcal{N}(C)$ satisfies $b\mathcal{N}(C) = \mathcal{N}(C)$.)

In fact, integral closure does not preserve the property of being Noetherian even for integral domains; see [**Nag62**, Appendix, Example E.5].

The following exercises, up to 19, give (a variant of) an example from [**Kap70**, Theorem 100] of an integral extension of Dedekind rings which is not finite. Our presentation here comes from [**Knaa**].

16. Let k be a field. Show that $k((t)) := \mathcal{F}(k[[t]])$ is transcendental over $k(t)$.

17. Let k be a field and $a \in k[[x]]$ an element transcendental over $k(x)$. Show that the ring $A := k[[x]] \cap k(x, a)$ is a Dedekind ring and in particular is integrally closed.

18. Let k be a field of characteristic p, $a \in k[[x]]$ an element transcendental over $k(x)$, and let $b = a^p$. Define the rings $A := k[[x]] \cap k(x, a)$ and $B := k[[x]] \cap k(x, b)$. Show that A is the integral closure of B inside $\mathcal{F}(A)$.

19. Let A, B be the rings of the previous exercise. Show that A is not a finitely generated B-module. (For any $n \in \mathbb{N}$, b can be written as $b = f_n^p + x^{pn} b_n$, where $f_n \in k[x]$ and $b_n \in k[[x]]$. Then b_n has a unique p-th root satisfying $b_n^{1/p} = x^{-n}(-f_n + a)$. If A was finitely generated, one could find $c \in A$ such that $cB \subset A[a]$. Show that this implies that x^n divides c in A, for all n, and find a contradiction.)

20. Prove the following strengthening of the Krull–Akizuki theorem. Let A be a reduced Noetherian ring, T its total ring of fractions, and B a ring such that $A \subset B \subsetneq T$. Assume that A is Noetherian and every nonzero prime ideal of A is maximal; then the same holds for B. (Consider the minimal primes of A and, for each such prime P, the map $B \to T \to \mathcal{F}(A/P)$.)

21. Let A be an integral domain, B its integral closure inside the algebraic closure of $k := \mathcal{F}(A)$. Prove that for any primes $P_1, \ldots, P_n \subset B$, the sum $P_1 + \cdots + P_n$ is either the whole of B or a prime ideal. This is a result of Artin from [**Art71**]. (Reduce to two primes P_1 and P_2. Assume $xy \in P_1 + P_2$ and define $z = y - x$, so that $x^2 + zx = a_1 + a_2$, with $a_i \in P_i$. Let t be a solution to $t^2 + zt = a_1$, and argue that either $t \in P_1$ or $t + z \in P_1$. Find a way to factor a_2, and check that in all cases either $x \in P_1 + P_2$ or $y \in P_1 + P_2$. This proof comes from [**HH92**].)

Chapter 6

Lattice Methods

This chapter is somewhat more specialized to the cases of interest in algebraic number theory. We are interested in rings that are obtained by taking the integral closure of $\mathbb{Z}$ inside a finite extension K of $\mathbb{Q}$. As an additive group, such a ring A is isomorphic to $\mathbb{Z}^n$, where n is the degree of K over $\mathbb{Q}$. Ideals can then be viewed as sublattices inside $\mathbb{Z}^n$, and this allows us to study them using geometric techniques. In particular, the *norm* of an ideal I—which measures how much I is sparser than A—will be a fundamental technical tool. While most of the chapter is devoted to this setting, some parts are valid over more general Dedekind rings.

In the first section, we study the properties of the integral closure of $\mathbb{Z}$ inside number fields; in particular their structure as an additive group. We also investigate what happens to prime ideals under extension. The norm of an ideal is the tool that we use here. We also generalize these results to Dedekind rings with a different technique. In the next section we study the case of an integral extension of Dedekind domains when the corresponding field extension is Galois.

We then go back to the special case of number fields, where we use more geometric techniques to prove bounds on the norm of ideals, from which we derive important results about the finiteness of the ideal class group and the finite generation of the unit group. Finally, we specialize these concepts to the interesting case of rings of integers inside cyclotomic extensions of $\mathbb{Q}$. This allows us to present a slick proof of the celebrated quadratic reciprocity formula of Gauss.

This chapter, together with the material needed from the previous ones (in particular the factorization theorem for Dedekind domains from Chapter

3), gives a quick introduction to the ideas and methods from algebraic number theory. It can be complemented with some parts of Chapter 7, which introduces the completion of rings and defines the so-called p-adic rings, giving a *local* analogue of the theory developed here. For more about number fields, [**Mar77**] is an excellent reference.

6.1. Additive structure of number rings

Let K be a number field, A its ring of integers (see Definition 5.4.3). By Theorem 5.4.1, A is a Dedekind ring. In particular, we can apply the results of Section 3.4, and factor every ideal I as a product of prime powers $I = P_1^{e_1} \cdots P_r^{e_r}$ in a unique way.

As a $\mathbb{Z}$-module, the structure of A is pretty simple.

Proposition 6.1.1. *Let $n = [K : \mathbb{Q}]$. Then A is a free $\mathbb{Z}$-module of rank n.*

Proof. By clearing denominators, we can find a $\mathbb{Q}$-basis for K given by elements $a_1, \ldots, a_n \in A$. Let $b_1, \ldots, b_n$ be the dual basis with respect to the trace (Definition 4.2.5). Then for every $a \in A$ we can write

$$a = \sum_{i=1}^{n} c_i b_i$$

for some $c_1, \ldots, c_n \in \mathbb{Q}$. But then $c_i = \operatorname{Tr}(a \cdot a_i) \in \mathbb{Z}$ by duality.

This shows that we have inclusions

$$\mathbb{Z}a_1 + \cdots + \mathbb{Z}a_n \subset A \subset \mathbb{Z}b_1 + \cdots + \mathbb{Z}b_n.$$

The latter inclusion implies that A is a free $\mathbb{Z}$-module of rank at most n, and the former shows that the rank is exactly n. □

Example 6.1.2. Let $K = \mathbb{Q}(\sqrt{m})$ be a quadratic number field; we can assume that m is squarefree. (Why?) Let us compute a basis for its ring of integers.

A number $\alpha = x + y\sqrt{m} \in K$ is an algebraic integer if and only if $\operatorname{Tr}(\alpha) = 2x$ and $\operatorname{N}(\alpha) = x^2 - my^2$ are in $\mathbb{Z}$. Let us write $x = a/2$ and $y = p/q$, where $a, p, q \in \mathbb{Z}$ and p and q are relatively prime. To have $\operatorname{N}(\alpha) \in \mathbb{Z}$ is equivalent to

$$a^2q^2 - 4mp^2 \equiv 0 \pmod{4q^2}.$$

This implies that q^2 divides $4m$, and since m is squarefree, q divides 2. So we can take $y = b/2$, with $b \in \mathbb{Z}$.

With this notation, α is integral over $\mathbb{Z}$ if and only if

$$a^2 \equiv mb^2 \pmod{4}.$$

Since m is squarefree, $m \equiv 1, 2$ or 3 modulo 4.

(a) If $m \equiv 2,3 \pmod 4$, a and b must both be even. In this case, $\alpha = s + t\sqrt{m}$, with $s, t \in \mathbb{Z}$, so a basis for $\mathcal{O}_K$ is $\{1, \sqrt{m}\}$

(b) If $m \equiv 1 \pmod 4$, a and b need just have the same parity. In this case, a basis for $\mathcal{O}_K$ is $\left\{1, \frac{1+\sqrt{m}}{2}\right\}$.

Generalizing the example, a natural question is to find a basis for the ring A of integers in K. In particular, one may ask whether it is possible to find some $a \in A$ such that $A = \mathbb{Z}[a]$. This is not true in general, but one can get bounds such as the following.

Proposition 6.1.3. *Let $a_1, \ldots, a_n \in A$ be a basis of K over $\mathbb{Q}$, $d = \operatorname{disc}(\{a_1, \ldots, a_n\})$. Then $dA \subset \mathbb{Z}[a_1, \ldots, a_n]$.*

Recall that the discriminant was defined in 4.2.6.

Proof. Let $b_1, \ldots, b_n$ be a basis of A, so that $\vec{a} = M\vec{b}$ for some matrix M with integer coefficients. Let $B = \mathbb{Z}[a_1, \ldots, a_n]$; then A/B is a finite abelian group of cardinality equal to $|\det M|$ by 1.5.9.

By Proposition 4.2.10,

$$d = \operatorname{disc}(\{a_1, \ldots, a_n\}) = m^2 \operatorname{disc}(\{b_1, \ldots, b_n\}),$$

where $m = \det M$. Since $m(A/B) = 0$, we have $mA \subset B$, and the result follows since $m|d$. $\square$

Definition 6.1.4. If A is the ring of integers in a number field K, we define its discriminant $\operatorname{disc}(A) := \operatorname{disc}(\{a_1, \ldots, a_n\})$, where $\{a_1, \ldots, a_n\}$ is any basis of A over $\mathbb{Z}$. This is well defined by 4.2.10, since any other basis can be obtained by a change of basis for an integer matrix with determinant ± 1. We sometimes denote $\operatorname{disc}(K) := \operatorname{disc}(A)$.

The discriminant has a natural geometric interpretation. First, recall that a *lattice* is a free abelian group $\mathcal{L} \cong \mathbb{Z}^n$ of finite rank with an embedding $i \colon \mathcal{L} \hookrightarrow \mathbb{R}^n$ such that the image of a $\mathbb{Z}$-basis for $\mathcal{L}$ is a basis of $\mathbb{R}^n$ as a vector space. This is equivalent to saying that $i(\mathcal{L})$ is discrete in the Euclidean topology (Exercise 8). Given a basis $g_1, \ldots, g_n$ of $\mathcal{L}$, the linear combinations $\sum_{j=1}^n a_j i(g_j)$ with $0 \leq a_j < 1$ form a parallelogram, called a *fundamental domain* for the lattice. The figure shows an example of 2-dimensional lattice, with generators v_1 and v_2, and two choices of a fundamental domain.

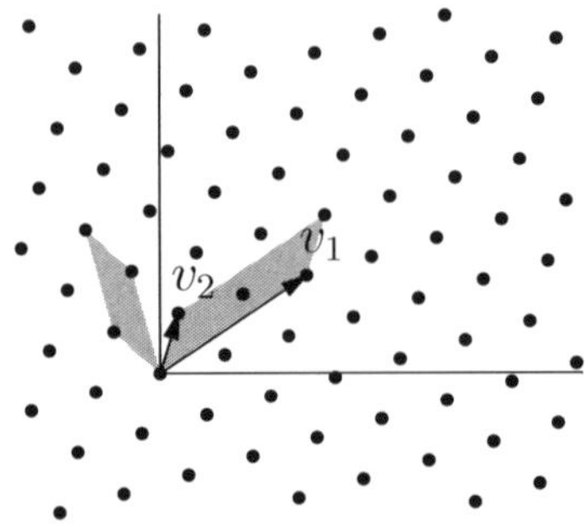

Definition 6.1.5. Let $i\colon \mathcal{L} \hookrightarrow \mathbb{R}^n$ be a lattice, $D \subset \mathbb{R}^n$ a fundamental domain with respect to the basis $\{g_1, \ldots, g_n\}$. The *volume* of D, denoted $\operatorname{vol}(D)$, is just its Lebesgue measure as a subset of $\mathbb{R}^n$. By elementary measure theory, $\operatorname{vol}(D) = |\det M|$, where M is a matrix having as columns the vectors $\{i(g_1), \ldots, i(g_n)\}$.

The *volume* of $\mathcal{L}$, denoted $\operatorname{vol}(\mathcal{L})$ is the volume of one of its fundamental domain. To see that this is well defined, notice that if we choose a different basis, the matrix for the change of basis has determinant ± 1, hence the volume of the fundamental domain does not change.

Let K be a number field with degree n over $\mathbb{Q}$, A its number ring. We have seen that—as a group—$A \cong \mathbb{Z}^n$. We can in fact view A as a lattice inside $\mathbb{R}^n$. To do so, let $K = \mathbb{Q}(\alpha)$, and say f is the minimal polynomial of α over $\mathbb{Q}$. Then f has n roots, of which, say, there are r real ones and s pairs of conjugate nonreal ones, so that $r + 2s = n$. By Corollary A.3.10, K admits n distinct embeddings into $\mathbb{C}$, each one determined by the image of α. For r such embeddings σ we have $\sigma(K) \subset \mathbb{R}$, so in fact the numbers r and s are independent of α.

Definition 6.1.6. The pair (r, s) is called the *signature* of the number field K.

The set of embeddings will be

$$\{\sigma_1, \ldots, \sigma_r, \tau_1, \overline{\tau_1}, \ldots, \tau_s, \overline{\tau_s}\},$$

where the σ_i are real. Together, they determine a homomorphism

$$\underline{\sigma}\colon K \to \mathbb{R}^r \times \mathbb{C}^s \tag{6.1.1}$$

given by $\underline{\sigma} = (\sigma_1, \ldots, \sigma_r, \tau_1, \ldots, \tau_s)$. As a real vector space, we can identify $\mathbb{R}^r \times \mathbb{C}^s$ with $\mathbb{R}^n$, and look at the image of A via $\underline{\sigma}$.

Proposition 6.1.7. *$\underline{\sigma}(A)$ is a lattice inside $\mathbb{R}^r \times \mathbb{C}^s$, whose volume is* $\operatorname{vol}(\underline{\sigma}(A)) = 2^{-s}\sqrt{|\operatorname{disc}(A)|}$.

Proof. The first assertion follows from the second one: as a group, $A \cong \mathbb{Z}^n$, and if $\underline{\sigma}(A)$ did not generate $\mathbb{R}^r \times \mathbb{C}^s$ as a real vector space, its volume would be 0.

Letting $\{\alpha_1, \ldots, \alpha_n\}$ be a $\mathbb{Z}$-basis of A, the image of A is generated by the columns of the matrix M given by

$$\begin{pmatrix} \sigma_1(\alpha_1) & \cdots & \sigma_r(\alpha_1) & \Re\tau_1(\alpha_1) & \Im\tau_1(\alpha_1) & \cdots & \Im\tau_s(\alpha_1) \\ \vdots & & \vdots & \vdots & \vdots & & \vdots \\ \sigma_1(\alpha_n) & \cdots & \sigma_r(\alpha_n) & \Re\tau_1(\alpha_n) & \Im\tau_1(\alpha_n) & \cdots & \Im\tau_s(\alpha_n) \end{pmatrix}.$$

The determinant of M is, up to sign, the volume of a fundamental domain. To compute it, we use column operations to replace the column $\Re\tau_k(\overrightarrow{\alpha})$ with $\tau_k(\overrightarrow{\alpha})$. Then, by multiplying the next column by $2i$ and subtraction, we can turn next column into $\overline{\tau_k(\overrightarrow{\alpha})}$.

The square of the determinant of the matrix thus obtained is $\operatorname{disc}(A)$, and our column operations have multiplied $\det M$ by $(2i)^s$, hence $|\det M| = 2^{-s}\sqrt{|\operatorname{disc}(A)|}$. □

There is an alternative way to look at the embedding $\underline{\sigma}$ in (6.1.1). Namely, let $K_{\mathbb{R}} = K \otimes_{\mathbb{Q}} \mathbb{R}$. Since K is a $\mathbb{Q}$-vector space of dimension n, $K_{\mathbb{R}}$ is a real vector space of the same dimension. More precisely, let $K = \mathbb{Q}(\alpha)$, and f the minimal polynomial of α. Then $K \cong \mathbb{Q}[x]/(f(x))$, so

$$K_{\mathbb{R}} \cong \frac{\mathbb{R}[x]}{(f(x))}.$$

By factoring f over $\mathbb{R}[x]$ as a product of r linear and s quadratic factors, we obtain an isomorphism $R[x]/(f) \cong \mathbb{R}^r \times \mathbb{C}^s$, and this gives another definition of the embedding

$$K \hookrightarrow K_{\mathbb{R}} \cong \mathbb{R}^r \times \mathbb{C}^s. \tag{6.1.2}$$

It is a simple verification that the two embeddings agree, for a suitable choice of the ordering of the factors of f (Exercise 11).

The discriminant can be used to bound the ring of integers of a composite field.

Proposition 6.1.8. *Let K, L be two number fields with $m = [K : \mathbb{Q}]$, $n = [L : \mathbb{Q}]$ and assume that $[KL : \mathbb{Q}] = m \cdot n$. Let A be the ring of integers of K, B that of L, and C that of KL. Setting $d = (\operatorname{disc}(A), \operatorname{disc}(B))$, we have*

$$C \subset \frac{1}{d} A \cdot B.$$

Proof. Clearly, $A \cdot B \subset C$. Let $\{a_1, \ldots, a_m\}$ be an integer basis of A and $\{b_1, \ldots, b_n\}$ an integer basis of B. The set $\{a_i \cdot b_j\}$ is a $\mathbb{Q}$-basis of KL.

C contains the product $A \cdot B$ as a subgroup of finite index, since both have rank mn. If c is this index, $c \cdot C \subset A \cdot B$. So, for $t \in C$ we can write

$$t = \sum_{i=1}^{m} \sum_{j=1}^{n} \frac{x_{ij}}{c} a_i b_j$$

for some $x_{ij} \in \mathbb{Z}$. We can assume that the GCD of all x_{ij} and c is 1.

Fix an embedding $L \subset \mathbb{C}$, and consider an embedding $\sigma_r : KL \to \mathbb{C}$ that fixes L. There are m of those, and their restrictions to K remain distinct. Hence

$$\begin{aligned}\sigma_r(t) &= \sum_{i=1}^{m} \sum_{j=1}^{n} \frac{x_{ij}}{c} \sigma_r(a_i) b_j \\ &= \sum_{i=1}^{m} \left(\sum_{j=1}^{n} \frac{x_{ij}}{c} b_j \right) \sigma_r(a_i) = \sum_{i=1}^{m} y_i \sigma_r(a_i).\end{aligned}$$

The terms y_i can be seen as the solution of a linear system defined by the matrix $(\sigma_r(a_i))_{r,i}$ and vector $(\sigma(t))_r$. By Cramer's rule,

$$y_i = \frac{\det M_i}{d_A},$$

where $d_A = \operatorname{disc}(\{a_1, \ldots, a_m\}) = \operatorname{disc}(A)$, and M_i is some cofactor matrix. In particular, $\det M_i \in C$, so $d_A y_i \in C$. Since $y_i \in L$, we have $d_A y_i \in B$, which can be expanded to $\frac{d_A x_{ij}}{c} \in \mathbb{Z}$.

So c divides $d_A x_{ij}$ for all i, j. Since c was chosen to be coprime with the x_{ij}, it follows that c divides d_A. Symmetrically, c divides d_B, hence $c|d$, which is the thesis. □

6.2. Prime extensions in number rings

Let $A \subset B$ be an inclusion of Dedekind domains, and let K be the fraction field of A and L that of B. Assume that B is integral over A. A key example is when K and L are number fields, with rings of integers A and B respectively.

We want to relate the factorization for ideals in A to that in B. Given an ideal $I \subset A$, we can consider the extension $I \cdot B$ and its factorization. We start with the case of prime ideals.

Remark 6.2.1. Let $P \subset A$ be a prime ideal. The ideal $P \cdot B$ is not trivial. In fact, since P is not the whole of A, there exists $\gamma \in K \backslash A$ such that $\gamma \cdot P \subset A$. This is clear if $P = (a)$ is principal: just take $\gamma = a^{-1}$. Otherwise, $P^{-1} \supset A$

is not the unit in the class group; then take any $\gamma \in P^{-1} \setminus A$. Assume that we can write

$$1 = p_1 b_1 + \cdots + p_k b_k$$

for some choice of $p_1, \dots, p_k \in P$ and $b_1, \dots, b_k \in B$.

Multiplying by γ we get that $\gamma \in B$ is integral over A. But we know that $\gamma \in K$, so we get a contradiction because A is integrally closed.

Since $P \cdot B$ is not trivial, there exists a factorization of the form

$$P \cdot B = Q_1^{e_1} \cdots Q_r^{e_r} \tag{6.2.1}$$

for some prime ideals Q_i of B. The condition that Q is a prime over P in B can be expressed in several equivalent ways.

Proposition 6.2.2. *Let $A \subset B$ be an integral extension of Dedekind rings, P a prime ideal of A and Q a prime ideal of B. The following conditions are equivalent:*

(i) *Q divides $P \cdot B$*

(ii) *$Q \supset P \cdot B$*

(iii) *$Q \supset P$*

(iv) *$Q \cap A = P$.*

Proof. The equivalence of the first three points follows immediately by the unique factorization for ideals in B. For the last point, it is clear that if $Q \cap A \supset P$, we must have equality because P is maximal. □

Definition 6.2.3. Let $A \subset B$ be an integral extension of Dedekind rings, P a prime ideal of A and Q a prime ideal of B over P.

(1) The degree of the field extension $A/P \subset B/Q$ is called the *degree of inertia* of Q over P, and is denoted $f(Q \mid P)$.

(2) The exponent e_i in the factorization (6.2.1) is called the *ramification index* of Q_i over P, and is denoted $e(Q_i \mid P)$.

(3) If $e(Q_i \mid P) = 1$ for all Q_i above P, we say that P is *unramified* in B; otherwise it is *ramified.*

So far we have been general, but now we restrict again to the case of rings of integers in number fields. In this case we can prove that the degree of inertia is always finite.

Remark 6.2.4. Let A be the integer ring in the number field K, $I \subset A$ any nonzero ideal. If $x \neq 0$ is any element of I, we have inclusions $x \cdot A \subset I \subset A$, which shows that, as an additive group, I is free of the same rank as A. In particular, if $P \subset A$ is prime, A/P is a finite field, and the degree of inertia is always finite for the case of number rings extensons.

Definition 6.2.5. Let $I \subset A$ be an ideal in a number ring. The index $A : I$ as additive groups—which is the same as the cardinality of A/I—is called the *norm* of I and denoted $\|I\|$.

Remark 6.2.6. By Proposition 6.1.7, it follows immediately that I embeds into $\mathbb{R}^r \times \mathbb{C}^s$ as a lattice of volume $2^{-s}\sqrt{|\mathrm{disc}(A)|} \cdot \|I\|$, where (r, s) is the signature of the number field.

The connection with the norm of elements is given by the following

Proposition 6.2.7. *If $I = (x)$ is principal, $\|I\| = |\mathrm{N}(x)|$, where we simply denote $\mathrm{N} = \mathrm{N}_{K/\mathbb{Q}}$.*

Proof. [**Con**, Theorem 7.2] Since I is a subgroup of A of the same rank, by Theorem 1.5.6 we can find a $\mathbb{Z}$-basis of A, say $\{r_1, \ldots, r_n\}$, such that $\{a_1r_1, \ldots, a_nr_n\}$ is a $\mathbb{Z}$-basis of I, for some integers $\{a_1, \ldots, a_n\}$. Another basis is of course $\{xr_1, \ldots, xr_n\}$, hence the linear map on K which sends a_ir_i to xr_i has determinant ± 1.

On the other hand, we have a commutative diagram of linear maps on K given by

$$\begin{array}{ccc} K & \xrightarrow{r_i \to a_ir_i} & K \\ & \searrow^{r_i \to xr_i} & \Big\downarrow {\scriptstyle a_ir_i \to xr_i}. \\ & & K \end{array}$$

Since the determinant of the vertical map is ± 1, the two other determinants are the same, up to sign. But the one above is $\|I\|$, and the one below is $\mathrm{N}(x)$. □

Remark 6.2.8. It follows that for any ideal I and $x \in I$, $\|I\|$ divides $\mathrm{N}(x)$, since (x) is a subgroup of I.

As it happens for the norm of numbers, the norm of ideals is multiplicative.

Proposition 6.2.9. *Let I, J be ideals of the number ring A. Then $\|I \cdot J\| = \|I\|\|J\|$.*

Proof. We use unique factorization and prove the equivalent fact that if $I = P_1^{m_1} \cdots P_k^{m_k}$, then $\|I\| = \|P_1\|^{m_1} \cdots \|P_k\|^{m_k}$.

First, let us treat the case of coprime ideals. If $I = S \cdot T$, with $S + T = (1)$, we have

$$\frac{A}{I} \cong \frac{A}{S} \times \frac{A}{T}$$

by the Chinese remainder theorem 1.2.12, hence $\|I\| = \|S\|\|T\|$.

This leaves with the case of a prime power, where we prove that $\|P^m\| = \|P\|^m$ by induction. Assume it holds for m and choose any $\alpha \in P^m \setminus P^{m+1}$. Consider the composition

$$\begin{aligned} A &\xrightarrow{\alpha\cdot} P^m \longrightarrow P^m/P^{m+1}, \\ x &\longmapsto \alpha \cdot x \end{aligned}$$

call it ϕ. Trivially, $P \subset \ker\phi$, and conversely if $x \in \ker\phi$, we must have $x \in P$ by unique factorization.

Since $P = \ker\phi$, we can descend ϕ to the injective map

$$\psi\colon A/P \to P^m/P^{m+1}.$$

By unique factorization, the ideal (α, P^{m+1}) is the largest common factor of (α) and P^{m+1}, which is P^m. This implies that ϕ is surjective, hence ψ is an isomorphism.

It follows that $A/P \cong P^m/P^{m+1}$, hence $\|P^{m+1}\| = \|P^m\|\|P\|$ as desired. □

We can use what we have proved so far about norms of ideals to get a special case of an important result.

Theorem 6.2.10. *Let L be a number field with ring of integers B, $p \in \mathbb{Z}$ a prime, $n = [L : \mathbb{Q}]$. Assume $p \cdot B = Q_i^{e_1} \cdots Q_r^{e_r}$. Then*

$$n = \sum_{i=1}^{r} e_i f_i,$$

where $f_i := f(Q_i \mid p)$.

We will generalize this in a moment to any extension of number fields, but this result will be needed as a starting point.

Proof. We compare the norms in the factorization of $p \cdot B$. For one, it is clear that $\|p \cdot B\| = p^n$, since B is free of rank n.

On the other hand, $\|Q_i\| = \#(B/Q_i)$, and B/Q_i is a field extension of $\mathbb{Z}/p\mathbb{Z}$ of degree f_i. Hence

$$\|p \cdot B\| = \|Q_1\|^{e_1} \cdots \|Q_r\|^{e_r} = (p^{f_1})^{e_1} \cdots (p^{f_r})^{e_r} = p^{\sum e_i f_i},$$

and the thesis follows by comparison. □

To study the more general case, we want to compare how the norm of ideals behaves under extension. First, we need

Lemma 6.2.11. *Let $K \subset L$ be number fields with integer rings A and B respectively. Let $P \subset A$ be a prime ideal, and $Q \subset B$ a prime over P—this gives an inclusion of fields $A/P \subset B/Q$. Then*

$$[B/Q : A/P] \leq [L : K].$$

Proof. Let $n = [L : K]$ and take elements $\overline{b_1}, \ldots, \overline{b_{n+1}} \in B/Q$ with representatives $b_1, \ldots, b_{n+1} \in B$. Then we have a linear dependence relation

$$a_1 b_1 + \cdots + a_{n+1} b_{n+1} = 0,$$

where can take $a_i \in A$ by clearing denominators. This gives a similar relation for the $\overline{b_i}$ over A/P, unless all $a_i \in P$. If this is the case, it is enough to find $\alpha \in K$ such that $\alpha a_i \in A$ for all i, but at least one $\alpha a_i \notin P$ and then multiply the equation by α.

Let $I = (a_1, \ldots, a_n)$ and choose an inverse ideal J of A, so that $I \cdot J = (\gamma)$ for some $\gamma \in A$. By unique factorization, $I \cdot J \not\subset (\gamma) \cdot P$, so we find $\beta \in J$ such that $\beta I \not\subset (\gamma) \cdot P$. Then it is easy to check that $\alpha := \frac{\beta}{\gamma}$ is the desired element. □

Proposition 6.2.12. *Let $K \subset L$ be number fields with integer rings A and B respectively. Let $I \subset A$ be an ideal. Then $\|I \cdot B\| = \|I\|^n$, where $n = [L : K]$.*

Proof. By the multiplicativity property 6.2.9, it is enough to prove the assertion for a prime ideal P. First, we consider the case where $K = \mathbb{Q}$; in this case, $P = (p)$. If $\{b_1, \ldots, b_n\}$ is a basis of B as a $\mathbb{Z}$-module, then a basis of $P \cdot B$ is $\{pb_1, \ldots, pb_n\}$, so $\|P \cdot B\| = p^n$.

For the general case, consider the inclusion $\mathbb{Q} \subset K \subset L$, and let $(p) = P \cap \mathbb{Z}$, so that we have also inclusions $(p) \subset P \subset P \cdot B$. Factor $(p) \cdot A = P_1^{e_1} \cdots P_r^{e_r}$, with inertia degrees $f_1, \ldots, f_r$, and $P = P_i$ for some i. By Theorem 6.2.10, we get

$$p^m = \|(p) \cdot A\| = p^{\sum e_i f_i},$$

where $m = [K : \mathbb{Q}]$.

For B, using 6.2.9 again, we get

$$p^{mn} = \|(p) \cdot B\| = \|P_1 B\|^{e_1} \cdots \|P_r B\|^{e_r}. \tag{6.2.2}$$

Moreover, $\|P_i \cdot B\| = \|P_i\|^{n_i}$ for some $n_i \leq n$ by Lemma 6.2.11. Combining with (6.2.2), we get

$$p^{mn} = \|P_1\|^{e_1 n_1} \cdots \|P_r\|^{e_r n_r} = p^{\sum e_i f_i n_i}.$$

Since all $n_i \leq n$, we must have equality for all i, hence $\|P_i \cdot B\| = \|P_i\|^n$. □

This allows us to extend 6.2.10 to the general case of extensions of number fields.

Theorem 6.2.13 (Inertia-ramification formula for number rings). *Let $K \subset L$ be number fields with ring of integers A and B respectively. Let $P \subset A$ be a prime ideal, $n = [L : K]$. Assume $P \cdot B = Q_i^{e_1} \cdots Q_r^{e_r}$. Then*

$$n = \sum_{i=1}^{r} e_i f_i,$$

where $f_i := f(Q_i \mid P)$.

Proof. By Propositions 6.2.12 and 6.2.9 we get

$$\begin{aligned} \|P\|^n &= \|P \cdot B\| = \|Q_1\|^{e_1} \cdots \|Q_r\|^{e_r} \\ &= \|P\|^{e_1 f_1} \cdots \|P\|^{e_r f_r} = \|P\|^{\sum e_i f_i}, \end{aligned}$$

and the result follows by comparing exponents. □

6.3. Prime extensions in Dedekind rings

We want to extend the results of the previous section to the more general case of Dedekind rings. The situtation is more subtle because we cannot assume the result of Section 6.1—in particular that the rings involved are additively free $\mathbb{Z}$-modules. For this reason, we will not use the ideal norm, instead we will prove the analogue of Theorem 6.2.13 with a different approach.

Let A be a Dedekind ring with fraction field K, L a finite separable extension of fields, and let B be the integral closure of A inside L. By Proposition 5.4.2, we know that B is finitely generated as an A-module, and in fact it is itself a Dedekind ring by Theorem 5.4.1.

Given a prime ideal $P \subset A$, we want to study the factorization of $P \cdot B$ as an ideal of B. Recall that Definition 6.2.3 is valid for general Dedekind rings, so we can talk about the index of ramification and the degree of inertia of a prime ideal $Q \subset B$ over P.

Proposition 6.3.1. *Let $K \subset L$ be a finite separable (Section A.3) extension of fields, $A \subset K$ a Dedekind ring with fraction field K, and $B \subset L$ its integral closure. For a prime $P \subset A$ we have*

$$\dim_{A/P} \frac{B}{P \cdot B} = [L : K],$$

where $\dim_{A/P}$ denotes dimension as a vector space over the field A/P.

Proof. Consider the multiplicative set $S = A \setminus P$, and let $A' = S^{-1}A = A_P$ and $B' = S^{-1}B$. Then, we have isomorphisms $A/P \cong A'/PA'$ and $B/P \cong B'/PB'$, so we can assume that A is local.

In this case, by unique factorization, all ideals are powers of P, which then must be principal. It follows that in fact, A is a principal ideal domain (actually, A is a discrete valuation ring, which we will encounter in Section 7.3). From the structure of finitely generated modules over principal ideal domains (Theorem 2.1.7) and the fact that B has no torsion, it follows that B is actually a free A-module. In this case, the thesis is clear. □

We can use this fact to generalize Theorem 6.2.13.

Theorem 6.3.2 (Inertia-ramification formula for Dedekind rings). *Let $K \subset L$ be a finite separable extension of fields, $A \subset K$ a Dedekind ring with fraction field K, and $B \subset L$ its integral closure. Let $P \subset A$ be a prime ideal, $n = [L : K]$. Assume $P \cdot B = Q_i^{e_1} \cdots Q_r^{e_r}$. Then*

$$n = \sum_{i=1}^{r} e_i f_i,$$

where $f_i := f(Q_i \mid P)$.

Proof. By the Chinese remainder theorem, we have

$$\frac{B}{P \cdot B} \cong \prod_{i=1}^{r} \frac{B}{Q_i^{e_i}}.$$

Moreover, $\dim_{A/P} \frac{Q_i^k}{Q_i^{k+1}} = \dim_{A/P} \frac{B}{Q_i}$ for all k (why?), hence

$$\dim_{A/P} \frac{B}{Q_i^{e_i}} = e_i \dim_{A/P} \frac{B}{Q_i}.$$

By Proposition 6.3.1 we have

$$[L : K] = \dim_{A/P} \frac{B}{P \cdot B} = \sum_{i=1}^{r} e_i \dim_{A/P} \frac{B}{Q_i} = \sum_{i=1}^{r} e_i f_i. \qquad \square$$

6.4. Galois extensions of Dedekind rings

Keeping the setting of the previous section, we investigate what happens when the extension $K \subset L$ is Galois (Section A.5). In this case, the Galois group $Gal(L/K)$ acts on B—because it preserves integral elements—hence it acts on ideals of B. In particular, given a prime ideal $P \subset A$, $Gal(L/K)$ must permute the prime ideals $Q_i \subset B$ that are over P.

Proposition 6.4.1. *Let $K \subset L$ be a finite Galois extension of fields, $A \subset K$ a Dedekind ring with fraction field K, and $B \subset L$ its integral closure. Let $P \subset A$ be a prime ideal, and $Q_1, \ldots, Q_r$ the primes of B over P. Then the action of $Gal(L/K)$ on $\{Q_1, \ldots, Q_r\}$ is transitive.*

Proof. Take any two primes over P, say Q_1 and Q_2. Assuming that there is no $\sigma \in G := Gal(L/K)$ such that $\sigma(Q_1) = Q_2$, we can solve the system

$$\begin{cases} x \equiv 0 \mod Q_2 \\ x \equiv 1 \mod \sigma(Q_1) \quad \text{for all } \sigma \in G \end{cases}$$

using the Chinese remainder theorem.

Let $y = \mathrm{N}_{L/K}(x) = \prod_{\sigma \in G} \sigma(x)$. On one hand, x itself is a factor in this product, hence $y \in Q_2$. On the other hand, $x \notin \sigma(Q_1)$ for every $\sigma \in G$, that is, no factor lies in Q_1. Since Q_1 is prime, $y \notin Q_1$. But this is a contradiction, since $y \in A \cap Q_2 = P$. □

From the proposition, it follows that $e = e(Q_i|P)$ and $f = f(Q_i|P)$ do not depend on i, hence the inertia-ramification formula simplifies to

$$n = [L : K] = r \cdot e \cdot f. \tag{6.4.1}$$

Definition 6.4.2. Let $K \subset L$ a finite Galois extension with Galois group G, $A \subset K$ a Dedekind ring with fraction field K, $B \subset L$ its integral closure. Let $P \subset A$ be a prime ideal, and Q a prime of B over P.

(1) The *decomposition group* of Q over P, denoted $D(Q|P)$ is
$$D(Q|P) = \{\sigma \in G \mid \sigma(Q) = Q\},$$
that is, the stabilizer of Q in G.

(2) The *inertia group* of Q over P, denoted $E(Q|P)$ is
$$E(Q|P) = \{\sigma \in G \mid \sigma(x) \equiv x \pmod{Q} \text{ for all } x \in B\}.$$

Remark 6.4.3. Since $D(Q|P)$ is the stabilizer of Q, and the action is transitive, we have $[G : D(Q|P)] = r$. Moreover, all subgroups $D(Q_i|P)$ are conjugate with each other: if $\tau(Q_i) = Q_j$,

$$D(Q_j|P) = \tau D(Q_i|P)\tau^{-1}.$$

Remark 6.4.4. It is immediate to check that $E(Q|P)$ is a subgroup of $D(Q|P)$. In fact, it is *normal* subgroup. To see this, notice that an automorphism $\sigma \in D(Q|P)$ induces a commutative diagram

$$\begin{array}{ccc} B & \xrightarrow{\sigma} & B \\ \downarrow & & \downarrow \\ B/Q & \xrightarrow{\overline{\sigma}} & B/Q, \end{array}$$

thereby inducing a homomorphism $\phi \colon D \to \mathrm{Aut}(B/Q)$.

In fact, A/P is a subfield of B/Q, and $\overline{\sigma} \in Gal(\frac{B}{Q}/\frac{A}{P})$. The kernel of ϕ is exactly $E(Q|P)$, which must then be normal in $D(Q|P)$. Moreover, since the cardinality of $Gal(\frac{B}{Q}/\frac{A}{P})$ is $f(Q|P)$, we have $[D(Q|P) : E(Q|P)] \leq f$.

By the Galois correspondence (Section A.5), we obtain a tower of extensions that sit between K and L:

$$\begin{array}{cc} L & \{e_G\} \\ | & | \\ L_E & E \\ | & | \\ L_D & D \\ | & | \\ K & G, \end{array}$$

where $D = D(Q|P)$, $E = E(Q|P)$, and on the vertical lines we denote the degree of the field extensions.

Proposition 6.4.5. *In the above setting, we have* $[L : L_E] = e$ *and* $[L_E : L_D] = f$.

Proof. Let us complete the diagram by showing the integral closure of A in L_E and L_D, as well as the restrictions of Q to these rings:

$$\begin{array}{cccc} Q & B & L & \{e_G\} \\ | & | & | & | \\ Q_E & B_E & L_E & E \\ | & | & | & | \\ Q_D & B_D & L_D & D \\ | & | & | & | \\ P & A & K & G. \end{array}$$

We know some numbers in this diagram: $[L_D : K] = r$ and $[L_E : L_D] \leq f$. We will compute some others and use the inertia-ramification (6.4.1) formula to conclude.

First, since $D = Gal(L/L_D)$, and the Galois group is transitive on primes over Q_D, we deduce that Q is the only prime of B over Q_D. By (6.4.1) we see that

$$e(Q|P) \cdot f(Q|P) = e(Q|Q_D) \cdot f(Q|Q_D),$$

and since there are obvious inequalities, we conclude that in fact $e(Q|Q_D) = e(Q|P)$ and $f(Q|Q_D) = f(Q|P)$.

The next step is to show that $f(Q|Q_E) = 1$, that is,

$$\frac{B}{Q} \cong \frac{B_E}{Q_E}.$$

Given $\alpha \in B$, the polynomial

$$g(x) = \prod_{\sigma \in E} (x - \sigma(\alpha))$$

is E-invariant, hence it belongs to $B_E[x]$. This gives a quotient polynomial $\overline{g}(x) \in B_E/Q_E[x]$.

By definition, $\sigma \in E$ implies that $\overline{\sigma(\alpha)} = \overline{\alpha}$ in B/Q, hence $\overline{g}(x)$ is a power of $(x - \overline{\alpha})$. Since B/Q is a separable extension of B_E/Q_E, the minimal polynomial of $\overline{\alpha}$ over B_E/Q_E must be $x - \overline{\alpha}$—that is, $\overline{\alpha} \in B_E/Q_E$. Since α was arbitrary, B/Q is the trivial extension of B_E/Q_E, as claimed.

At this point, since inertia degrees are clearly multiplicative in extensions, we find that $f(Q_E|Q_D) = f(Q|Q_D) = f$. By the inequality we already know we conlude that $[L_E : L_D] = f$ and now we can fill in all remaining degrees in the diagram, which becomes

$$\begin{array}{rcccc}
 & Q & B & L & \{e_G\} \\
{\scriptstyle f=r=1} & \Big| & \Big| & \Big|{\scriptstyle e} & \Big| \\
 & Q_E & B_E & L_E & E \\
{\scriptstyle e=r=1} & \Big| & \Big| & \Big|{\scriptstyle f} & \Big| \\
 & Q_D & B_D & L_D & D \\
{\scriptstyle e=f=1} & \Big| & \Big| & \Big|{\scriptstyle r} & \Big| \\
 & P & A & K & G.
\end{array}$$

□

The fields L_D and L_E, that we have defined in terms of subgroups of the Galois groups, enjoy an intrinsic characterization. To see this, let F be any intermediate field, $K \subset F \subset L$. If B_F is the integral closure of A in F, and $Q_F = Q \cap B_F$, by the definitions it follows immediately that $D(Q|Q_F) = D(Q|P) \cap H$ and $E(Q|Q_F) = E(Q|P) \cap H$, where $H = Gal(L/F)$. Since the intersection of Galois groups corresponds to the composite of the relative fields, we get the following diagram of field extensions:

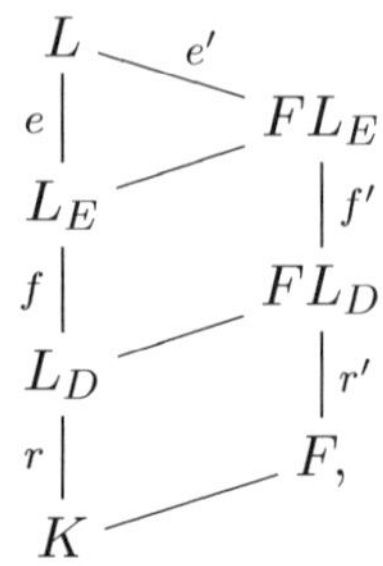

where r', f', and e' are the number of factors, degree of inertia and ramification index of Q_F in B.

Proposition 6.4.6. *Fix a finite Galois extension $K \subset L$ with Galois group G, $A \subset K$ a Dedekind ring with fraction field K, $B \subset L$ its integral closure. Given a prime $P \subset A$ and a prime $Q \subset B$ over it, let L_D and L_E be the subfields of L associated to $D(Q|P)$ and $E(Q|P)$ respectively.*

(i) *L_D is the biggest subfield $F \subset L$ for which $e(Q_F|P) = 1$ and $f(Q_F|P) = 1$.*

(ii) *L_D is the smallest subfield $F \subset L$ such that Q is the only prime over Q_F.*

(iii) *L_E is the biggest subfield $F \subset L$ for which $e(Q_F|P) = 1$.*

(iv) *L_E is the smallest subfield $F \subset L$ such that Q is totally ramified over Q_F, that is, $Q_F \cdot B = Q^{[L:F]}$.*

Proof. (i) In the diagram above, assume $e = e' = 1$ and $f = f' = 1$. Then $[L : FL_D] = [L : F]$, which means $F \subset L_D$.

(ii) Let $H = Gal(L/F)$. If Q is the only prime over Q_F, every $\sigma \in H$ preserves Q, which means $H \subset D(Q|P)$ and the thesis follows from the Galois correspondence.

(iii) In the diagram above, assume $e = e' = 1$. Then $[L : FL_E] = [L : L_E]$, which means $F \subset L_E$.

(iv) If Q is totally ramified over Q_F we must have $f' = r' = 1$, which means $FL_E = F$, or $L_E \subset F$.

□

Corollary 6.4.7. *Let A be a Dedekind ring with fraction field K, and let L_1, L_2 be two finite separable extensions of K. Let P be a prime of A. If P splits completely (resp., is unramified) in L_1 and L_2, then the same holds in L_1L_2.*

Proof. Fix a finite Galois extension L of K containing both L_1 and L_2. The hypothesis is equivalent to $L_1, L_2 \subset L_{D(Q_i|P)}$ for each prime Q_i over P (resp., $L_1, L_2 \subset L_E$), hence the same holds for the composite field L_1L_2. $\square$

Since the normal closure of a field L over K is the composite of $\sigma(L)$ for all embeddings σ of L into $\overline{K}$, we also get

Corollary 6.4.8. *Let A be a Dedekind ring with fraction field K, and L a finite separable extension of K. Let P be a prime of A. If A splits completely (resp., is unramified) in L, then the same holds in the normal closure of L.*

6.5. Discriminant and ramification

As it turns out, the discriminant of a number field is strongly related to the properties of ramification of primes in that field. In fact, we have the following result:

Theorem 6.5.1. *Let K be a number field with ring of integers A. The prime number p is ramified in A if, and only if, it divides* $\operatorname{disc}(A)$.

We will see in Corollary 6.7.4 that the discriminant of a number field is always bigger than 1, so ramified primes always exist.

Since the proof is quite long, we split it in two.

Proof of forward implication. Assume that p is ramified, say $p \cdot A = P_1^{e_1} \cdots P_r^{e_r}$ with $e_1 > 1$. We can then write $p \cdot A = P_1 \cdot I$ for some ideal I divisible by all primes over p.

Fix an integer basis $\{\alpha_1, \dots, \alpha_n\}$ of A, and choose $\alpha \in I \setminus p \cdot A$. Write

$$\alpha = m_1\alpha_1 + \cdots + m_n\alpha_n,$$

where p does not divide all m_i, say m_1 is not multiple of p. The basis $\{\alpha, \alpha_2, \dots, \alpha_n\}$ satisfies

$$\operatorname{disc}(\{\alpha, \alpha_2, \dots, \alpha_n\}) = m_1^2 \operatorname{disc}(A),$$

so if we can prove that p divides $\operatorname{disc}(\{\alpha, \alpha_2, \dots, \alpha_n\})$, we are done. Denote $\alpha_1' = \alpha$ and $\alpha_i' = \alpha_i$ for $i > 1$.

Now let L be the normal closure of K, with ring of integers B. Since $\alpha \in I$, it is inside every prime of A over p, hence inside every prime of B over p. The above discriminant is then

$$\operatorname{disc}(\{\alpha, \alpha_2, \dots, \alpha_n\}) = (\det(\sigma_i(\alpha_j')))^2,$$

where $\sigma_1, \dots, \sigma_n$ are the complex embeddings of K. Since $\alpha_1' = \alpha$ belongs to every prime Q of B over p, the same holds for each of its conjugates $\sigma_i(\alpha)$. It follows that $\operatorname{disc}(\{\alpha, \alpha_2, \dots, \alpha_n\})$ belongs to every prime Q of B over p, and since it is an integer number, $\operatorname{disc}(\{\alpha, \alpha_2, \dots, \alpha_n\}) \in Q \cap \mathbb{Z} = (p)$. $\square$

Proof of backward implication. We use the characterization of discriminant given by (4.2.1), so given a basis $\{\alpha_1, \dots, \alpha_n\}$ of A we have

$$\operatorname{disc}(A) = \det(\operatorname{Tr}(\alpha_i\alpha_j))_{ij}.$$

The matrix $(\operatorname{Tr}(\alpha_i\alpha_j))_{ij}$ has integer coefficients, and p dividing its determinant implies that its rows are linearly dependent in $\mathbb{Z}/p\mathbb{Z}$. So there exist integer numbers $m_1, \dots, m_n$, not all divisible by p, such that

$$\sum_{i=1}^{n} m_i(\operatorname{Tr}(\alpha_i\alpha_1), \dots, \operatorname{Tr}(\alpha_i\alpha_n)) \equiv 0 \mod p.$$

In turn, this implies that $\operatorname{Tr}(\alpha x) \equiv 0 \mod p$ for all $x \in A$, where

$$\alpha := \sum_{i=1}^{n} m_i\alpha_i.$$

Notice that by construction $\alpha \notin pA$, since the linear combination is not trivial.

Now assume by contradiction that $p \cdot A = P_1 \cdots P_r$ is unramified. Since $\alpha \notin pA$, there is some i such that $\alpha \notin P_i$. If L is the normal closure of K, with ring of integers B, and Q is a prime of B above P_i, we have $\alpha \notin Q$, but by Proposition A.7.5 $\operatorname{Tr}(\alpha x) \in Q$ for all $x \in B$.

Factor $P_iB = Q_1 \cdots Q_r$, with $Q = Q_1$, and choose $\beta \in B$ satisfying $\beta \notin Q$ and $\beta \in Q_i$ for all $i \geq 2$. Our choice of α guarantees that

$$\sum_{\sigma \in Gal(L/\mathbb{Q})} \sigma(\alpha\beta y) = \operatorname{Tr}(\alpha\beta y) \in Q$$

for all $y \in B$. On the other hand for every $\sigma \in Gal(L/\mathbb{Q})$ such that $\sigma \notin D(Q|P_i)$, we have $\sigma^{-1}(Q) = Q_i$ for some $i \geq 2$, hence $\sigma(\alpha\beta y) \in Q$ for all $y \in B$. By difference, we find

$$\sum_{\sigma \in D(Q|P_i)} \sigma(\alpha\beta y) \in Q \tag{6.5.1}$$

for all $y \in B$.

Since the prime (p) is unramified, we have an isomorphism $D(Q|P_i) \cong Gal((B/Q)/(A/P_i))$, and then (6.5.1) implies that $\overline{\alpha\beta} = 0$ in B/Q. This means that $\alpha\beta \in Q$, which is a contradiction since neither α nor β belongs to Q. □

Example 6.5.2. Continuing Example 6.1.2, let us compute the discriminant of $K = \mathbb{Q}(\sqrt{m})$, with m squarefree. When $m \equiv 1 \pmod 4$, a basis for $\mathcal{O}_K$

is just $\left\{1, \frac{1+\sqrt{m}}{2}\right\}$, hence

$$\operatorname{disc}(K) = \left(\det\begin{pmatrix} 1 & 1 \\ \frac{1+\sqrt{m}}{2} & \frac{1-\sqrt{m}}{2} \end{pmatrix}\right)^2 = m,$$

so the only ramified primes are divisors of m.

When $m \equiv 2, 3 \pmod 4$,

$$\operatorname{disc}(K) = \left(\det\begin{pmatrix} 1 & 1 \\ \sqrt{m} & -\sqrt{m} \end{pmatrix}\right)^2 = 4m,$$

so the ramified primes are divisors of m, as well as 2.

6.6. Computing prime factorizations

At this point, we know a great deal about factorization of primes in number rings. It remains to find a way to actually compute them.

In order to do this, let $K \subset L$ be number fields, with integer rings A and B respectively. Take any $\alpha \in B$ such that $L = K(\alpha)$. In general, $A[\alpha]$ is a subring of B, but since they have the same rank as $\mathbb{Z}$-modules (why?), the index $m := [B : A[\alpha]]$ is finite. We will be able to provide a factorization in B for all primes $P \subset A$ that are coprime with mA.

Theorem 6.6.1 (Kummer). *Let $K \subset L$ be number fields, with integer rings A and B respectively. Choose some $\alpha \in B$ such that $L = K(\alpha)$; let $g(x) \in A[x]$ be its minimal polynomial over K and denote $m := [B : A[\alpha]]$.*

Let $P \subset A$ be a prime such that $P \cap \mathbb{Z} = (p)$, for some p that does not divide m. Factor

$$\overline{g}(x) = \overline{g_1}^{e_1}(x) \cdots \overline{g_r}^{e_r}(x) \text{ in } (A/P)[x],$$

where the $\overline{g_i}$ are irreducible monic polynomials. Then $P \cdot B = Q_1^{e_1} \cdots Q_r^{e_r}$, where $Q_i := (P, g_i(\alpha))$ (as an ideal of B).

Proof. Since A/P is a field, and $\overline{g_i}$ is irreducible in $(A/P)[x]$, $\frac{(A/P)[x]}{(\overline{g_i}(x))}$ is again a field, hence the ideal $(P, g_i(x))$ is maximal inside $A[x]$.

Now consider the composition

$$\psi_i \colon A[x] \xrightarrow{f \to f(\alpha)} A[\alpha] \longrightarrow B \longrightarrow B/Q_i.$$

The evaluation map is clearly surjective. The map $A[\alpha] \to B/Q_i$ is surjective as well. In fact, $A[\alpha]$ and Q_i are both subgroups of full rank in B—the first has index m, while the second has index a power of p by Proposition 6.2.12.

It follows that they must generate B, hence ψ_i is surjective. Since both P and g_i are in its kernel, we derive a surjective homomorphism

$$\frac{A[x]}{(P, g_i(x))} \to \frac{B}{Q_i}. \tag{6.6.1}$$

Since the former is a field, B/Q_i is a field or trivial, which means that Q_i is either prime or the whole B.

Now let us show that $Q_i + Q_j = B$ for $i \neq j$. To see this, since $\overline{g_i}$ and $\overline{g_j}$ are coprime in $(A/P)[x]$, we can write

$$h_i g_i + h_j g_j \equiv 1 \mod P$$

for some $h_i, h_j \in A[x]$. Evaluating at α, we get to write

$$h_i(\alpha)g_i(\alpha) + h_j(\alpha)g_j(\alpha) + f(\alpha) = 1$$

for some $f(x) \in P[x]$, which shows that P, $g_i(\alpha)$ and $g_j(\alpha)$ generate B, as desired.

We now prove that $P \cdot B$ divides $Q_1^{e_1} \cdots Q_r^{e_r}$, or equivalently

$$Q_1^{e_1} \cdots Q_r^{e_r} \subset P \cdot B.$$

In fact, $Q_i^{e_i} \subset (P, g_i(\alpha)^{e_i})$, hence

$$Q_1^{e_1} \cdots Q_r^{e_r} \subset (P, g_1(\alpha)^{e_1} \cdots g_r(\alpha)^{e_r}).$$

But $g_1(\alpha)^{e_1} \cdots g_r(\alpha)^{e_r} = g(\alpha) + r(\alpha) = r(\alpha)$ for some $r(x) \in P[x]$, hence the ideal simplifies to $P \cdot B$.

It follows that $P \cdot B = Q_1^{d_1} \cdots Q_r^{d_r}$ for some $d_i \leq e_i$ (omit Q_i if it is the whole B). We want to prove we have equality for each i. By Theorem 6.2.13 we get that

$$n := [L : K] = \sum_{i=1}^{r} d_i f_i,$$

where $f_i = f(Q_i|P) = \deg \overline{g_i}$, the last equality in virtue of the isomorphism (6.6.1) (we have omitted terms where B/Q_i is trivial).

On the other hand,

$$n = \deg g = \deg \overline{g} = \sum_{i=1}^{r} e_i f_i.$$

This means that we have $d_i = e_i$ for all i, and no term can be omitted (and hence the case where $Q_i = B$ never arises), proving the theorem. $\square$

Example 6.6.2. By Example 6.5.2, we know the ramified primes of $K = \mathbb{Q}(\sqrt{m})$, with m squarefree. With Kummer's theorem, we can be more explicit. Let us assume that $m \equiv 2, 3 \pmod 4$, so that a basis for $\mathcal{O}_K$ is $\{1, \sqrt{m}\}$.

The minimal polynomial for $\sqrt{m}$ is just $f(x) = x^2 - m$. Let us reduce this modulo a prime p.

(a) If p divides m, $\overline{f}(x) = x^2$, and by Kummer's theorem $p\mathcal{O}_K = (p, \sqrt{m})^2$.

(b) If $p = 2$, $\overline{f}(x) = x^2$ or $(x+1)^2$ according to whether $m \equiv 2$ or $3 \pmod 4$, and correspondingly

$$2\mathcal{O}_K = \begin{cases} (2, \sqrt{m})^2 \\ (2, \sqrt{m}+1)^2. \end{cases}$$

(c) If p does not divide m and is not 2, $f(x)$ has distinct roots modulo p. If m is a square modulo p, say $m \equiv c^2$, then

$$p\mathcal{O}_K = (p, \sqrt{m} - c)(p, \sqrt{m} + c).$$

Otherwise, f is irreducible modulo p, and in this case $p\mathcal{O}_K = (p)$ is irreducible as well.

The case where $m \equiv 1 \pmod 4$ is Exercise 1.

6.7. Geometry of ideal lattices

The results that we have proved so far in the chapter are mostly valid over general Dedekind rings, although in the case of number rings we have been able to give more explicit proofs based on the properties of the discriminant and the ideal norms. In this section, we make more explicit use of the embedding of a number ring as a lattice in $\mathbb{R}^r \times \mathbb{C}^s$ to derive some finiteness results that are actually specific to the case of number fields. Namely, if A is the ring of integers in a number field K, the multiplicative group A^* is finitely generated, and the class group $G(A)$ is finite.

Remark 6.7.1. The two groups are related by a simple exact sequence. Namely, given a nonzero $a \in A$, factor the ideal (a) as a product of primes

$$(a) = P_1^{e_1} \cdots P_r^{e_r}.$$

This defines a monoid homomorphism $A \setminus 0 \to \bigoplus_P \mathbb{Z}$, where the sum ranges over all primes P of A, which sends a to a vector with component e_i at P_i. By adding inverses, we get a homomorphism $\phi\colon K^* \to \bigoplus_P \mathbb{Z}$, and by definition we have $\ker \phi = A^*$ and $\operatorname{coker} \phi = G(A)$. Putting all together we get an exact sequence

$$0 \longrightarrow A^* \longrightarrow K^* \longrightarrow \bigoplus_P \mathbb{Z} \longrightarrow G(A) \longrightarrow 0.$$

The main technical tool of this section is the following geometric theorem of Minkowski. We will use this result—together with the computation of the volume of the embedding of a number ring—to construct nontrivial integer algebraic numbers with some control on their size.

Theorem 6.7.2 (Minkowski's convex body theorem). *Let $\mathcal{L} \subset \mathbb{R}^n$ be a lattice and let $B \subset \mathbb{R}^n$ be a convex, Lebesgue measurable subset, symmetric with respect to the origin (that is, $B = -B$). If $\operatorname{vol}(B) > 2^n \operatorname{vol}(\mathcal{L})$, then there exists a nonzero $\lambda \in \mathcal{L} \cap B$. If moreover B is compact, the same holds under the weaker assumption $\operatorname{vol}(B) \geq 2^n \operatorname{vol}(\mathcal{L})$.*

Proof. Let D be a fundamental domain for $\mathcal{L}$, so that

$$\mathbb{R}^n = \overset{\circ}{\bigcup_{\lambda \in \mathcal{L}}} (\lambda + D),$$

where the dot denotes that the union is disjoint. Denote by $\frac{1}{2}B$ a rescaling of B by $1/2$, so that $\operatorname{vol}(\frac{1}{2}B) > \operatorname{vol}(D)$. We can write

$$\frac{1}{2}B = \overset{\circ}{\bigcup_{\lambda \in \mathcal{L}}} \left(\frac{1}{2}B \cap (\lambda + D) \right),$$

hence

$$\operatorname{vol}\left(\frac{B}{2}\right) = \sum_{\lambda \in \mathcal{L}} \operatorname{vol}\left(\frac{1}{2}B \cap (\lambda + D)\right) = \sum_{\lambda \in \mathcal{L}} \operatorname{vol}\left(\left(\frac{1}{2}B - \lambda\right) \cap D\right).$$

This implies that the sets $(\frac{1}{2}B - \lambda) \cap D$ cannot be disjoint, otherwise $\operatorname{vol}(\frac{1}{2}B) \leq \operatorname{vol}(D)$. So we find λ_1, λ_2 such that

$$\left(\frac{1}{2}B - \lambda_1\right) \cap \left(\frac{1}{2}B - \lambda_2\right) \neq \emptyset,$$

that is, there are $b_1, b_2 \in B$ such that $b_1/2 - \lambda_1 = b_2/2 - \lambda_2$. Then $b := (b_1 - b_2)/2 = \lambda_2 - \lambda_1 \in \mathcal{L}$, and moreover $b \in B$ since B is convex and symmetric.

When B is compact and $\operatorname{vol}(B) \geq 2^n \operatorname{vol}(\mathcal{L})$, we can apply the above results for $(1+\epsilon)B$ for all $\epsilon > 0$, and find a nonzero element in $(1+\epsilon)B \cap \mathcal{L}$. The intersection $(1+\epsilon)B \cap \mathcal{L}$ is actually finite, since B is compact and $\mathcal{L}$ is discrete, so the elements we find for various ϵ will converge to a nonzero element in $B \cap \mathcal{L}$. □

To apply the above result, let K be a number field, A its ring of integers, and $I \subset A$ an ideal. If K admits r real and $2s$ nonreal embeddings into $\mathbb{C}$, we have an embedding $\underline{\sigma} : A \to \mathbb{R}^r \times \mathbb{C}^s$ which exhibits A as a lattice in $\mathbb{R}^n$, for $n = r + 2s$. By Proposition 6.1.7 and Remark 6.2.6 we know that

$$\operatorname{vol}(A) = 2^{-s}\sqrt{|\operatorname{disc}(A)|}$$
$$\operatorname{vol}(I) = 2^{-s}\sqrt{|\operatorname{disc}(A)|} \cdot \|I\|$$

for an ideal $I \subset A$. By applying Minkowski's theorem for various choices of convex bodies, we can produce nontrivial algebraic integers satisfying various constraints.

Proposition 6.7.3. *Let K be a number field of degree n over $\mathbb{Q}$, A its ring of integers, $I \subset A$ an ideal. Let r, s be the signature of K, so that $n = r+2s$. Then I contains an element x of norm*

$$|\mathrm{N}(x)| \leq \lambda \|I\|,$$

where

$$\lambda = \left(\frac{4}{\pi}\right)^s \frac{n!}{n^n}\sqrt{|\mathrm{disc}(A)|}. \tag{6.7.1}$$

Proof. Choose the convex body $B_t \subset \mathbb{R}^r \times \mathbb{C}^s$ defined by

$$B_t := \left\{ (y_1, \dots, y_r, z_1, \dots, z_s) \mid \sum_{i=1}^r |y_i| + 2\sum_{j=1}^s |z_j| \leq t \right\}.$$

We choose t to have $\mathrm{vol}(B_t) = 2^n \mathrm{vol}(I)$, in order to apply Minkowski. A straightforward but long computation (see Exercise 9) gives

$$\mathrm{vol}(B_t) = 2^r \left(\frac{\pi}{2}\right)^s \frac{t^n}{n!}, \tag{6.7.2}$$

hence we choose t so that

$$t^n = \left(\frac{4}{\pi}\right)^s n!\sqrt{|\mathrm{disc}(A)|} \cdot \|I\|. \tag{6.7.3}$$

For this value of t, we find $x \in I$ such that $\underline{\sigma}(x) \in B_t$. If $\sigma_1, \dots, \sigma_r$ are the real embeddings and $\tau_1, \dots, \tau_s$ are the nonreal embeddings of K,

$$|\mathrm{N}(x)| = \prod_{i=1}^r |\sigma_i(x)| \cdot \prod_{j=1}^s |\tau_j(x)|^2,$$

so $|\mathrm{N}(x)| \leq \frac{t^n}{n^n}$ by the arithmetic-geometric inequality. Substituting our choice of t^n from (6.7.3) gives the desired result. □

Corollary 6.7.4. *Let K be a number field of degree $n \geq 2$ over $\mathbb{Q}$. Then*

$$|\mathrm{disc}(K)| \geq \frac{\pi^2}{4}\left(\frac{81\pi}{64}\right)^{n-2}.$$

In particular, $|\mathrm{disc}(K)| > 1$ unless $K = \mathbb{Q}$.

You will prove in Exercise 13 that for any d there exists only finitely many number fields with discriminant bounded by d.

Proof. Taking $I = A$ in Proposition 6.7.3, we find an $x \in A$ with $|\mathrm{N}(x)| \leq \lambda$, in particular $\lambda \geq 1$. This simplifies to

$$|\mathrm{disc}(A)| \geq \left(\frac{\pi}{4}\right)^{2s} \frac{n^{2n}}{(n!)^2} \geq \left(\frac{\pi}{4}\right)^{n} \frac{n^{2n}}{(n!)^2}.$$

Calling a_n the right-hand side, so that $a_2 = \pi^2/4$, we find

$$\frac{a_{n+1}}{a_n} = \frac{\pi}{4}\left(1 + \frac{1}{n}\right)^{2n} \geq \frac{81\pi}{64},$$

for $n \geq 2$, and the thesis follows by induction. □

Corollary 6.7.5. *Let $C \in G(A)$ be any class of ideals. Then there exists $I \in C$ such that $\|I\| \leq \lambda$, where λ is defined in* (6.7.1).

Proof. Let $J \in C^{-1}$ be any integer ideal. By Proposition 6.7.3 we find $x \in J$ with $|\mathrm{N}(x)| \leq \lambda\|J\|$. Then there exists $I \in C$ such that $(x) = I \cdot J$. By Proposition 6.2.9 we get

$$|\mathrm{N}(x)| = \|(x)\| = \|I\| \cdot \|J\|,$$

hence $\|I\| \leq \lambda$. □

Proposition 6.7.3 also implies the following fundamental result.

Theorem 6.7.6. *Let K be a number field, A its ring of integers. The class group $G(A)$ is finite.*

Proof. Let $C \in G(A)$ be any class of ideals. By Corollary 6.7.5, there is an integer ideal $I \in C$ such that $\|I\| \leq \lambda$. This bounds the norm of each prime P in the factorization of I by Proposition 6.2.9. So there are finitely many choices for $(p) = P \cap \mathbb{Z}$, and in turn finitely many choices for P. □

Definition 6.7.7. Let K be a number field with ring of integers A. The cardinality of $G(A)$ is called the *class number* of K, and denoted by $h(K)$.

We will see in the exercises some easy cases in which $h(K)$ can be computed explicitly. Minkowski's theorem also implies another fundamental finiteness result.

Theorem 6.7.8. *Let K be a number field of degree n over $\mathbb{Q}$, A its ring of integers, $I \subset A$ an ideal. Let (r, s) be the signature of K, so that $n = r + 2s$. The group of units of A is finitely generated—more precisely,*

$$A^* \cong \mathbb{Z}^\rho \times U,$$

where $\rho = r + s - 1$ and U is a finite group, namely U is the group of roots of unity in K.

As it turns out, the proof is rather elementary, but the delicate part is the equality $\rho = r + s - 1$. We need a lemma, which is useful in itself:

Lemma 6.7.9. *Let A be a number ring, $x \in A$. Then $x \in A^*$ if, and only if,* $\mathrm{N}(x) = \pm 1$.

Proof. Assume $\mathrm{N}(x) = \pm 1$, so that x satisfies the polynomial

$$x^n + a_{n-1}x^{n-1} + \cdots + a_1 x \pm 1 = 0$$

for some $a_1, \ldots, a_{n-1} \in \mathbb{Z}$. Then

$$x \cdot (x^{n-1} + a_{n-1}x^{n-2} + \cdots + a_1) = \pm 1,$$

hence x is invertible. The other implication is trivial. □

Proof of Theorem 6.7.8. Since we work inside the multiplicative group A^*, we consider a modification of our standard embedding $\underline{\sigma}$. Let $\sigma_1, \ldots, \sigma_r$ be the real embeddings of K and $\tau_1, \ldots, \tau_s$ the nonreal ones. We define a homomorphism

$$\begin{aligned} \log \underline{\sigma} \colon\ A^* &\longrightarrow \mathbb{R}^{r+s} \\ x &\longmapsto (\log(|\sigma_1(x)|), \ldots, \log(|\tau_s(x)|)) \end{aligned}$$

and notice that $\mathrm{N}(x) = \pm 1$ if, and only if, $\log \underline{\sigma}(x)$ lies in the hyperplane H defined by the equation

$$\sum_{i=1}^{r} y_i + 2 \sum_{j=1}^{s} z_j = 0.$$

Notice that $\log \underline{\sigma}$ is a group homomorphism, with kernel

$$\tilde{K} := \ker \log \underline{\sigma} = \{x \in A^* \mid |\sigma_i(x)| = |\tau_j(x)| = 1 \text{ for all } i, j\}.$$

The coefficients of the minimal polynomial of any element $x \in \tilde{K}$ are bounded, which implies that $\tilde{K}$ is finite. Since $\tilde{K}$ is a multiplicative finite subgroup of $\mathbb{C}^*$, it must be cyclic (Proposition A.2.2)—hence $\tilde{K}$ is the group of roots of unity in K.

It remains to study the image $S := \log \underline{\sigma}(A^*) \subset H$. By the same reasoning, for each compact $W \subset H$, the intersection $S \cap W$ is finite. This implies that S is a discrete subgroup of H, hence it is a free abelian group of rank $\rho \leq r + s - 1$. We conclude that we have an exact sequence

$$0 \longrightarrow U \longrightarrow A^* \longrightarrow \mathbb{Z}^\rho \longrightarrow 0,$$

which splits since $\mathbb{Z}^\rho$ is free, and this implies the theorem, save for the equality $\rho = r + s - 1$.

To conclude, we must prove that S generates H. Given any choice of $\lambda_1, \ldots, \lambda_r, \mu_1, \ldots, \mu_s > 0$, consider the convex body $B_{\lambda,\mu} \subset \mathbb{R}^r \times \mathbb{C}^s$ defined by

$$B_{\lambda,\mu} = \{(y_1, \ldots, y_r, z_1, \ldots, z_s) \mid |y_i| \leq \lambda_i, |z_j| \leq \mu_j\}.$$

Then

$$\operatorname{vol}(B_{\lambda,\mu}) = \prod_{i=1}^{r}(2\lambda_i) \prod_{j=1}^{s}(\pi\mu_j^2).$$

In order to apply Minkowski, we choose the λ_i and μ_j so that

$$\lambda_1 \cdots \lambda_r \mu_1^2 \cdots \mu_s^2 = T \geq 2^n \frac{1}{(2\pi)^s}\sqrt{|\operatorname{disc}(A)|}. \tag{6.7.4}$$

Then we have

$$\operatorname{vol}(B_{\lambda,\mu}) \geq 2^{n-s}\sqrt{|\operatorname{disc}(A)|} = 2^n \operatorname{vol}(A).$$

By Theorem 6.7.2 we find $x_{\lambda,\mu} \in A$, $x \neq 0$, such that $\underline{\sigma}(x) \in B_{\lambda,\mu}$. Moreover, $|\mathrm{N}(x_{\lambda,\mu})| \leq T$ by construction (and hence $T \geq 1$). Notice that

$$\frac{\lambda_i}{T} \leq |\sigma_i(x_{\lambda,\mu})| \leq \lambda_i,$$

and similarly

$$\frac{\mu_j}{T} \leq |\tau_j(x_{\lambda,\mu})| \leq \mu_j.$$

By taking logarithms and rearranging things a bit, these inequalities become

$$\begin{aligned} 0 &\leq \log \lambda_i - \log|\sigma_i(x_{\lambda,\mu})| \leq \log T \\ 0 &\leq \log \mu_j - \log|\tau_j(x_{\lambda,\mu})| \leq \log T. \end{aligned} \tag{6.7.5}$$

The projection on the first $r+s-1$ coordinates is an isomorphism between H and $\mathbb{R}^{r+s-1}$. If we are able to show that there is no nontrivial linear form on H that vanishes on S, we are done. Now take any linear form over $\mathbb{R}^{r+s-1}$, say

$$F(y_1, \ldots, y_r, z_1, \ldots, z_{s-1}) = \sum_{i=1}^{r} c_i y_i + \sum_{j=1}^{s-1} d_j z_j.$$

By multiplying (6.7.5) by c_i and d_j and summing, using the triangular inequality we get

$$\begin{aligned} &\left|\sum_{i=1}^{r} c_i \log \lambda_i + \sum_{j=1}^{s-1} d_j \log \mu_j - F(\log \underline{\sigma}(x_{\lambda,\mu}))\right| \\ &\qquad \leq \left(\sum_{i=1}^{r}|c_i| + \sum_{j=1}^{s-1}|d_j|\right)\log T. \end{aligned} \tag{6.7.6}$$

Take any $B > (\sum |c_i| + \sum |d_j|) \log T$. Given $h \in \mathbb{N}$, choose λ_i and μ_j to satisfy

$$\sum_{i=1}^{r} c_i \log \lambda_i + \sum_{j=1}^{s-1} d_j \log \mu_j = 2Bh,$$

apart from μ_s, which is chosen to satisfy (6.7.4). The corresponding $x_{\lambda,\mu}$ is denoted x_h. Equation (6.7.6) simplifies to

$$|F(\log \underline{\sigma}(x_h)) - 2Bh| < B,$$

so

$$(2h-1)B < F(\log \underline{\sigma}(x_h)) < (2h+1)B.$$

This forces all the values obtained for x_h to be distinct, as the choice of h varies. On the other hand, they all satisfy $|\mathrm{N}(x_h)| \leq T$.

Since the ideals of limited norm are finite in number, we must have $(x_h) = (x_k)$ for some $h \neq k$, so $x_h = \xi x_k$ for some $\xi \in A^*$. From the fact that

$$F(\log \underline{\sigma}(x_h)) \neq F(\log \underline{\sigma}(x_k))$$

we find that $F(\log \underline{\sigma}(\xi)) \neq 0$, showing that F does not vanish identically on S and concluding the proof that $\rho = r + s - 1$. □

Another beautiful application of Minkowski's theorem is in Exercise 13.

6.8. Cyclotomic rings

We now give an example of application of the previous theory to a special but very interesting case, namely the cyclotomic fields. Recall that the *cyclotomic polynomials* $\phi_n(x)$ are defined inductively by the rule

$$x^n - 1 = \phi_n(x) \cdot \prod_{d|n} \phi_d(x), \tag{6.8.1}$$

starting from $\phi_1(x) = x - 1$. They are constructed in such a way that the roots of $\phi_n(x)$ are exactly the primitive nth roots of 1.

Remark 6.8.1. Using (6.8.1) it is immediate to prove by induction that $\phi_n(x)$ is monic with coefficients in $\mathbb{Z}$. Just note that the quotient of two monic polynomials in $\mathbb{Z}[x]$ is again monic and with integer coefficients.

Let ζ_n be a primitive nth root of unity. All other nth primitive roots of unity are powers of ζ_n, so the field $\mathbb{Q}(\zeta_n)$ is the splitting field of ϕ_n. In particular the field $\mathbb{Q}(\zeta_n)$ is a Galois extension of $\mathbb{Q}$.

Definition 6.8.2. A field of the form $\mathbb{Q}(\zeta_n)$ is called a *cyclotomic field.*

The following result is well known.

Proposition 6.8.3. *The polynomials $\phi_n(x)$ are irreducible.*

Proof. It is equivalent to say that all primitive nth roots of unity are conjugate. For this it is enough to show that if p is a prime not dividing n, ζ_n and ζ_n^p are conjugate.

Let $f(x) \in \mathbb{Z}[x]$ be the minimal polynomial of ζ_n, and write

$$\phi_n(x) = f(x)g(x);$$

by Gauss's lemma we can take both $f, g \in \mathbb{Z}[x]$.

Now assume by contradiction that $f(\zeta_n^p) \neq 0$, hence $g(\zeta_n^p) = 0$. So ζ_n is also a root of

$$h(x) := g(x^p),$$

which means $f(x)|g(x^p)$. Reducing this modulo p we find

$$\overline{f}(x)|\overline{g}(x^p) = \overline{g}(x)^p.$$

This implies that $\phi_n(x)$ has some repeated root in a finite extension of $\mathbb{F}_p$. But $x^n - 1$ does not have repeated roots modulo p, since it is relatively prime to its derivative nx^{n-1}, a contradiction. □

Corollary 6.8.4. *The Galois group of $\mathbb{Q}(\zeta_n)$ over $\mathbb{Q}$ is isomorphic to $(\mathbb{Z}/n\mathbb{Z})^*$.*

Proof. For any $a \in (\mathbb{Z}/n\mathbb{Z})^*$ we can define an automorphism τ_a of $\mathbb{Q}(\zeta_n)$ fixing $\mathbb{Q}$ by the requirement that

$$\tau_a(\zeta_n) = \zeta_n^a.$$

Such an automorphism exists because ζ_n and ζ_n^a are conjugate, and it is clearly uniquely determined since ζ_n generates the field.

It is clear that

$$\tau_a \circ \tau_b = \tau_{ab},$$

and this gives an injective homomorphism

$$(\mathbb{Z}/n\mathbb{Z})^* \to Gal(\mathbb{Q}(\zeta_n)/\mathbb{Q}).$$

Since both groups have cardinality $\phi(n)$, this is an isomorphism. □

We can also compute the ring of integers of $\mathbb{Q}(\zeta_n)$ by using discriminants.

Lemma 6.8.5. *The discriminant* $\operatorname{disc}(\mathbb{Z}[\zeta_n])$ *divides a power of* n.

Proof. Let $N = \mathrm{N}_{\mathbb{Q}(\zeta_n)/\mathbb{Q}}$ be the norm. By 4.2.11 we need to compute $N(\phi_n'(\zeta_n))$.

Write $x^n - 1 = \phi_n(x)g(x)$. Taking derivatives and evaluating at ζ_n we get

$$n\zeta_n^{-1} = n\zeta_n^{n-1} = \phi_n'(\zeta_n)g(\zeta_n).$$

Taking the norm and using the fact that $N(\zeta_n) = 1$ we get

$$N(n) = N(\phi_n'(\zeta_n))N(g(\zeta_n)), \tag{6.8.2}$$

and the left-hand side is a power of n. □

Lemma 6.8.6. *Let $n = p^k$ be a power of p, a prime integer. Then $\phi_n(1) = p$.*

Proof. For every n, denote

$$f_n(x) = \frac{x^n - 1}{x - 1} = 1 + x + \cdots + x^{n-1},$$

so that $f_n(1) = n$. Then for $k \geq 1$ we get

$$\phi_{p^k}(1) = \frac{f_{p^k}(1)}{f_{p^{k-1}}(1)} = p.$$

□

Theorem 6.8.7. *The ring of integers of $\mathbb{Q}(\zeta_n)$ is $\mathbb{Z}[\zeta_n]$.*

Proof. Let A be the ring of integers of $\mathbb{Q}(\zeta_n)$, so that $\mathbb{Z}[\zeta_n] \subset A$.

First, assume that $n = p^a$ is the power of a prime p, so that $\operatorname{disc}(\mathbb{Z}[\zeta_n])$ is a power of p as well by the above lemma. By Proposition 6.1.3, we have the inclusion

$$A \subset \frac{1}{p^c}\mathbb{Z}[\zeta_n] = \frac{1}{p^c}\mathbb{Z}[1 - \zeta_n]$$

for some $c \geq 0$.

Assume by contradiction that $\mathbb{Z}[\zeta_n] \subsetneq A$, and let $x \in A$ not in $\mathbb{Z}[\zeta_n]$. Write

$$x = \frac{m_0 + m_1(1 - \zeta_n) + \cdots + m_{t-1}(1 - \zeta_n)^{t-1}}{p^c},$$

where t is the order of $(\mathbb{Z}/n\mathbb{Z})^*$, i.e., the degree of $\phi_n(x)$. We can change x with $x \cdot p^s$ for some s, so we can assume $c = 1$. Also we can remove the first terms if p divides some m_i, so assume that

$$x = \frac{m_i(1 - \zeta_n)^i + \cdots + m_{t-1}(1 - \zeta_n)^{t-1}}{p}$$

with $p \nmid m_i$.

We can also remove higher order terms, but this is trickier. Since

$$\phi_n(x) = \prod_{k=1,(k,p)=1}^{n} (x - \zeta_n^k),$$

using Lemma 6.8.6 we have

$$p = \phi_n(1) = \prod_{k=1,(k,p)=1}^{n} (1 - \zeta_n^k).$$

Each term $1-\zeta_n^k$ is a multiple of $1-\zeta_n$, hence $(1-\zeta_n)^t$ divides p in $\mathbb{Z}[\zeta_n]$. Since $i<t$, this implies that $y := \frac{p}{(1-\zeta_n)^{i+1}} \in \mathbb{Z}[\zeta_n]$, hence $xy \in A$. By expanding this we get

$$\frac{m_i}{1-\zeta_n} + m_{i+1} + \cdots + m_{t-1}(1-\zeta_n)^{t-1-i} \in A,$$

hence $\frac{m_i}{1-\zeta_n} \in A$, with $p \not| m_i$.

Write this as $m_i = (1-\zeta_n)\cdot\alpha$ and take norms to find $m_i^r = N(1-\zeta_n)N(\alpha) = pN(\alpha)$, which is a contradiction since p does not divide m_i. This proves the thesis for n a prime power.

For the general case, write $n = p_1^{a_1}\cdots p_k^{a_k}$. We can apply the above for each field $\mathbb{Q}(\zeta_{p_i^{a_i}})$. To conclude, use Proposition 6.1.8 and the fact that the discriminant of $\mathbb{Z}[\zeta_{p_i^{a_i}}]$ divides a power of p_i, so all these discriminants are prime with each other. □

We now focus in more detail on the ring $\mathbb{Z}[\zeta_p]$, where p is an odd prime. First, we can compute the discriminant exactly. Namely, $x^p-1=(x-1)\phi_p(x)$, so (6.8.2) simplifies to

$$p^{p-1} = \mathrm{N}(\phi_p'(\zeta_p))\,\mathrm{N}(\zeta_p - 1).$$

The minimal polynomial of $\zeta_p - 1$ is $\phi_p(x+1)$, and its constant term is p, so $\mathrm{N}(\zeta_p-1)=p$. Since

$$\operatorname{disc}(\mathbb{Z}[\zeta_p]) = (-1)^{\binom{p-1}{2}}\,\mathrm{N}(\phi_p'(\zeta_p)),$$

we get

$$\operatorname{disc}(\mathbb{Z}[\zeta_p]) = \pm p^{p-2},$$

where the sign is + for $p \equiv 1 \pmod 4$ and − otherwise. Since the discriminant is defined as the square of the determinant of a matrix with entries in $\mathbb{Z}[\zeta_p]$, we conclude that $\sqrt{\pm p} \in \mathbb{Z}[\zeta_p]$. In particular, this shows the inclusion $\mathbb{Q}(\sqrt{\pm p}) \subset \mathbb{Q}(\zeta_p)$.

Let us understand the splitting of a prime $q \neq p$ in $\mathbb{Z}[\zeta_p]$. By Kummer's theorem 6.6.1, we can do this by computing the factorization of $\overline{\phi_p}$ inside $(\mathbb{Z}/q\mathbb{Z})[x]$.

The finite extensions of $\mathbb{F}_q = \mathbb{Z}/q\mathbb{Z}$ are the fields $\mathbb{F}_{q^k}$, while the roots of ϕ_p are primitive p-th roots of unity. In fact, the roots of $\overline{\phi_p}$ over $\mathbb{F}_q$ are distinct, since $\overline{\phi_p}'$ is not 0. The group $\mathbb{F}_{q^k}^*$ is cyclic, so the roots of $\overline{\phi_p}$ lie in $\mathbb{F}_{q^k}$ if and only if q^k-1 is multiple of p. In this case, since all the roots are inside the same extension of $\mathbb{F}_q$ of degree k, the factorization of $\overline{\phi_p}$ is

$$\overline{\phi_p} \equiv \overline{g_1}\cdots\overline{g_r},$$

where $r = \frac{p-1}{k}$. The corresponding factorization of q looks like

$$q\mathbb{Z}[\zeta_p] = Q_1 \cdots Q_r,$$

where $e(Q_i|q) = 1$ and $f(Q_i|q) = k$.

In this case $E(Q_i|q)$ is trivial, but $D(Q_i|q)$ is not, and corresponds to a subfield $\mathbb{Q}(\zeta_p)_D$ of $\mathbb{Q}(\zeta_p)$ of degree $r = \frac{p-1}{k}$ over $\mathbb{Q}$. By Corollary 6.8.4, the Galois group of $\mathbb{Q}(\zeta_p)$ over $\mathbb{Q}$ is cyclic, so there is a subgroup for each d that divides $p-1$. Let us denote the by $F(d)$ the subfield of $\mathbb{Q}(\zeta_p)$ of degree d over $\mathbb{Q}$. By comparison, we get the following useful result.

Theorem 6.8.8. *Let p be an odd prime number, $q \neq p$ another prime, and d a divisor of $p-1$. Then q splits completely inside $F(d)$ if and only if q is a d-th power $\bmod p$.*

Proof. The group $(\mathbb{Z}/p\mathbb{Z})^*$ is cyclic, so q is a d-th power if and only if $q^{(p-1)/d} \equiv 1 \pmod p$. The order of q in $(\mathbb{Z}/p\mathbb{Z})^*$ is exactly $f(Q_i|q) = k$, so this amounts to $(p-1)/d$ being a multiple of k. Equivalently, this is the same as $r = (p-1)/k$ being a multiple of d, or $F(d)$ being contained in $\mathbb{Q}(\zeta_p)_D$. The conclusion follows by Proposition 6.4.6. □

In fact, we know that $F(2) = \mathbb{Q}(\sqrt{\pm p})$, where the sign is positive if $p \equiv 1 \pmod 4$ and negative otherwise. We can use this to obtain a slick proof of the celebrated quadratic reciprocity formula. To introduce that, we first need a

Definition 6.8.9. Let $a, p \in \mathbb{Z}$ with p prime. The *Legendre symbol* of a and p is defined by

$$\left(\frac{a}{p}\right) = a^{\frac{p-1}{2}} \pmod p = \begin{cases} 0 \text{ if } a \text{ is multiple of } p \\ 1 \text{ if } a \text{ is a square} \quad \pmod p \\ -1 \text{ otherwise.} \end{cases}$$

It is immediate to check that

$$\left(\frac{ab}{p}\right) = \left(\frac{a}{p}\right)\left(\frac{b}{p}\right),$$

which can be used to reduce the computation of a Legendre symbol to the case where a is a prime number. The following famous result allows us to simplify the computation more.

Theorem 6.8.10 (Quadratic reciprocity theorem)**.** *Let p and q be two different primes. If both are odd, then*

$$\left(\frac{p}{q}\right) = (-1)^{\left(\frac{p-1}{2}\right)\left(\frac{q-1}{2}\right)}\left(\frac{q}{p}\right).$$

For the prime 2,

$$\left(\frac{2}{p}\right) = (-1)^{\frac{p^2-1}{8}}.$$

Proof. Let us consider the odd case, and assume that $p \equiv 1 \pmod 4$, in which case we aim to prove that $\left(\frac{p}{q}\right) = \left(\frac{q}{p}\right)$. Let us consider the field $F(2) \subset \mathbb{Q}(\zeta_p)$; as we have verified above this is just $\mathbb{Q}(\sqrt{p})$.

By Theorem 6.8.8, $\left(\frac{q}{p}\right) = 1$ if and only if q splits completely inside $\mathbb{Z}[\sqrt{p}]$. By Kummer's theorem 6.6.1, this happens if and only if the polynomial $x^2 - p$ factorizes in $\mathbb{F}_q$, which happens exactly when $\left(\frac{p}{q}\right) = 1$.

The other cases are handled similarly. □

6.9. Exercises

1. Compute the factorization of a prime p in a quadratic number field $K = \mathbb{Q}(\sqrt{m})$, where $m \equiv 1 \pmod 4$ is squarefree.

2. Show that $\mathbb{Z}[\sqrt{-3}]$ does not have unique factorization for ideals by considering the ideal $I = (2, 1+\sqrt{-3})$ and proving that $I^2 = 2I$, but $I \neq (2)$.

3. Show that $\mathbb{Z}[\sqrt{7}]$ is a principal ideal domain, as follows. By Corollary 6.7.5, we know that every class contains an ideal of norm at most $\lambda = 2.6457\ldots$. It is thus enough to show that primes of norm at most 2 are principal. If P is such a prime, show that $P \cap \mathbb{Z}$ is (2). Factor 2 explicitly using Kummer's theorem.

You can do the same computation for other small quadratic fields.

4. Show in the same way that $\mathbb{Z}[\sqrt{-163}]$ is a principal ideal domain. In fact, it is known that the only quadratic fields $\mathbb{Z}[\sqrt{-m}]$, $m > 0$ of class number one appear for $m = 1, 2, 3, 7, 11, 19, 43, 67$ and 163 (Heegner–Stark theorem, [**Sta69**]).

5. Prove that $A = \mathbb{Z}[\frac{1+\sqrt{-19}}{2}]$ is a principal ideal domain, but is not Euclidean. (For the last part, let $\alpha \in A$ be an nonunit of minimal Euclidean norm—what are the possibilities for $A/(\alpha)$?)

6. Let α be a solution of $f(x) = x^3 + x^2 - 2x + 8 = 0$, and let $K = \mathbb{Q}(\alpha)$ (K is known as the *Dedekind field*). Compute $\operatorname{disc}(f) = -2^2 \cdot 503$. Let $\beta = 4/\alpha$; prove that β is integral, and check that the discriminant of $\mathbb{Z}[\alpha, \beta]$ is -503. Deduce that $\mathbb{Z}[\alpha, \beta]$ is the ring of integers of K, and that $\mathbb{Z}[\alpha]$ has index 2 inside it.

7. Let K be the Dedekind field from Exercise 6, A its ring of integers. Show that 2 is unramified in A, and that there exists no surjective homomorphism from A to one of the finite fields $\mathbb{F}_4$ or $\mathbb{F}_8$. Conclude that 2 factors in A as a product of 3 different primes.

Deduce from this that A is not of the form $\mathbb{Z}[\gamma]$ for any $\gamma \in A$, and in fact for any $\gamma \in A$ the index of $\mathbb{Z}[\gamma]$ in A must be even.

Verify the same thing directly by computing $\operatorname{disc}(\mathbb{Z}[\gamma])$ for a generic $\gamma \in \mathbb{Z}[\alpha, \beta]$ and showing that it must be even.

8. Let $\mathcal{L} \subset \mathbb{R}^n$ be a subgroup which is abstractly isomorphic to $\mathbb{Z}^n$. Show that $\mathcal{L}$ is discrete in the Euclidean topology if and only if $\mathcal{L}$ generates $\mathbb{R}^n$ as a vector space.

9. Compute the volume of B_t, proving (6.7.2) (use induction on both r and s).

10. Use Eisenstein's criterion to give another proof that $\phi_p(x)$ is irreducible, when p is prime.

11. Check that the two embeddings defined in (6.1.1) and (6.1.2) agree, when using the same ordering for the factors of f.

12. Finish the proof of Theorem 6.8.10 by analyzing in detail the other cases.

13. Prove Hermite's theorem: for any d there exist only finitely many number fields K with discriminant less than d. (One can assume that the degree of K over $\mathbb{Q}$ is bounded. Choose a convex body in $\mathbb{R}^r \times \mathbb{C}^s$ where each coordinate is bounded by a small constant, except for one bounded by a big constant times $\operatorname{disc}(K)$. Use this to get an algebraic integer α in K; prove that in fact $K = \mathbb{Q}(\alpha)$ and use the fact that the coefficients for the minimal polynomial of α are bounded.)

The following exercises, up to Exercise 16 discuss the Frobenius element of a Galois number field extension.

14. Let L/K be a Galois extension of number fields, $P \subset \mathcal{O}_K$ be a prime which is not ramified in $\mathcal{O}_L$. Let Q be a prime of $\mathcal{O}_L$ over P. Show that there is a distinct *Frobenius element* $\Phi(Q|P) \in Gal(L/K)$ such that

$$\Phi(Q|P)(x) \equiv x^f \pmod{Q}$$

for all $x \in \mathcal{O}_L$, where $f = f(Q|P)$.

15. Continuing the previous exercise, show that if Q' is another prime over P, say $Q' = \sigma(Q)$ for some $\sigma \in Gal(L/K)$, the Frobenius elements are related by

$$\Phi(Q'|P) = \sigma\Phi(Q|P)\sigma^{-1}.$$

In particular, if L/K is an abelian extension, $\Phi(Q|P) = \Phi(P)$ only depends on P.

16. Let $L = \mathbb{Q}(\zeta_m)$ be a cyclotomic field, where we can take $m \not\equiv 2 \pmod 4$. A prime $p \in \mathbb{Z}$ is unramified in L if and only if p does not divide m. The Galois group $Gal(\mathbb{Q}(\zeta_m)/\mathbb{Q})$ is isomorphic to $(\mathbb{Z}/m\mathbb{Z})^*$, and acts by sending ζ_m to some power ζ_m^k.

Show that the Frobenius element $\Phi(p)$ is the element of the Galois group defined by $\Phi(p)(\zeta_m) = \zeta_m^p$.

17. The following example is taken from [**Cla65**]. Let K be a quadratic number field with class number bigger then 1 - for concreteness, take $K = \mathbb{Q}[\sqrt{-5}]$. Choose a prime p that factors as $(p) = Q_1 Q_2$ in $\mathcal{O}_K$, where Q_1 and Q_2 are principal (in our case, $p = 29$ will do). Writing $Q_1 = (q_1)$, let S be the multiplicative system generated by q_1, and $A = S^{-1}\mathcal{O}_K$. Prove that A is not the integral closure of a principal ideal domain in K. (A cannot be the integral closure of a subring of $\mathbb{Q}$ since $Q_1 \cdot A$ is invertible, but $Q_2 \cdot A$ is not. Prove that the class group of A does not become trivial.)

18. Prove that the equation $x^2 - 2y^2 = 1$ has infinitely many integer solutions. (Notice that $x^2 - 2y^2$ is a norm in a suitable number ring). More generally, show that the same holds for the equation

$$x^2 - dy^2 = 1, \tag{6.9.1}$$

where d is a squarefree integer, that is not congruent to 1 modulo 4. Can you modify your argument to make it work for the case where $d \equiv 1 \pmod 4$?. Equation (6.9.1) is known as *Pell's equation.*

19. Give an alternative proof of quadratic reciprocity following Eisenstein. Let S be the set of even integers s such that $2 \le s \le p-1$. Denote by $r(s)$ the remainder of the division of qs by p. Prove that $s \to (-1)^{r(s)} r(s) \pmod p$ is a permutation of S, and deduce from this that $\left(\frac{q}{p}\right) = (-1)^{\sum_{s\in S} \lfloor \frac{qs}{p} \rfloor}$.

This formula tells us that q is a square modulo p if and only if the number of points with integer coordinates in a certain triangle is even. Use symmetry to conclude quadratic reciprocity from this.

This is just another one of a plethora of different proofs that have been found since Gauss (see for example the book [**Bau15**]).

The following exercises (up to Exercise 22) lead to a proof of a special case of Fermat's last theorem, usually denoted by Case 1 for regular primes: there is no integer solution to $a^p + b^p = c^p$, where p does not divide abc. The case where p divides abc (Case 2) is much harder; see [**Tha99**]. The general case (when p is not regular) is the celebrated Wiles and Taylor theorem; see [**CSS97**].

20. Let p be an integer prime and ζ_p a primitive p^r-th root of unity, for some $r \ge 1$. Show that $p\mathbb{Z}[\zeta_p] = (1 - \zeta_p)^n$, where $n = \phi(p^r)$, and deduce that $1 - \zeta_p$ is prime.

21. Let p be a prime, $u \in \mathbb{Z}[\zeta_p]$ a unit. Prove that $u/\overline{u}$ is a power of ζ_p, where $\bar{\cdot}$ denotes complex conjugation.

22. Let $p \geq 5$ be a prime, and assume that p does not divide $h(\mathbb{Q}(\zeta_p))$ (p is called a *regular prime*). Show that there is no integer solution to $a^p + b^p = c^p$, where p does not divide abc. (Assuming there is a solution, show that $a + \zeta_p b = u \cdot t^p$ for $u, t \in \mathbb{Z}[\zeta_p]$ with u invertible, then use the previous exercise.)

The following exercises (up to Exercise 27) discuss some aspects of the geometry of lattices in more detail. For a detailed treatment of these topics, see [**MG02**] or [**Sim10**].

23. Let $\mathcal{L}$ be a lattice in $\mathbb{R}^n$, and denote by $\lambda_1(\mathcal{L})$ the minimum length of a nonzero vector in $\mathcal{L}$. Prove the result of Minkowski that $\lambda_1(\mathcal{L}) < \sqrt{n}(\operatorname{vol}\mathcal{L})^{\frac{1}{n}}$. This is a theoretical bound, but actually finding a short vector inside a lattice is a difficult computational problem, which we tackle in the next exercises.

24. Let $\mathcal{L}$ be a lattice in $\mathbb{R}^n$, and $\{b_1, \ldots, b_n\}$ a basis of $\mathcal{L}$. Denote by $\{b_1^*, \ldots, b_n^*\}$ the basis of $\mathbb{R}^n$ obtained by Gram–Schmidt orthogonalization, so

$$b_k^* = b_k - \sum_{i<k} u_{i,k} b_i,$$

where $u_{i,k} = \frac{\langle b_k, b_i^* \rangle}{\langle b_i^*, b_i^* \rangle}$. Prove that $\lambda_1(\mathcal{L}) \geq \min \|b_i^*\|$.

25. Keep the notation of the previous exercises. We say that the basis is *LLL-reduced* if $\|u_{i,k}\| \leq \frac{1}{2}$ for all $i < k$, and

$$\|b_{i+1}^* + u_{i,i+1} b_i^*\|^2 \geq \frac{3}{4}\|b_i^*\|^2 \tag{6.9.2}$$

for all i.

The following algorithm was introduced in [**LLL82**]. It alternates a *reduction step* and a *swap step*. The reduction step looks like an approximate Gram–Schmidt: for all $i = 2, \ldots, n$ and for all $j = i-1, \ldots, 1$, replace b_i by $b_i - \lceil u_{i,j} \rfloor b_j$, where $\lceil \cdot \rfloor$ denotes rounding by the nearest integer. The swap step consists in looking at violations of (6.9.2): if b_i and b_{i+1} violate the condition, swap them and start again from the beginning.

Prove that the LLL algorithm always terminates and produces an LLL-reduced basis for a lattice. (The quantity $F(b_1, \ldots, b_n) := \prod_i \|b_i\|^2$ always decreases in the swap step.)

26. With the notation of the previous exercise, prove that if the basis $\{b_1, \ldots, b_n\}$ is LLL-reduced, then

$$\|b_1\| \leq 2^{\frac{n-1}{2}} \lambda_1(\mathcal{L}),$$

thus obtaining an algorithmic way to produce a not too long vector in $\mathcal{L}$.

27. The LLL algorithm can be used to find algebraic equations with small coefficients satisfied by a number, given a good enough approximation of the number.

Let α be some algebraic number of degree n, and $\overline{\alpha}$ some approximation of α. Choose some big constant K, and consider the vectors

$$\begin{aligned} b_0 &= (1, \ldots, 0, K) \\ b_1 &= (0, 1, \ldots, 0, K\overline{\alpha}) \\ &\vdots \\ b_n &= (0, 0, \ldots, 1, K\overline{\alpha}^n). \end{aligned}$$

Let $\mathcal{L}$ be the lattice generated by $\{b_0, \ldots, b_n\}$. Using the LLL algorithm, we can find a short vector $v \in \mathcal{L}$. But such a vector must have the form

$$v = \left(a_0, \ldots, a_n, K \cdot \sum_{i=0}^{n} a_i \overline{\alpha}^i \right).$$

In order for it to be short, the polynomial $\sum_{i=0}^{n} a_i \overline{\alpha}^i$ must be approximately 0, and we can hope that the equation is satisfied exactly by α.

Try this in practice using $\overline{\alpha} = 1.6180339887$ and $K = 10^4$: you should recover the equation $x^2 - x + 1$ for the golden ratio.

Chapter 7

Metric and Topological Methods

In this chapter we use topological methods to investigate rings. We start with the notion of absolute value on a field, which is analogue to the notion of norm in the theory of Banach spaces, and we show that the metric completion of a field endowed with an absolute value is again a field. The important example of the p-adic absolute values give rise to the so-called p-adic fields.

In the sequel our treatment becomes increasingly algebraic. First we consider the notion of valuation, which is strictly connected to that of absolute value. We exhibit a general result of existence for valuations on a field. As a particular case we study the so-called discrete valuation rings (or DVR), which turn out to be rings obtained by localization of Dedekind rings.

Finally, in Section 7.5 we show that the approach which uses the notion of metric was not necessary to define the completion, and we can replace it by purely algebraic methods. The choice of an ideal I of the ring A determines a topology on A, and we can define the completion of A with respect to this topology using the machinery of inverse limits, which we also introduce. In the rest of the chapter, we study the properties of this topology and the relation between a ring and its completion.

Both the metric and the algebraic approach allow us to give a definition of the p-adic integers $\mathbb{Z}_p$. Section 7.7 ends the chapter by showing yet another way to build $\mathbb{Z}_p$, this time using explicit algebraic formulas. These can be generalized to any ring, giving rise the construction of rings of Witt vectors.

7.1. Absolute values

Definition 7.1.1. Let k be a field. A function

$$|\cdot|: k \to \mathbb{R}$$

is called an *absolute value* if it satisfies:

(1) $|a| \geq 0$ and $|a| = 0$ if and only if $a = 0$;
(2) $|ab| = |a|\,|b|$;
(3) $|a+b| \leq |a| + |b|$

for all choices of $a, b \in k$. As usual, condition 3. is called the *triangular inequality*.

Remark 7.1.2. This notion is clearly akin to that of a norm in functional analysis. A norm is required to be homogeneous, and that only makes sense for real (or complex) vector spaces. By changing that into condition 2. we obtain a notion which is better suited to study abstract fields.

We will give examples soon, but before doing that let us introduce an important distiction.

Definition 7.1.3. The absolute value $|\cdot|$ is called *nonarchimedean* if it satisfies the stronger inequality

3'. $|a+b| \leq \max\{\,|a|\,, |b|\,\}$

for all $a, b \in k$, and *archimedean* otherwise.

Example 7.1.4.

(a) The standard absolute value $|\cdot|_{st}$ on $\mathbb{R}$ is an absolute value in our sense, and it restricts to an absolute value on every subfield of $\mathbb{R}$.
(b) Similarly we have the Euclidean absolute value $|\cdot|_{st}$ on $\mathbb{C}$.
(c) On every field, we have the trivial absolute value defined by $|a|_{tr} = 1$ for all $a \neq 0$. This is the only absolute value on a finite field. A quick way to see this is the following.

    The multiplicative group $\mathbb{F}_q^*$ is cyclic; if a is a generator we must have $|a^n| = |a|^n$ for any absolute value $|\cdot|$. Taking $n = q-1$ we find $|a|^{q-1} = |1| = 1$; since $|a|$ is real and positive we then have $|a| = 1$.

(d) Here comes the real motivation for this notion. Fix a prime number $p \in \mathbb{Z}$; for all $a \in \mathbb{Z} \setminus \{0\}$ denote $v_p(a) = r$ if p^r divides a but p^{r+1} does not. Note that

$$v_p(a \cdot b) = v_p(a) + v_p(b), \tag{7.1.1}$$

hence it makes sense to extend this to $\mathbb{Q}^*$ by the rule

$$v_p\left(\frac{a}{b}\right) := v_p(a) - v_p(b),$$

and (7.1.1) still holds. Also note that

$$v_p(a+b) \geq \min\bigl\{v_p(a), v_p(b)\bigr\}. \tag{7.1.2}$$

These relations look like the logarithm of those defining an absolute value. Indeed we can get an absolute value $|\cdot|_p$ on $\mathbb{Q}$ defined by

$$|a|_p := p^{-v_p(a)}.$$

Note that $|\cdot|_p$ is nonarchimedean.

The function v_p is called the *p-adic valuation* and $|\cdot|_p$ is called the *p-adic absolute value.* We will come back on valuations in next section.

(e) By the same token, let A be a Dedekind ring with fraction field k. Fix a prime P of A; for $a \in A$ we let $v_P(a)$ be the exponent of P in the factorization of (a). This satisfies properties analogue to (7.1.1) and (7.1.2), hence we can extend it to k^* by the rule

$$v_P\left(\frac{a}{b}\right) := v_P(a) - v_P(b).$$

Letting q be the norm of P, we define the nonarchimedean absolute value

$$|a|_P := q^{-v_P(a)}.$$

Again we call v_P the *P-adic valuation* and $|\cdot|_P$ the *P-adic absolute value.*

An absolute value on a field allows us to define a metric on it. Given an absolute value $|\cdot|$ on the field k define the distance

$$d(a,b) := |a-b|$$

for $a, b \in k$. The function d turns k into a metric space; in turn this induces a topology on k, such that the field operations are continuous.

It may happen that two different absolute values dominate each other; in this case they are equivalent for our purposes.

Definition 7.1.5. Let $|\cdot|_1, |\cdot|_2$ be two absolute values on k. We say that $|\cdot|_1$ is *equivalent* to $|\cdot|_2$ if there exist a real number $e > 0$ such that

$$|a|_2 = |a|_1^e$$

for all $a \in k$. An equivalence class of absolute values on k is called a *place* of k.

It is clear that two equivalent absolute values determine the same topology on k; it may come as a surprise that the converse is true.

Proposition 7.1.6. *Let $|\cdot|_1, |\cdot|_2$ be two absolute values on k. If the topologies induced on k by $|\cdot|_1$ and $|\cdot|_2$ are the same, the two absolute values are equivalent.*

Proof. First note that for any absolute $|\cdot|$ and any $a \in k$ we have $|a| < 1$ if and only if $a^n \to 0$. Since this is a topological property, $|a|_1 < 1$ if, and only if, $|a|_2 < 1$. Passing to the reciprocal we find $|a|_1 > 1$ if, and only if, $|a|_2 > 1$. It follows that for any $a \in k$ we can write

$$|a|_2 = |a|_1^e$$

for some $e > 0$, depending on a.

Now assume

$$|a|_2 = |a|_1^e \quad \text{and} \quad |b|_2 = |b|_1^f\,;$$

we shall prove that $e = f$. If this is not the case, let

$$c := \frac{\log |b|_1}{\log |a|_1}.$$

We can assume that $e < f$ and $c > 0$. Then $ce < cf$, so we can find a rational number r/s such that

$$ce < \frac{r}{s}e < cf.$$

This gives the two inequalities

$$cs < r \quad \text{and} \quad re < csf.$$

From the first one we derive

$$|b|_1^s = |a|_1^{cs} < |a|_1^r\,,$$

so

$$\left|\frac{a^r}{b^s}\right|_1 > 1.$$

From the second

$$|a|_2^r = |a|_1^{re} < |a|_1^{csf} = |b|_1^{sf} = |b|_2^s\,,$$

hence

$$\left|\frac{a^r}{b^s}\right|_2 < 1.$$

This is a contradiction. □

Remark 7.1.7. The second part of the proof may seem tricky, but it is simpler than it looks. Once one looks for a number of the form a^r/b^s which violates the conclusion of the first part, it is just a matter of spelling out the needed costraints.

Remark 7.1.8. The function $|\cdot|^e$ may fail to be an absolute value, even if $|\cdot|$ is. For instance if $|\cdot|_{st}$ is the standard absolute value on $\mathbb{R}$, $|\cdot|_{st}^2$ does not satisfy the triangular inequality. Still, $|\cdot|^e$ is an absolute value when $|\cdot|$ is nonarchimedean.

The main reason we consider absolute values is that we can use them to costruct new rings or fields by completion of old. Let k be a field endowed with an absolute value $|\cdot|$, and let $A \subset k$ a subring. Then the metric completion $\widehat{A}$ of A is itself a ring, with the ring operations extended from A by continuity. To make sense of the preceding sentence, let us recall some definitions.

Definition 7.1.9. Let (X, d) be a metric space. A sequenxe (x_n) of elements of X is called *Cauchy* if for any $\epsilon > 0$ we can find N such that

$$d(x_m, x_n) < \epsilon$$

for all $m, n \geq N$. The metric space X is called *complete* if every Cauchy sequence has a limit in X.

It is trivial to check that, conversely, a converging sequence is always Cauchy. If (X, d) is any metric space, there is a construction of a *complete* metric space $(\widehat{X}, \widehat{d})$ together with an embedding $X \hookrightarrow \widehat{X}$ such that $\widehat{d}$ restricts to d on X and X is dense in $\widehat{X}$. Moreover such $\widehat{X}$ is unique up to isometry, and is called the *completion* of X. We give the standard construction of $\widehat{X}$, leaving uniqueness to the reader.

Consider the set Y of Cauchy sequences of elements of X. We call two sequences (x_n) and (y_n) equivalent if

$$d(x_n, y_n) \to 0.$$

We define $\widehat{X}$ as the quotient of Y by this equivalence relation. The equivalence class of a sequence is denoted $[x_n]$. On $\widehat{X}$ we define the distance

$$\widehat{d}([x_n], [y_n]) := \lim_{n\to\infty} d(x_n, y_n).$$

Such a limit exists in $\mathbb{R}$, since both x_n and y_n are Cauchy and d is uniformly continous, so $(d(x_n, y_n))$ is a Cauchy sequence in $\mathbb{R}$. Moreover it is immediate to see that $\widehat{d}$ does not depend on the choice of the representatives.

We claim that $\widehat{X}$ is a complete metric space with a dense isometric embedding of X. The embedding of X is given by the function

$$\iota\colon\ X \longrightarrow \widehat{X},$$

$$x \longmapsto [x].$$

This is clearly an isometry, and moreover the image is dense. Indeed for any Cauchy sequence $[x_n] \in \widehat{X}$ choose N such that $d(x_n, x_m) < \epsilon$ for $n, m \geq N$. Then

$$\widehat{d}([x_n], [x_N]) = \lim_{n\to\infty} d(x_n, x_N) \leq \epsilon.$$

Finally, we check that $\widehat{X}$ is complete. Let $(x(m))$ be a Cauchy sequence in $\widehat{X}$ and write $x(m) = [x(m)_n]$. One can define the diagonal sequence y by $y_n = x(n)_n$. Using the triangular inequality, it is easy to check that

$$\lim_{m\to\infty} (x(m)) = [y_n]$$

in $\widehat{X}$.

We now come back to the setting of absolute values. Fix an absolute value $|\cdot|$ on the field k. For a subring $A \subset k$ let $\widehat{A}$ be the completion. The ring operations on A extend to $\widehat{A}$, by putting

$$[a_n] + [b_n] := [a_n + b_n]$$
$$[a_n] \cdot [b_n] := [a_n \cdot b_n].$$

It is immediate to check that the sequences $(a_n + b_n)$ and $(a_n \cdot b_n)$ are indeed Cauchy, and that the operations are well defined. Note that this is determined by the requirement that the operations should be continous on the whole $\widehat{A}$. The ring axioms for the operations are valid on A which is dense, hence hold on all $\widehat{A}$ by continuity. This gives $\widehat{A}$ the structure of a ring.

If moreover A is itself a field—for instance $A = k$—we also define

$$1/[a_n] := [1/a_n]$$

for $[a_n] \neq 0$. In this case it is slightly less obvious that $1/a_n$ is Cauchy. Since $[a_n] \neq 0$, $a_n \not\to 0$, which implies that $|a_n| > \epsilon$ for some $\epsilon > 0$ and infinitely many n. Since moreover a_n is Cauchy, we have $|a_n| > \epsilon/2$ eventually. For any $\delta > 0$ choose N such that $|a_n - a_m| < \delta$ for $n, m \geq N$. Then

$$\left|\frac{1}{a_n} - \frac{1}{a_m}\right| = \left|\frac{a_m - a_n}{a_n a_m}\right| < \frac{4\delta}{\epsilon^2},$$

proving that the sequence $(1/a_n)$ is Cauchy. We deduce that the completion of a field is itself a field.

Note that the completion $\widehat{k}$ inherits an absolute value, again obtained extending by continuity the absolute value $|\cdot|$ on k. We shall denote this absolute value by the same symbol.

Remark 7.1.10. Once one has constructed $\widehat{k}$, $\widehat{A}$ is simply the closure of A inside $\widehat{k}$. So, it is enough to consider completions of fields.

Example 7.1.11.

(a) Consider the Euclidean absolute value $|\cdot|_{st}$ on $\mathbb{Q}$. The completion of $\mathbb{Q}$ is then $\mathbb{R}$. If we complete $\mathbb{Q}(i)$ with respect to the Euclidean absolute value, we obtain $\mathbb{C}$.

(b) The completion of $\mathbb{Q}$ with respect to the p-adic absolute value $|\cdot|_p$ is called the *field of p-adic numbers* and is denoted by $\mathbb{Q}_p$. It contains the ring $\mathbb{Z}_p$ of *p-adic integers* as the completion of $\mathbb{Z}$.

We claim that $\mathbb{Z}_p$ is local; its maximal ideal is $p\mathbb{Z}_p$, or equivalently the closure of (p) inside $\mathbb{Z}_p$. To see this, we try to understand what a p-adic integer looks like. Choose any set D of representatives for $\mathbb{Z}/(p)$, for instance

$$D = \{0, 1, \ldots, p-1\}.$$

Let $a \in \mathbb{Z}_p$; then a is the limit of a sequence (a_n) of integers which is Cauchy for $|\cdot|_p$. Up to passing to a subsequence, we can assume that if $m, n \geq N$ then $a_m - a_n$ is divisible by p^N. In particular the class of a_n modulo p^N is independent of $n \geq N$; we shall call $\alpha_N \in D$ the representative for such class.

It follows that

$$a_n - \sum_{i=0}^{n} \alpha_i p^i$$

is divisible by p^{n+1}, so the sequence $\left(\sum_{i=0}^{n} \alpha_i p^i\right)$ is equivalent to (a_n). We conclude that we can write

$$a = \sum_{i=0}^{\infty} \alpha_i p^i := \lim_{n\to\infty} \sum_{i=0}^{n} \alpha_i p^i. \tag{7.1.3}$$

This is completely analogous to writing a real number between 0 and 1 in terms of powers of $1/10$, and we can think of the coefficients α_n as the p-adic digits of a (note that the argument above shows that the α_i are uniquely determined). The numbers having an expansion with a finite number of digits are exactly the usual integers.

The function

$$\mathbb{Z}_p \to \mathbb{Z}/(p^n)$$

defined by sending a to the finite sum $\sum_{i=0}^{n-1} \alpha_i p^i$ is easily seen to be a surjective ring homomorphism extending the projection from $\mathbb{Z}$. It follows that p^n generates a nontrivial ideal in $\mathbb{Z}_p$, and in particular $p\mathbb{Z}_p$ is maximal.

We can also verify that any p-adic integer $a \in \mathbb{Z}_p \setminus p\mathbb{Z}_p$ is invertible. Indeed, consider the element $1/a \in \mathbb{Q}_p$. For any $n \in \mathbb{N}$ we can use density to find some $b_n \in \mathbb{Q}$ such that

$$|b_n - 1/a|_p < p^{-n};$$

in particular, the denominator of b_n is not divisible by p and we can find some $c_n \in \mathbb{Z}$ such that

$$|c_n - b_n|_p < p^{-n}.$$

It follows that (c_n) is a sequence of integers which converges to $1/a$ in $\mathbb{Q}_p$, in particular $1/a \in \mathbb{Z}_p$. This completes the proof of our claim that $\mathbb{Z}_p$ is local with maximal ideal $p\mathbb{Z}_p$. Moreover the universal property of the localization shows that we have inclusions

$$\mathbb{Z} \subset \mathbb{Z}_{(p)} \subset \mathbb{Z}_p.$$

Since $\mathbb{Z}_p$ has projections to $\mathbb{Z}/(p^n)$ for any n but no other homomorphisms, it is particularly well-suited for studying congruences modulo powers of p. We shall make this more precise in Section 7.5, when we treat the completion of a ring as an inverse limit.

(c) Let k be a number field with ring of integers A. For a prime P of A consider the absolute value $|\cdot|_P$ of k. Then the completion of k is a field k_P, containing the completion A_P of A. By the same arguments as above, A_P is a local ring with maximal ideal PA_P, and every element of A_P can be written as

$$a = \sum_{i=0}^{\infty} a_i,$$

where $a_i \in P^i \setminus P^{i-1}$. The only difference is the lack of a preferred system of representatives for the digits.

On p-adic numbers, one can mimic Newton's method to solve analytic equations. A possible formulation is the following, but see Theorem 7.6.1 for a generalization.

Theorem 7.1.12 (Hensel's lemma for p-adics). *Let $f \in \mathbb{Z}[x]$ be a polynomial, $\overline{f}(x) \in \mathbb{Z}/p\mathbb{Z}[x]$ its residue modulo the prime p. Assume $\overline{f}$ has a simple* *root $\alpha \in \mathbb{Z}/p\mathbb{Z}$—that is $\overline{f}'(\alpha) \not\equiv 0$. Then α can be lifted uniquely to a p-adic root of f in $\mathbb{Z}_p$.*

Proof. The idea is to lift α to roots α_k of f modulo p^k. By induction, assume that we find $\alpha_k \in \mathbb{Z}$ such that

$$f(\alpha_k) \equiv 0 \mod p^k$$

and

$$f'(\alpha_k) \not\equiv 0 \mod p,$$

the case $k = 1$ being the hypothesis.

Expand

$$f(\alpha_k + t \cdot p^k) = \sum_{i=0}^{n} c_i t^i p^{ik}$$

as a polynomial in $t \cdot p^k$. By construction $c_0 = f(\alpha_k)$ and $c_1 = f'(\alpha_k)$, so modulo p^{k+1} we have

$$f(\alpha_k + t \cdot p^k) \equiv f(\alpha_k) + f'(\alpha_k) \cdot t \cdot p^k \mod p^{k+1}.$$

Since p^k divides $f(\alpha_k)$, we can solve for t in $\mathbb{Z}/p\mathbb{Z}$ by

$$t = -\frac{f(\alpha_k)}{f'(\alpha_k)p^k}. \tag{7.1.4}$$

With this choice of t, we get our desired lift α_{k+1}. The sequence $\{\alpha_k\}$ converges in $\mathbb{Z}_p$ to a root of f that lifts α. □

Notice the resemblance of (7.1.4) with the equations used in the Newton method to approximate roots of real functions.

We shall generalize the completion construction in Section 7.5. In the final part of this section we classify all places on number fields. This consists of two, pretty independent, steps: first we classify places on $\mathbb{Q}$, and then we study the ways one can extend absolute values from one field to a bigger one. It will turn out that the only places are those we already met in the examples.

Theorem 7.1.13 (Ostrowski). *Every nontrivial absolute value on $\mathbb{Q}$ is equivalent either to the Euclidean absolute value or to a p-adic one.*

Proof. Let $|\cdot|$ be an absolute value on $\mathbb{Q}$. Given $m, n \in \mathbb{N}$ with $n > 1$ we write m in base n as

$$m = a_0 + a_1 n + \cdots + a_r n^r$$

with $a_i \in \{0, 1, \ldots, n-1\}$. By the triangular inequality

$$|m| \le \sum_{i=0}^{r} |a_i|\, N^i,$$

where $N = \max\{1, |n|\}$. Combining this with the easy estimates $r \le \frac{\log m}{\log n}$ and $|a_i| \le a_i \le n$ we find

$$|m| \le (1+r)nN^r \le \left(1 + \frac{\log m}{\log n}\right) nN^{\frac{\log m}{\log n}}. \tag{7.1.5}$$

This is true for all $m \in \mathbb{N}$, so we can substitute m^k for m in (7.1.5). After taking the k-th root and letting $k \to \infty$ we find

$$|m| \le N^{\frac{\log m}{\log n}}. \tag{7.1.6}$$

Two cases now arise

(i) Assume $|n| > 1$ for all $n \ge 2$. In this case $N = |n|$ and (7.1.6) becomes

$$|m|^{\frac{1}{\log m}} \le |n|^{\frac{1}{\log n}}.$$

Since m, n were arbitrary, by symmetry we must have the reverse inequality, so in fact equality holds. In other words,

$$|n|^{\frac{1}{\log n}} = c$$

is the same constant $c > 0$ for all $n \ge 2$. Then

$$|n| = c^{\log n} = n^{\log c}$$

for all $n \ge 2$. It follows that $|\cdot|$ is equivalent to $|\cdot|_{st}$.

(ii) Otherwise there is some $n \ge 2$ such that $|n| \le 1$. Then $N = 1$ and from (7.1.6) we find $|m| \le 1$ for all $m \in \mathbb{N}$.

It is immediate to check that the set

$$\{n \in \mathbb{Z} \mid |n| < 1\}$$

is a prime ideal, hence it is generated by some prime $p \in \mathbb{N}$. In this case, $|\cdot|$ is equivalent to $|\cdot|_p$. □

Finally we discuss how to extend absolute values. The situation is easier for complete fields, and is completely analogous to the fact that all norms on $\mathbb{R}^n$ are equivalent.

Proposition 7.1.14. *Let k be a field complete with respect to the absolute value $|\cdot|$ and let E/k be a finite extension. Then there is at most one extension of $|\cdot|$ to E.*

Proof. We abuse notation and denote by $|\cdot|$ an extension of $|\cdot|$ to E. Choose a basis $e_1, \ldots, e_n$ of E over k; for $v \in E$ we can write

$$v = v_1 e_1 + \cdots v_n e_n.$$

Triangular inequality yields

$$|v| \leq \sum_{i=1}^{n} |v_i|\,|e_i| \leq M \sum_{i=1}^{n} |v_i|\,, \tag{7.1.7}$$

where $M = \max\{|e_i|\}$.

Note that the function

$$\|v\| := \sum_{i=1}^{n} |v_i|$$

may fail to be an absolute value, but still allows us to define a distance on E by $d(v, w) := \|v - w\|$. Equation (7.1.7) then shows that $|\cdot|$ is continous with respect to this distance.

The sphere

$$S := \{v \in E \mid \|e\| = 1\}$$

is compact since $|\cdot|$ is complete on k. It follows that $|\cdot|$ assumes a minimum m on S, so

$$m\,\|v\| \leq |v| \leq M\,\|v\|$$

for $v \in S$; by homogeneity the same inequality holds for all $v \in E^*$. In particular $|\cdot|$ and $\|\cdot\|$ induce the same topology on E. So all extensions of $|\cdot|$ to E are equivalent. $\square$

To generalize Ostrowski's theorem, let k be a number field with an absolute value $|\cdot|$. The restriction of $|\cdot|$ to $\mathbb{Q}$ is either $|\cdot|_p$ for some prime p or the standard absolute value.

Let us consider the first case. The completion $\widehat{k}$ is a finite extension of $\mathbb{Q}_p$—namely, if $k = \mathbb{Q}(\alpha)$, then $\widehat{k} = \mathbb{Q}_p(\alpha)$. More explicitly, if $g(x)$ is the minimal polynomial for α, and

$$g(x) = g_1(x) \cdots g_d(x)$$

over $\mathbb{Q}_p$, then $\widehat{k} \cong \mathbb{Q}_p[x]/(g_i)$ for some i. By Proposition 7.1.14, the absolute value on $\widehat{k}$ is uniquely determined by its restriction to $\mathbb{Q}_p$.

In fact, we know how to extend the absolute value $|\cdot|_p$ from $\mathbb{Q}_p$ to $\widehat{k}$. The ring $\mathbb{Z}_p$ is a Dedekind domain with a single prime ideal (in the terminology of Section 7.3 it is a DVR), so the integral closure A of $\mathbb{Z}_p$ inside $\widehat{k}$ is a Dedekind ring. If $\widehat{P}$ is any prime of A over $p\mathbb{Z}_p$, we can define a $\widehat{P}$-adic absolute value on $\widehat{k}$ which will extend $|\cdot|_p$. By uniqueness, this must agree with the absolute value of $\widehat{k}$ (and in fact, $\widehat{P}$ is the only prime above $p\mathbb{Z}_p$).

It follows that the absolute value on k is the P-adic absolute value, where $P = \widehat{P} \cap \mathcal{O}_k$.

In a similar way, we can prove that the only archimedean absolute values on k are obtained by embeddings into $\mathbb{C}$. Our conclusion is the following classification.

Theorem 7.1.15 (Ostrowski). *Every nontrivial absolute value on a number field k is equivalent either to the Euclidean absolute value for some embedding $k \hookrightarrow \mathbb{C}$, or to a P-adic one.*

7.2. Valuations and valuation rings

Recall that we have defined the p-adic absolute value by taking the exponential of the function v_p defined by $v_p(a) = r$ if p^r divides a but p^{r+1} does not. In this section we consider generalizations of this function.

Notice that we can see v_p as a group homomorphism $\mathbb{Q}^* \to \mathbb{Z}$, and that (7.1.2) implies that if $v_p(a) \in \mathbb{N}$ and $v_p(b) \in \mathbb{N}$, then $v_p(a+b) \in \mathbb{N}$.

Definition 7.2.1. An *ordered group* G (written additively) is an abelian group endowed with a subset P closed under the group operation (the *positive* elements) such that for every $g \in G$, either $g \in P$ or $-g \in P$.

An ordered group has an order compatible with its group structure. Namely, define $g \preccurlyeq h$ if $h - g \in P$. The fact that P is closed under the group operation implies that $\prec$ is transitive, and the last requirement in the definition implies that $\prec$ is a total order.

Example 7.2.2. Taking $G = \mathbb{Z}$ and $P = \mathbb{N}$ we see that $\mathbb{Z}$ is an ordered group with the usual order.

Definition 7.2.3. A *valuation* on the field k is a group homomorphism $v : k^* \to G$, where G is an ordered group, such that

$$v(a+b) \succcurlyeq \min\{v(a), v(b)\}. \tag{7.2.1}$$

Valuations and absolute values are strictly related concepts, as shown in the following remark.

Remark 7.2.4. When the target group of v is $\mathbb{Z}$ with the standard ordering, one can define a nonarchimedean absolute value by taking $|x| = a^{v(x)}$ for any real $a > 1$. In general, there exist valuations with target groups different from $\mathbb{Z}$, as well as archimedean absolute values, so neither concept is more general than the other.

Associated to a valuation v, there is a ring

$$A = \{a \in k | v(a) \succcurlyeq 0\} \cup \{0\}.$$

Notice that if $a \in k$ is a nonzero element, either $a \in A$, or $1/a \in A$. We will see that this property alone characterizes rings of this form, and in fact it is enough to recover the group G and the valuation from the ring alone.

Definition 7.2.5. Let A be an integral domain with field of fractions k. We say that A is a *valuation ring* if for all $a \in k^*$ either $a \in A$ or $1/a \in A$.

The set $\mathcal{M}$ of elements $a \in A$ such that $1/a \notin A$ is an ideal. In fact, let $x \in A$ and $a \in \mathcal{M}$. Then $ax \in \mathcal{M}$. If this was not the case, $1/(ax) \in A$, hence $1/a = x/(ax) \in A$, a contradiction. Similarly take $a, b \in \mathcal{M}$—then we can assume by symmetry that $a/b \in A$, so

$$a + b = \left(1 + \frac{a}{b}\right) \cdot b \in \mathcal{M}$$

by the previous point. Since $\mathcal{M}$ consist exactly of nonunit elements, it follows that it is the only maximal ideal. We conclude:

Proposition 7.2.6. *Let A be a valuation ring. Then A is local, with maximal ideal*

$$\mathcal{M} = \{a \in A | 1/a \notin A\} = A \setminus A^*.$$

Knowing this, we can construct the valuation from a valuation ring as follows. Let A^* be the group of units of A. We know that, if A arises from a valuation v, $v(A^*) = 0$. So we can just take the abelian group $G = k^*/A^*$, with positive elements P the image of A. It is clear that P is closed under the group operation, hence G is an ordered group.

The valuation v is just the projection $k^* \to k^*/A^*$. It remains to check condition (7.2.1). This amounts to saying that given $a, b \in k$, either $(a + b)/a \in A$ or $(a + b)/b \in A$. But

$$\frac{a+b}{a} = 1 + \frac{b}{a}; \frac{a+b}{b} = 1 + \frac{a}{b},$$

so this follows from the definition of a valuation ring.

We summarize the discussion so far.

Theorem 7.2.7. *Let A be a valuation ring with fraction field k. Then there exists a valuation v on k such that*

$$A = \{x \in k | v(x) \succcurlyeq 0\}. \tag{7.2.2}$$

Vice versa, given a valuation v on a field k, define A by (7.2.2). Then A is a valuation ring.

In the rest of the section, we make some important connections between valuation rings and integral closure.

Proposition 7.2.8. *Let A be a valuation ring. Then A is integrally closed.*

Proof. Let k be the field of fractions of A and $x \in k$ an element integral over A so that

$$x^n + a_{n-1}x^{n-1} + \cdots + a_0 = 0$$

for some $a_0, \ldots, a_{n-1} \in A$. Rewrite this as

$$x = -\left(a_{n-1} + \cdots + \frac{a_0}{x^{n-1}}\right).$$

Since A is a valuation ring, either $x \in A$ or $1/x \in A$. In the latter case the above equation shows that $x \in A$ as well. □

We now turn our attention to a way to actually construct valuation rings for a field k.

Theorem 7.2.9. *Let k and K be two fields, with K algebraically closed, and consider homomorphisms $f\colon A \to K$, where A is a subring of k. Assume (f, A) is maximal among such pairs. This means that there is no pair (g, B) of a ring $B \subset k$ containing A and a homomorphism $g\colon B \to K$ which restricts to f on A. Then A is a valuation ring.*

Notice that maximal pairs of this kind always exist thanks to Zorn's lemma.

Example 7.2.10. When $k = \mathbb{Q}$ and K is the algebraic closure of $\mathbb{Z}/p\mathbb{Z}$, the maximal ring is the localization $\mathbb{Z}_{(p)}$. In fact the projection $\mathbb{Z} \to \mathbb{Z}/p\mathbb{Z}$ can be extended to any element $a/b \in \mathbb{Q}$ such that b is not divisible by p. Unsurprisingly, the associated valuation is the p-adic valuation.

Before getting to the theorem, we need a couple of intermediate results.

Lemma 7.2.11. *Let $f\colon A \to K$ be a maximal homomorphism as in Theorem 7.2.9. Then A is a local ring with maximal ideal* $\ker f$.

Proof. It is enough to prove that if $a \in A$ and $f(a) \neq 0$, f can be extended to $A[1/a]$. By maximality, it will follow that $1/a \in A$.

Consider the homomorphism $g : A[x] \to K$ sending x to $1/f(a)$. We want to check that this descends to a homomorphism defined on $A[1/a]$. In other words, let $\overline{g} : A[x] \to A[1/a]$ be the evaluation morphism; we need to check that $\ker \overline{g} \subset \ker g$.

For this, let $p(x) \in A[x]$, and assume that $p(1/a) = 0$. If

$$p(x) = p_0 + p_1 x + \cdots + p_r x^r$$

with $p_0, \ldots, p_r \in A$, then

$$g(p) = f(p_0) + \frac{f(p_1)}{f(a)} + \cdots + \frac{f(p_r)}{f(a)^r},$$

hence we can multiply by $f(a)^r$ to get

$$f(a)^r g(p) = f(a^r p(1/a)) = 0.$$

□

Lemma 7.2.12. *Let $f\colon A \to K$ be a maximal homomorphism as in Theorem 7.2.9 and $a \in k$ a nonzero element. Let $\mathcal{M} \subset A$ be the maximal ideal. Then either $\mathcal{M}[a] \subsetneq A[a]$ or $\mathcal{M}[1/a] \subsetneq A[1/a]$.*

Proof. Assume by contradiction that $\mathcal{M}[a]$ generates $A[a]$ and the same holds for $1/a$. Then we have equations

$$\begin{aligned} 1 &= r_0 + r_1 a + \cdots + r_m a^m \\ 1 &= s_0 + s_1/a + \cdots + s_n/a^n, \end{aligned}$$

where all the $r_i, s_j \in \mathcal{M}$ and we can take m, n minimal. By symmetry, assume $m \geq n$—then we rewrite the second equation as

$$(1 - s_0)a^n = s_1 a^{n-1} + \cdots + s_n.$$

Notice that $1 - s_0$ is invertible, so we get

$$a^n = t_1 a^{n-1} + \cdots + t_n$$

for some $t_i \in \mathcal{M}$. This allows us to write a^m as a combination of powers of a of lower degree, hence in the first equation m cannot be minimal. □

Proof of Theorem 7.2.9. Let $a \in k$; we want to show that either a or $1/a$ belongs to A. Let $\mathcal{M}$ be the maximal ideal of A. By Lemma 7.2.12 and symmetry, we can assume that $\mathcal{M}[a] \subsetneq A[a]$, so $\mathcal{M}[a]$ is contained in some maximal ideal $\mathcal{M}'$ of $A[a]$.

Since $\mathcal{M}$ is maximal, $\mathcal{M}' \cap A = \mathcal{M}$, and we get an embedding of fields $L = A/\mathcal{M} \subset L' = A[a]/\mathcal{M}'$. Notice that $L' = L(a)$, and a is algebraic over L. Since L comes equipped with an embedding into K, we can extend this to an embedding of L' into K.

By composition, this gives an extension of f to the ring $A[a]$. By maximality, it follows that $a \in A$. □

Corollary 7.2.13. *Let k be a field $A \subset k$ a ring. The integral closure of A in k is the intersection of all valuation rings of k which contain A.*

Proof. One inclusion is clear by Proposition 7.2.8. For the other one let $a \in k$ and assume that a is not integral over A. In particular, a is not in $B = A[1/a]$. This means that $1/a$ is not invertible in B, hence it is contained in a maximal ideal $\mathcal{M}$.

Let K be an algebraic closure of $B/\mathcal{M}$. We have a homomorphism $f\colon B \to K$ sending $1/a$ to 0. In particular, f cannot be extended to $B[a]$.

On the other hand, it can be extended to a valuation ring $C \supset B$ by Zorn's lemma and Theorem 7.2.9. This shows that $a \notin C$, which gives the reverse inclusion. □

7.3. Discrete valuation rings

A *discrete valuation ring*, or *DVR*, is a valuation ring A, where the underlying valuation v has target the ordered group $\mathbb{Z}$. These rings happen to have a particularly simple structure.

First, we already know that they are local rings with maximal ideal $\mathcal{M} = \{a \in A \mid v(a) \geq 1\}$. More generally, observe that for every $n \in \mathbb{N}$, we can define the ideal $I_n = \{a \in A \mid v(a) \geq n\}$, so that $I_0 = A$ and $I_1 = \mathcal{M}$.

Let $a, b \in A$ be two elements such that $v(a) = v(b)$. Then $v(a/b) = 0$, so that a/b is an invertible element of A. It follows that a and b generate the same ideal. Similarly, if $v(a) < v(b)$, b lies in the ideal generated by a. Hence, we can easily describe all ideals of A.

Let t be the minimal valuation of a noninvertible element, say $t = v(a)$. Let b be any other noninvertible element, and t' be the remainder of $v(b)$ by t, so that $v(b) = q * t + t'$. If $t' \neq 0$, we can produce a ring element of valuation strictly between 0 and t, namely b/a^q. It follows that $t' = 0$, hence $v(b)$ is a multiple of $v(a)$.

By rescaling, we can assume that $t = 1$. The ideal I_k contains a^q for $q \geq k$. It follows that these ideals are all distinct, and in fact $I_k = (a^k) = \mathcal{M}^k$.

This tells us that all ideals are powers of the maximal ideals, and all elements are powers of a single generator, up to invertible elements. We summarize the discussion so far.

Proposition 7.3.1. *Let A be a discrete valuation ring. Then A is local with a principal maximal ideal $\mathcal{M}$, and all its nonzero ideals are powers of $\mathcal{M}$.*

We can characterize discrete valuation rings via the following result.

Proposition 7.3.2. *Let A be a local integral domain with maximal ideal $\mathcal{M} \neq 0$. Assume that A is Noetherian and integrally closed, and that $\mathcal{M}$ is the only nonzero prime ideal. Then A is a discrete valuation ring.*

Proof. By Theorem 5.1.19, A is a Dedekind ring. Since $\mathcal{M}$ is the only nonzero prime, all nonzero ideals are powers of $\mathcal{M}$, and these powers are distinct by unique factorization.

We can then define a valuation on A as follows. Let $a \in A$ be any element; if $(a) = \mathcal{M}^k$, define $v(a) = k$. It is easy to check that v is a well-defined function, and we can extend it to the field of fractions of A by the rule $v(a/b) = v(a) - v(b)$. The extension becomes a discrete valuation.

To check that A is the valuation ring of v, assume that $v(a/b) \geq 0$. Then $v(a) \geq v(b)$ hence $(a) = \mathcal{M}^k$, $(b) = \mathcal{M}^h$ for some $k \geq h$. We conclude that $a/b \in A$, hence A is the valuation ring of v. □

We can reformulate this result by saying that every local Dedekind ring is a discrete valuation ring. There is also a converse:

Proposition 7.3.3. *A discrete valuation ring is a local Dedekind ring.*

Proof. A discrete valuation ring is integrally closed by 7.2.8. Moreover, it is Noetherian and has a single nonzero prime ideal, both by 7.3.1. □

Remark 7.3.4. The localization of a Dedekind ring is again a Dedekind ring. This is because being Noetherian or being integrally closed are properties that are preserved under localization (2.2.12) and 5.1.18 and because of the correspondence between prime ideals in a localization. Thus, the above can be rephrased by saying that a localization of a Dedekind ring is a discrete valuation ring.

The valuation can also be expressed explicitly. Namely, if P is a prime ideal of a Dedekind ring A, for every element $a \in A$ we can factorize the ideal $(a) = P^t \cdot P_1^{t_1} \cdots P_k^{t_k}$. The valuation is obtained by setting $v(a) = t$ and extending by the valuation rule, hence it essentially encodes the multiplicity of the factor P at a.

7.4. Direct and inverse limits

In Section 7.1 we gave the definition of the completion of an integral domain with respect to an absolute value on its fraction field. We want generalize this construction to other rings, as well as express it in algebraic terms. To do so, we introduce the machinery of *limits*.

Definition 7.4.1. Let $I \neq \emptyset$ be a set with a partial order $\leq$. I is said to be *directed* if for any $i, j \in I$ we find $k \in I$ such that $i \leq k$ and $j \leq k$.

Definition 7.4.2. Let $\{A_i\}$ be a family of groups (or modules over a fixed ring, or rings) over a directed set of indices I. For every pair $i < j$ assume given a homomorphism $\{f_{ij} : A_i \to A_j\}$, and assume that these are *compatible*, in the sense that for every $i < j < k$ the diagram

$$\begin{array}{ccc} A_i & \xrightarrow{f_{ij}} & A_j \\ & {\scriptstyle f_{ik}}\searrow \quad \swarrow{\scriptstyle f_{jk}} & \\ & A_k & \end{array}$$

commutes. This datum is called a *direct system*.

Definition 7.4.3. Let $\{A_i\}_{i\in I}$, $\{f_{ij}\}$ be a direct system and let A be a group (module, ring) with given homomorphisms $g_i : A_i \to A$, which are *compatible*, in the sense that for every $i < j$ the diagram

$$\begin{array}{ccc} A_i & \xrightarrow{\quad f_{ij} \quad} & A_j \\ & {\scriptstyle g_i}\searrow \quad \swarrow{\scriptstyle g_j} & \\ & A & \end{array}$$

commutes.

A is called the *direct limit* of the system—denoted $\varinjlim A_i$—if it enjoys the following universal property. For every other group (module, ring) B, equipped with compatible homomorphisms $h_i : A_i \to B$, there exists a unique homomorphism $h : A \to B$ such that $h_i = h \circ g_i$ for all i.

Remark 7.4.4. The direct limit of a system of groups (or modules or rings) is unique (up to a uniquely determined homomorphism) due to its universal property. In fact, given two direct limits A and A' (with their homomorphisms), the universal property determines two homomorpshisms $A \to A'$ and $A' \to A$, and again it follows from the uniqueness requirement of the universal property that these morphisms are inverse to each other.

Let us consider how to construct the direct limit of a system of groups. Assuming it exists, for every element x in some A_i, we have its image $g_i(x) \in A$. Moreover, if we have $f_{ij} : A_i \to A_j$, the image $g_i(x)$ has to be the same as $g_j(f_{ij}(x))$.

Since there are no other conditions, the idea is to construct the universal object by just putting together these requirements. Namely, we start from the disjoint union U of all A_i, and we introduce the equivalence relation $\sim$ generated by the following: elements $a_i \in A_i$ and $a_j \in A_j$ are equivalent if $a_j = f_{ij}(a_i)$.

The resulting quotient is equipped with a group operation. Namely, let $a_i \in A_i$ and $a_j \in A_j$ be any two elements of U. Then, we can find k bigger than both i and j. The images $f_{ik}(a_i) \sim a_i$ and $f_{jk}(a_j) \sim a_j$ belong to the same group A_k, so we can use the operation there to combine them and get a new element of U.

It is easy to check that this gives a well-defined operation on the quotient $A := U/\sim$ that satisfies the group axioms. If moreover the A_i have additional structure, such as being modules over a ring R or rings themselves, the same construction gives A the structure of a module or ring.

The natural inclusion followed by the quotient gives maps $A_i \to A$, and the universal property is easily checked.

Example 7.4.5.

(a) All this may seem very abstract, but in fact it is something that every child learns to do when summing fractions. Fractions with a fixed denominator d are easy to sum: they form an additive group A_d which is isomorphic to $\mathbb{Z}$—for instance $\frac{1}{7} + \frac{3}{7} = \frac{4}{7}$.

There are homomorphism $A_a \to A_b$ given by multiplication by k whenever $b = k \cdot a$. Hence we have a direct system where the index set is $\mathbb{N}_+$ and the ordering relation is divisibility.

In order to sum fractions with different denominators, one has to consider the equivalence relation induced by these maps. To sum, say, $\frac{3}{7}$ and $\frac{2}{5}$, one has to consider the equivalent elements $\frac{15}{35}$ and $\frac{14}{35}$ of A_{35} to get the result $\frac{29}{35}$. This is exactly the construction we have outlined above, which in particular shows that $\mathbb{Q}$ is the direct limit of the system of the A_d.

(b) When the maps of the system are inclusions, the direct limit is just the union of the A_i. For an example of this, consider the construction of the algebraic closure of a field k. Finite extensions can be constructed explicitly as quotients of the form $k[x]/(f)$, where f is some irreducible polynomial. To get the algebraic closure, one has to take the direct limit of the system of finite extensions.

The above examples should make the notion of direct limit more natural. The notion of inverse limit is obtained by essentially reversing all the arrows. Unfortunately, examples of inverse limits are not as natural.

Definition 7.4.6. Let $\{A_i\}$ be a family of groups (or modules over a fixed ring, or rings) over a directed set of indices I. For every pair $i < j$ assume given a homomorphism $\{f_{ij} : A_j \to A_i\}$, and assume that these are compatible. This datum is called an *inverse system*.

Definition 7.4.7. Let $\{A_i\}_{i \in I}$, $\{f_{ij}\}$ be an inverse system and let A be a group (module, ring) with given homomorphisms $g_i : A \to A_i$, which are *compatible*, in the sense that for every $i < j$ the diagram

$$\begin{array}{ccc} & A & \\ {\scriptstyle g_j}\swarrow & & \searrow{\scriptstyle g_i} \\ A_j & \xrightarrow{\quad f_{ij}\quad} & A_i \end{array}$$

is commutative.

A is called the *inverse limit* of the system—denoted $\varprojlim A_i$—if it enjoys the following universal property. For every other group (module over R, ring) B, equipped with compatible homomorphisms $h_i : B \to A_i$, there exists a unique homomorphism $h : B \to A$ such that $h_i = g_i \circ h$ for all i.

As in the case of direct limits, there is a form of uniqueness, up to a uniquely determined isomorphism. To show the existence, though, we must follow a different way, since the arrows go in the reverse direction, and there is no way to push elements $a_i \in A_i$ to the inverse limit.

Instead, assuming $A = \varprojlim A_i$ exists, for each $a \in A$ we get a compatible system of elements $a_i = g_i(a) \in A_i$.

Definition 7.4.8. Let $\{A_i\}_{i \in I}$, $\{f_{ij}\}$ be an inverse system. We say that the collection of elements (a_i), where $a_i \in A_i$, is *compatible* (or *coherent*) if $f_{ij}(a_j) = a_i$ for all $i < j$.

The idea is to *define* the inverse limit A as the set of compatible system of elements $\{a_i \in A_i\}$. The group operation can be applied separately on each component, and it is easy to check that this gives a groups structure on A. When the A_i are modules over a ring, or ring themselves, the inverse limit inherits the same structure, again by applying operations component by component.

Given a group B equipped with compatible homomorphisms $h_i : B \to A_i$, for any $b \in B$ we get a compatible sequence $(g_i(b))$. This defines a homomorphism $h : B \to A$ such that $h_i = g_i \circ h$ for all i. This is enough to show that A is the inverse limit of the A_i, as desired.

Example 7.4.9. The most prominent example of inverse limit is the construction of p-adic numbers. Recall from Section 7.1 that these can be constructed by considering the p-adic absolute value on $\mathbb{Q}$ and taking the completion $\mathbb{Q}_p$. The closure $\mathbb{Z}_p$ of $\mathbb{Z}$ is the ring of p-adic integers—each such number $a \in \mathbb{Z}_p$ has a series expansion

$$a = \sum_{i=0}^{\infty} \alpha_i p^i,$$

which converges in the p-adic topology.

There is a homomorphism $\mathbb{Z}_p \to \mathbb{Z}/(p^n)$ defined by sending a to the finite sum $\sum_{i=0}^{n-1} \alpha_i p^i$. These homomorphisms make $\mathbb{Z}_p$ into the inverse limit of the system of rings $\{\mathbb{Z}/(p^n)\}_{n \in \mathbb{N}}$, equipped with the natural projections $\mathbb{Z}/(p^{n+1}) \to \mathbb{Z}/(p^n)$.

This is easily proved by making use of the explicit description of the inverse limit. Namely, given a compatible sequence $\{a_n \in \mathbb{Z}/(p^n)\}$, we can write

$$a_n = \sum_{i=0}^{n-1} \alpha_{i,n} p^i,$$

for some coefficients $\alpha_{i,n}$ that are determined uniquely modulo p. The compatibility condition ensures that the coefficients $\alpha_{i,n} = \alpha_i$ are in fact independent from n. This allows us to define the element

$$a = \sum_{i=0}^{\infty} \alpha_i p^i \in \mathbb{Z}_p.$$

This gives a homomorphism from the inverse limit of the system to $\mathbb{Z}_p$. The inverse homomorphism is guaranteed by the universal property.

A similar construction shows that for a Dedekind ring A and a prime ideal P, the ring of P-adic integers A_P is the inverse limit of the system of rings $\{A/P^i\}_{i\in\mathbb{N}}$, equipped with the natural projections.

In the next section we are going to generalize this construction and define the completion of a ring in purely algebraic terms.

Proposition 7.4.10. *Let $\{A_i\}$, $\{B_i\}$ and $\{C_i\}$ be inverse systems over the same index set I and assume we have exact sequences $0 \to A_i \to B_i \to C_i$ for every $i \in I$, compatible with the system maps (a left-exact sequence of inverse systems). Then we get an exact sequence*

$$0 \longrightarrow \varprojlim A_i \longrightarrow \varprojlim B_i \longrightarrow \varprojlim C_i$$

of the inverse limits.

If moreover we have short exact sequences $0 \to A_i \to B_i \to C_i \to 0$, the index set is $\mathbb{N}$, and the maps f_{ij} for A_i are surjective, the sequence

$$0 \longrightarrow \varprojlim A_i \longrightarrow \varprojlim B_i \longrightarrow \varprojlim C_i \longrightarrow 0$$

is exact.

Proof. The maps $\varprojlim A_i \to \varprojlim B_i$ and $\varprojlim B_i \to \varprojlim C_i$ are defined by the universal property, and in fact are just restrictions of the product maps $\prod A_i \to \prod B_i$ and $\prod B_i \to \prod C_i$. This already makes clear that $\varprojlim A_i \to \varprojlim B_i$ is injective, since the product map is. Moreover, it also implies that the composition $\varprojlim A_i \to \varprojlim C_i$ is 0.

Let (b_i) be a compatible set that maps to 0 in $\varprojlim C_i$. We can then lift the set uniquely to a set $(a_i) \in \prod A_i$, so we only need to show that this lifted set is itself compatible. To do so, take $i > j$, so that we have a map $f_{ij} : A_i \to A_j$. The equality $f_{ij}(a_i) = a_j$ can be checked by taking the images in B_j, since $A_j \to B_j$ is injective, but there it follows because (b_i) is a compatible system.

For the last assertion, assume exactness at C_i and that maps $A_i \to A_j$ are surjective. Given compatible set (c_i), we can lift it to $(b_i) \in \prod B_i$ as above. The elements b_i are determined up to the image of some $a_i \in A_i$.

Call $\alpha_i : A_i \to B_i$ the given map. We need to find some collection $a_i \in A_i$ such that $b_i' := b_i + \alpha_i(a_i)$ is a compatible system. Calling g_{ij} the maps for B_i, this amounts to saying that

$$g_{ij}(b_j + \alpha_j(a_j)) = b_i + \alpha_i(a_i),$$

or equivalently

$$g_{ij}(b_j) - b_i = \alpha_i(f_{ij}(a_j) - a_i).$$

The left side is in the image of α_i by exactness, call it $\alpha_i(a_i')$, hence we need to solve

$$f_{ij}(a_j) = a_i' + a_i.$$

If f_{ij} is surjective and the index set is $\mathbb{N}$, this can be solved inductively. □

7.5. Completion of rings and modules

In this section, we reformulate the notion of completing a field (or a subring of it) with respect to an absolute value in terms of inverse limits. In doing so we will obtain a notion that is useful in a more general setting.

For a start, we notice that the operation of constructing Cauchy sequences only requires that we are able to take the difference of two elements and tell whether an element is close to 0. The following is thus the natural setting to define Cauchy completions:

Definition 7.5.1. Let G be a group endowed with a structure of topological space. We say that G is a *topological group* if the group operations are continous.

In the following, we will only work with abelian topological groups and use additive notation.

Remark 7.5.2. Since translations in the group are continuous, the topology is determined by the neighborhoods of 0, and in fact G is Hausdorff if, and only if, 0 is closed in G.

In general, the closure of 0 is a subgroup of G (why?), and the quotient of G by this subgroup, with the induced topology, is Hausdorff.

Assuming that G is *first countable*, that is, each point has a countable fundamental system of neighborhoods, we can define convergence in terms of sequences as usual.

Definition 7.5.3. A sequence $(g_n) \subset G$ *converges* to $g \in G$ if for every neighborhood U of g there exists N such that $g_n \in U$ for all $n \geq N$. A sequence (g_n) is called *Cauchy* if for every neighborhood U of 0 there exists N such that $g_m - g_n \in U$ for all $m, n \geq N$.

It is easily seen that each convergent sequence is Cauchy—if the converse holds, the group is called *complete*. The methods of 7.1 are easily translated in this more general setting, allowing us to construct the *completion* of a topological group G.

Definition 7.5.4. Let G be an abelian, first countable, topological group. Let $\widehat{G}$ be the set of Cauchy sequences in G modulo the equivalence relation $\sim$, where we say that $(g_n) \sim (h_n)$ if $(g_n - h_n)$ converges to 0. Then $\widehat{G}$ has a natural structure of topological group and is called the *completion* of G.

Remark 7.5.5. There is a natural map $G \to \widehat{G}$ that sends g to the constant sequence $g_n = g$, but unlike the case of metric spaces, this map is not necessarily injective. The kernel is the set of elements equivalent to 0, that is, the intersection of all neighborhoods of 0. In particular this map is injective if, and only if, G is Hausdorff.

This is a simple generalization of something we have already seen: the main twist of this section is to express this is as an inverse limit.

Proposition 7.5.6. *Let G be an abelian topological group, and assume that $0 \in G$ has a fundamental system of neighborhoods G_n which are* subgroups. *Then $\widehat{G}$ is the inverse limit of the system G/G_n.*

Proof. Assume, as we may, that the groups G_n are in fact nested: $G \supset G_1 \supset G_2 \supset \cdots \supset G_n \supset \cdots$. The inverse limit of the inverse system given by G/G_n is explicitly constructed as the set of coherent sequences $x_n \in G/G_n$. If we lift x_n to some $g_n \in G$, the coherence condition tells us exactly that the sequence (g_n) is Cauchy. The different choices in lifting x_n differ exactly by a convergent sequence. □

Remark 7.5.7. This reconciles the two different constructions of p-adic numbers that we have seen in 7.1.11 and 7.4.9. Not all examples of completions are inverse limits, though. In particular the condition that 0 has a fundamental system of neighborhoods that are subgroups rules out the case of archimedean absolute values.

Remark 7.5.8. This proposition opens the way to studying filtrations of groups topologically. Given a filtration on *any* group G, say

$$G \supset G_1 \supset \cdots \supset G_n \supset \cdots,$$

we can endow G with the topology where neighborhoods of 0 are generated by the G_n, and then link its completion to the inverse limit of the quotients G/G_n.

By Proposition 7.4.10 we immediately obtain

Proposition 7.5.9. *Let $0 \to H \to G \to G/H \to 0$ be an exact sequence of topological abelian groups, and assume G has the topology generated by a filtration $\{G_n\}$. If we give H and G/H the induced topologies, then we have an exact sequence of completions*

$$0 \longrightarrow \widehat{H} \longrightarrow \widehat{G} \longrightarrow \widehat{G/H} \longrightarrow 0\,.$$

Corollary 7.5.10. *Completion is idempotent, that is, $\widehat{\widehat{G}} \cong \widehat{G}$.*

Proof. By the inclusion $G_n \hookrightarrow G$ we get

$$0 \longrightarrow \widehat{G_n} \longrightarrow \widehat{G} \longrightarrow \widehat{G/G_n} \longrightarrow 0\,.$$

But the topology on G/G_n is discrete, so G/G_n equals its completion, and we deduce

$$\frac{G}{G_n} \cong \frac{\widehat{G}}{\widehat{G_n}}.$$

The thesis follows by passing to inverse limits. □

In the rest of the section we will specialize to the case where $G = M$ is a filtered A-module.

Definition 7.5.11. Let I be an ideal of the ring A, M an A-module with a filtration $M \supset M_1 \supset \cdots \supset M_n \supset \cdots$. We say that this is an I-filtration if $I \cdot M_n \subset M_{n+1}$. The I-filtration is called *stable* if $I \cdot M_n = M_{n+1}$ for n large enough.

Remark 7.5.12. Given two different filtrations $\{M_i\}$ and $\{M_i'\}$, both of which are I-stable, we have some N such that $M_{N+i} \subset M_i'$, and vice versa. We say that they have *bounded difference*. In particular, they determine the same topology on M. In the particular case where $M_n = I^n \cdot M$, this is called the I-adic topology.

Remark 7.5.13. The most obvious example of a stable I-filtration is of course $M_n = I^n \cdot M$, but things are easier if we keep a little more generality.

Recall from Section 1.7 that associated to a filtration

$$M \supset M_1 \supset \cdots \supset M_n \supset \cdots$$

we have a graded module $\mathrm{Gr}(M)$. If this is an I-filtration, $\mathrm{Gr}(M)$ is in fact a module over the graded ring $\mathrm{Gr}_I(A)$.

A related but different construction arises by taking the graded ring

$$B_I(A) = \textstyle\bigoplus_{n=0}^{\infty} I^n$$

and graded module

$$B(M) = \textstyle\bigoplus_{n=0}^{\infty} M_n.$$

In general, these behave worse than the usual associated graded ring and module, where we take quotients on each homogeneous component. Still, they are useful for the following result, which is the cornerstone of the theory that we are going to develop next.

Proposition 7.5.14. *Let A be a Noetherian ring, M a finitely generated A-module with an I-filtration $\{M_n\}$. Then $\{M_n\}$ is stable if, and only if, $B(M)$ is finitely generated over $B_I(A)$.*

Proof. First, notice that $B_I(A)$ is generated by I as an A-algebra. Since A is Noetherian, I is finitely generated, so $B_I(A)$ is itself Noetherian by Hilbert's basis theorem 2.3.3.

Consider the sum $\bigoplus_{i=0}^{n} M_i \subset B(M)$. This is not an $B_I(A)$-submodule, because it is not closed under multiplication by I. The generated submodule includes all terms of the form $I^k \cdot M_n$, hence it is

$$B(M)_n := M_0 \oplus \cdots M_n \oplus I \cdot M_n \oplus I^2 \cdot M_n \oplus \cdots .$$

The union of all $B(M)_n$ is $B(M)$. If the chain $\{B(M)_n\}$ stabilizes, $B(M)$ equals a member of this chain, hence it is finitely generated. Vice versa, if $B(M)$ is finitely generated over $B_I(A)$, it is Noetherian, so this chain must stabilize.

But it is clear from the definition that the chain stabilizes if, and only if, the filtration $\{M_n\}$ is I-stable. □

Corollary 7.5.15 (Artin–Rees lemma). *Let A be a Noetherian ring, $I \subset A$ an ideal and $M \supset M_1 \supset \cdots \supset M_n \supset \cdots$ a stable I-filtration of the A-module M. If M is finitely generated, for each submodule $N \subset M$, the $\{N \cap M_n\}$ form a stable I-filtration of N.*

Proof. It is clear that $\{N \cap M_n\}$ is an I-filtration, and stability follows from Proposition 7.5.14 and the fact that $B(N) \subset B(M)$, hence it is Noetherian if $B(M)$ is. □

Remark 7.5.16. As a particular case, taking $M_n = I^n \cdot M$, we obtain that $N \cap (I^n \cdot M)$ is stable. In particular, it has bounded difference from $I^n \cdot N$, and it induces the I-topology on N.

Corollary 7.5.17. *Let A be a Noetherian ring, $I \subset A$ an ideal and M_1, M_2, M_3 finitely generated modules over A. If*

$$0 \longrightarrow M_1 \longrightarrow M_2 \longrightarrow M_3 \longrightarrow 0$$

is exact, then

$$0 \longrightarrow \widehat{M_1} \longrightarrow \widehat{M_2} \longrightarrow \widehat{M_3} \longrightarrow 0$$

is exact, where $\widehat{M_i}$ is the completion of M_i with respect to the I-adic topology.

Proof. This is just a rephrasing of 7.5.9 combined with the above remark that the I-adic topology on M_1 agrees with the topology induced by the inclusion in M_2. □

Corollary 7.5.18. *Let A be a Noetherian ring, $I \subset A$ an ideal. Then, for all $n \geq 0$, the extension $I^n \cdot \widehat{A}$ is the completion of I^n, and moreover $I^n/I^{n+1} \cong \widehat{I^n}/\widehat{I^{n+1}}$.*

Proof. This is just the previous corollary in the case where $M_2 = A$ and $M_1 = I^n$. In this case, $M_3 = A/I^n$ has the discrete topology, hence $\widehat{M_3} \cong M_3$ and the thesis follows by a little diagram chasing. The second assertion follows by considering $M_1 = I^{n+1}$ inside $M_2 = I^n$. □

We are now in a position to understand the kernel of the map $M \to \widehat{M}$. By construction, this is the intersection $\bigcap_k I^k \cdot M$, but we have a more explicit description.

Theorem 7.5.19. *Let A be a Noetherian ring, $I \subset A$ an ideal and M a finitely generated module over A, endowed with the I-adic topology. The kernel of $M \to \widehat{M}$ is*

$$\{m \in M \mid (1-x) \cdot m = 0 \text{ for some } x \in I\}.$$

Proof. Let $K = \bigcap_k I^k \cdot M$ be the kernel of $M \to \widehat{M}$. This is the intersection of all neighborhoods of 0, hence it has the discrete topology. By the Artin–Rees lemma 7.5.15, this is the I-adic topology on K, so it follows that $I \cdot K = K$. By 5.1.8 we find $x \in I$ such that $(1-x) \cdot K = 0$.

For the converse, assume $(1-x) \cdot m = 0$, that is, $m = xm$. Then $m = x^k m$ for all k, so $m \in \bigcap_k I^k \cdot M = K$. □

Remark 7.5.20. The identity $I \cdot K = K$ may seem trivial, but in fact it is not, since products and intersections do not commute! This is the crucial step where Artin–Rees enters the proof.

Given a Noetherian ring A with an ideal I, let $\widehat{A}$ be its completion. For an element $x \in \widehat{I}$, the series $1 + x + \cdots + x^n + \cdots$ is well defined and converges in $\widehat{A}$. To see this, consider the truncated sums $s_k := \sum_{i=0}^k x^i$. If $h > k \geq N$, the difference $s_k - s_h \in \widehat{I}^N$, so that the sequence $\{s_k\}$ defines an element of the inverse limit $\varprojlim \widehat{A}/\widehat{I}^k$, which is just the I-completion of $\widehat{A}$, and $\widehat{A}$ is complete.

If we denote this limit by $s = \sum_{i=0}^{\infty} x^i$, it follows that $s \cdot (1-x) = 1$ (why?), so that $1 - x$ is invertible inside $\widehat{A}$. By 1.1.19, $\widehat{I}$ is contained in the Jacobson radical of $\widehat{A}$. We use this fact to give a few examples of completions.

Example 7.5.21.

(a) Let A be a local Artinian ring with maximal ideal $\mathcal{M}$. Then by 2.4.4, $\mathcal{M}^k = 0$ for k big enough, hence A is complete with respect to the $\mathcal{M}$-adic topology.

(b) Let $I \subset A$ an ideal. Then $S = 1 + I$ is a multiplicative set and we can take the localization $S^{-1}A$. Since every element of S is invertible inside $\widehat{A}$, there is natural map $S^{-1}A \to \widehat{A}$. If A is Noetherian, Theorem 7.5.19 implies that this map is injective, so that we can see $S^{-1}A$ as a subring of the completion.

(c) [**Ces**] Let $\mathcal{M} \subset A$ be maximal ideal, and assume that A is Noetherian. For the $\mathcal{M}$ completion $\widehat{A}$ we know by 7.5.18 that $\widehat{A}/\widehat{\mathcal{M}} \cong A/\mathcal{M}$ is a field, so $\widehat{\mathcal{M}}$ is maximal. Moreover we have just seen that it is contained inside $\mathcal{J}(\widehat{A})$, so $\widehat{A}$ is local.

Moreover, consider the completion of the localization $A_{\mathcal{M}}$. This is just the inverse limit of quotients $A_{\mathcal{M}}/(\mathcal{M}A_{\mathcal{M}})^k$. Since localization is exact, this is just the localization $(A/\mathcal{M}^k)_{\mathcal{M}}$. But $A/\mathcal{M}^k$ is already local, so this is in fact $A/\mathcal{M}^k$, and the limit of these is $\widehat{A}$.

It follows that $\widehat{A}$ is the completion of $A_{\mathcal{M}}$ as well, and in particular we have again a natural map $A_{\mathcal{M}} \to \widehat{A}$.

(d) Let k be a field and $A = k[x_1, \dots, x_n]$, with the ideal $\mathcal{M} = \{f \in A \mid f(0) = 0\}$. The completion of A can be identified with the formal power series ring $k[[x_1, \dots, x_n]]$.

To see this, notice that I^k is the ideal of polynomial vanishing at least of order k in 0, so A/I^k can be identified with the module of polynomials of degree less than k. Each compatible sequence of such polynomials defines a unique power series, showing that $\widehat{A} \cong k[[x_1, \dots, x_n]]$.

Actually the link between power series rings and completions allows us to prove that completion preserves the Noetherian property.

Theorem 7.5.22. *Let A be a Noetherian ring, $I \subset A$ an ideal. Then the completion $\widehat{A}$ with respect to the I-adic topology is also Noetherian.*

Proof. First, we establish that $A[[x]]$, the ring of power series on A, is Noetherian. This is the content of Exercise 4 in Chapter 2. We will not spoil it completely, but the proof is essentially the same as Theorem 2.3.3—just replace the ideal generated by leading coefficients by the ideal generated by the lowest-degree coefficients. The proof goes on mostly unchanged. By induction—using the fact that $A[[x, y]]$ is isomorphic to $A[[x]][[y]]$—it follows that $A[[x_1, \dots, x_n]]$ is Noetherian for all n.

Now let $a_1, \ldots, a_n$ be generators for I. We show that there is a well-defined surjective homomorphism

$$\begin{aligned} ev_a\colon\ A[[x_1, \ldots, x_n]] &\longrightarrow \widehat{A} \\ x_i &\longmapsto a_i \end{aligned}$$

defined by evualuating a power series in $a_1, \ldots, a_n$, which implies that $\widehat{A}$ is Noetherian.

To show that ev_a is well defined, take any power series p and let p_k be the polynomial obtained by truncating up to degree k. Then if $h, k \geq N$, we have $p_h(\underline{a}) - p_k(\underline{a}) \in I^N$, which shows that the $p_k(\underline{a})$ converge to a well-defined element in $\widehat{A}$.

To show that ev_a is surjective, recall that any element $b \in \widehat{A}$ is defined by a compatible sequence $\{\overline{b_k} \in A/I^k\}$. Choose representatives b_k of $\overline{b_k}$. If we let $c_k := b_{k+1} - b_k$ and $c_0 := b_0$, then by compatibility $c_k \in I^k$, and

$$b = \sum_{k=0}^{\infty} c_k.$$

Recursively write c_k as a homogeneous polynomial of degree k in $a_1, \ldots, a_n$—plus possibly an element of I^{k+1}, which we incorporate into c_{k+1}. This gives the desired expression of b as a power series in $a_1, \ldots, a_n$, showing that ev_a is indeed surjective. □

In the end of the section, we derive a few consequences of the Artin–Rees lemma, and in particular of the strictly related Theorem 7.5.19. We try to focus on some applications of the theory developed so far that do not mention the machinery of completion.

The first result follows from the fact that for an element x in the Jacobson radical, $1 + x$ is invertible.

Corollary 7.5.23. *Let A be a Noetherian ring, $I \subset \mathcal{J}(A)$ an ideal and M a finitely generated A-module. Then $\bigcap_{n\geq 0} I^n \cdot M = 0$—in other words, the I-adic topology is Hausdorff on M.*

Specializing to the case where $M = A$ we get the famous

Corollary 7.5.24 (Krull intersection theorem)**.** *Let A be a Noetherian ring, $I \subset \mathcal{J}(A)$ an ideal. Then $\bigcap_{n\geq 0} I^n = 0$.*

Remark 7.5.25. In our terminology, a module M over a ring is complete with respect to the I topology if every Cauchy sequence converges. Other authors require the more stringent condition that the map $c : M \to \widehat{M}$ is an isomorphism. The difference lies in the fact that the kernel of c may be nontrivial, or in other words, that M may not be Hausdorff. The above results imply that the two definitions coincide in many cases of interest.

Another application is the following:

Proposition 7.5.26. *Let A be a Noetherian ring, $P \subset A$ a prime ideal. Then the kernel of the localization map $A \to A_P$ is the intersection of all P-primary ideals.*

Proof. The ring A_P is local, hence applying the preceding corollary we see that the intersection of all ideals $P^n \cdot A_P$ is 0. This is the same as the intersection of all P-primary ideals of A_P, since every P-primary ideal sits among P and some power P^k.

Pulling back this to A and using 3.2.25, we get the thesis. □

7.6. Hensel's lemma

In this section we study various generalizations of Hensel's lemma for p-adics (Theorem 7.1.12) to arbitrary complete local rings. Since in this case there is not a preferred generator for the maximal ideal, we cannot translate its proof directly. Rather, we prove the following more general statement.

Theorem 7.6.1 (Hensel's lemma, strong form). *Let A be a Noetherian local ring, complete with respect to its maximal ideal $\mathcal{M}$. Denote $k = A/\mathcal{M}$. Let $f \in A[x]$ be a monic polynomial, $\overline{f}(x) \in k[x]$ its residue modulo $\mathcal{M}$. Assume $\overline{f}$ factorizes as $\overline{f} = G \cdot H$ in $k[x]$, where G and H are coprime. Then we have a factorization $f = g \cdot h$ in $A[X]$, where g and h are monic of the same degrees of G and H, and $\overline{g} = G$ and $\overline{h} = H$. Moreover, g and h are uniquely determined.*

The following, more familiar form, follows immediately by taking $G(x) = x - \alpha$:

Corollary 7.6.2 (Hensel's lemma, weak form). *Let A be a Noetherian local ring, complete with respect to its maximal ideal $\mathcal{M}$. Denote $k = A/\mathcal{M}$. Let $f \in A[x]$ be a monic polynomial, $\overline{f}(x) \in k[x]$ its residue modulo $\mathcal{M}$. Assume $\overline{f}$ has a simple root α in k. Then α lifts uniquely to a root of f in A.*

Proof. We show by induction that there exist monic polynomials $g_k, h_k \in A[x]$ such that

$$f \equiv g_k \cdot h_k \mod \mathcal{M}^k[x],$$

such that $\overline{g_k} = G$ and $\overline{h_k} = H$. Moreover, such polynomials are uniquely determined modulo $\mathcal{M}^k[x]$. The case $k = 1$ is the hypothesis of the theorem.

Having produced g_k and h_k we look for polynomials of the form

$$g_{k+1} = g_k + s$$
$$h_{k+1} = h_k + t,$$

where $\deg s < \deg g_k$ and $\deg t < \deg h_k$.

To find the right s and t, consider the difference

$$d := f - g_k h_k \in \mathcal{M}^k[x].$$

Using the fact that G and H are coprime, we can write $1 \equiv a g_k + b h_k \mod \mathcal{M}[x]$ for suitable $a, b \in A[x]$. Multiplying by d we almost get what we want:

$$d \equiv (d \cdot a) g_k + (d \cdot b) h_k \mod \mathcal{M}^{k+1}[x],$$

so if we let $g^\circ_{k+1} := g_k + d \cdot b$ and $h^\circ_{k+1} := h_k + d \cdot a$ we have

$$g^\circ_{k+1} \equiv g_k \mod \mathcal{M}^k[x],$$
$$h^\circ_{k+1} \equiv h_k \mod \mathcal{M}^k[x],$$

and

$$g^\circ_{k+1} h^\circ_{k+1} \equiv g_k \cdot h_k + d \equiv f \mod \mathcal{M}^{k+1}[x]. \tag{7.6.1}$$

The only issue with this choice is that $d \cdot a$ and $d \cdot b$ may have large degree. To remedy this, use the fact that g_k is monic to perform the division of $d \cdot b$ by g_k, so that

$$d \cdot b = l_k \cdot g_k + s,$$
$$d \cdot a = m_k \cdot h_k + t.$$

With this choice of s and t, the relation (7.6.1) is valid for g_{k+1} and h_{k+1}—that is, $g_{k+1} h_{k+1} \equiv f \mod \mathcal{M}^{k+1}[x]$. Moreover, s has smaller degree than g_k, so that g_{k+1} is monic, and similarly for h_{k+1}.

Uniqueness can be proved inductively, in a similar fashion. Since by construction $g_{k+1} - g_k \in \mathcal{M}^k[x]$, and all g_k have the same degree, the coefficients of the $\{g_k\}$ converge in A. This defines a limit polynomial g, monic of the same degree as G. Symmetrically, we get a limit polynomial h, monic of the same degree of H.

The equality $f = g \cdot h$ follows by passing to the limit. More precisely, the coefficients of $f - g \cdot h$ belong to $\mathcal{M}^k$ for all k. But $\bigcap_k \mathcal{M}^k$ is 0 by Krull intersection theorem 7.5.24. □

Local rings that satisfy the conclusion of Hensel's lemma are called Hensel rings, and are not necessarily complete (see [**Ray70**] for a detailed treatment, or Exercise 23). The following results have a flavor similar to

Hensel's lemma, in that they allow to lift a condition on a module of the form M/I to the I-adic completion of M.

Proposition 7.6.3. *Let A be a Noetherian ring, $I \subset A$ an ideal, M, N two A-modules with a map $\phi\colon M \to N$. If the induced map $M/I \to N/I$ is surjective, then the induced map $\widehat{M} \to \widehat{N}$ is surjective, where $\widehat{\cdot}$ denotes completion with respect to the I-adic topology.*

Proof. By induction, the map $M/I^k \to N/I^k$ is surjective for all k. In fact, assume this holds for k, and take $\overline{n} \in N/I^{k+1}$. Then we can write

$$n = \phi(m) + \sum_{j=1}^{r} a_j n_j,$$

for some $m \in M$, $a_j \in I^k$ and $n_j \in N$. Since $M/I \to N/I$ is surjective, we can write $n_j = \phi(m_j) + x_j$, where $m_j \in m$ and $x_j \in I \cdot N$, so

$$n \equiv \phi\left(m + \sum_{j=1}^{r} a_j m_j\right) \pmod{I^{k+1} \cdot N},$$

which means that $M/I^{k+1} \to N/I^{k+1}$ is surjective as well.

It follows that there are exact sequences

$$0 \longrightarrow \frac{L_k}{I^k M} \longrightarrow \frac{M}{I^k M} \longrightarrow \frac{N}{I^k N} \longrightarrow 0\,.$$

where $L_k = \phi^{-1}(I^k N)$. By a similar reasoning, the map $L_{k+1}/I^{k+1}M \to L_k/I^k M$ is surjective (check this!). Using Proposition 7.4.10, we get an exact sequence of the inverse limits, so $\widehat{M} \to \widehat{N}$ is surjective. □

Corollary 7.6.4. *Let A be a Noetherian ring, complete with respect to the I-adic topology, M an A-module. If $\bigcap_{n=1}^{\infty} I^n \cdot M = 0$ and M/IM is finitely generated over A/I, then M is finitely generated over A.*

Proof. Let $m_1, \ldots, m_k \in M$ such that their classes in M/I generate M/I as an A/I-module, and let $N = \langle m_1, \ldots, m_k \rangle_A$. By the previous Proposition, the map $\widehat{N} \to \widehat{M}$ is surjective. On the other hand, N is a quotient of A^k, hence it is complete. (Why?) Since $\bigcap_{n=1}^{\infty} I^n \cdot M = 0$, the map $M \to \widehat{M}$ is an inclusion, and we conclude that $N \cong M \cong \widehat{M}$. □

7.7. Witt vectors

In the course of this chapter, we have seen two different construction of p-adic numbers. The first one starts with the p-adic absolute value on $\mathbb{Q}$ and defines $\mathbb{Q}_p$ as the Cauchy completion of $\mathbb{Q}$ with respect to this absolute value. The p-adic integers $\mathbb{Z}_p$ can be seen either as the topological closure of $\mathbb{Z}$ inside $\mathbb{Q}_p$, or alternatively as the integral closure of $\mathbb{Z}$ in the same field. The second approach is more algebraic in nature and constructs the p-adic integers $\mathbb{Z}_p$ as the inverse limit of the system of quotients $\mathbb{Z}/p^k\mathbb{Z}$, and $\mathbb{Q}_p$ as the fraction field of $\mathbb{Z}_p$. Both points of view are fruitful, and their generalizations lead to the study of general absolute values and valuations in one direction, and to I-adic completions in the other one.

In this section, we are going to introduce a third approach to study p-adic numbers, which will turn out to be crucial to prove Cohen's structure theorem in Section 10.6. The idea is that every $a \in \mathbb{Z}_p$ has an expansion into digits

$$a = \sum_{i=0}^{\infty} \alpha_i p^i,$$

where the coefficients α_i are taken in $\{0, \ldots, p-1\}$. One should be able to *define* p-adic numbers this way, and perform algebraic operations using this system of digits.

Predictably, this plan is bound to fail due to the issue of carries. This is very similar to trying to define addition or multiplication of real numbers using digits: the formulas become very complicated, and keeping track of carries in an organized way is tricky. The insight of Witt [**Wit36**] was that one *can* perform this plan and find simple, universal formulas, *provided* one chooses a smarter set of representatives for the digits.

Let $\pi\colon \mathbb{Z}_p \to \mathbb{Z}/p\mathbb{Z}$ be the projection. A choice of "digits" essentially amounts to a section of π, that is, a function $\sigma\colon \mathbb{Z}/p\mathbb{Z} \to \mathbb{Z}_p$ such that $\pi \circ \sigma = \mathrm{id}$. Witt noticed that computations are simplified if one chooses a particular σ. Namely, each element $a \in \mathbb{Z}/p\mathbb{Z}$ satisfies $a^p = a$, so by Hensel's lemma there exists a unique section $\tau\colon \mathbb{Z}/p\mathbb{Z} \to \mathbb{Z}_p$ such that $\tau(a)^p = \tau(a)$ for all $a \in \mathbb{Z}/p\mathbb{Z}$. The section τ is called the *Teichmüller character* and $\tau(a)$ is called the *Teichmüller representative* of a. Clearly, τ satisfies $\tau(ab) = \tau(a)\tau(b)$. One can find suitable formulas for p-adic numbers when expressed using Teichmüller representatives.

Instead of pursuing this and building p-adic numbers out of $\mathbb{Z}/p\mathbb{Z}$ digits, we are going to define a universal construction that works over any ring. This will lead us to the definition of p-adic Witt vectors. Our presentation will follow parts of [**Haz09**], which has a much more extensive treatment. Witt vectors satisfy many universal properties, and accordingly one can construct them in various different ways. For possible alternative approaches, see [**Rab14**] or [**Hes05**].

The following remark is meant to give some motivation for the construction of the Witt polynomials. If it is more confusing than inspiring on first read, one can safely skip it and come back to read it after the rest of the section.

Remark 7.7.1. Fix a prime number p. Given a ring A, we are going to define another ring $W_p(A)$ together with a projection $\pi\colon W_p(A) \to A$—when $A = \mathbb{Z}/p\mathbb{Z}$, we want to recover $\mathbb{Z}_p$. Mimicking this special case, we want to be able to write elements of $W_p(A)$ as infinite sums

$$a = \sum_{i=0}^{\infty} \sigma(a_i)p^i,$$

for some $a_i \in A$, where $\sigma\colon A \to W_p(A)$ is a section. This representation is going to induce a set bijection

$$\lambda_n\colon A^n \to W_p(A)/p^n W_p(A)$$

given by

$$\lambda_n(a_0, \ldots, a_{n-1}) := \sum_{i=0}^{n-1} \sigma(a_i)p^i.$$

Having fixed such a bijection, we get an induced ring structure on A^n, and the question becomes how to express addition and multiplication in such coordinates.

Unfortunately, λ_n depends on the choice of the section σ, so it is unlikely that we can find nice universal formulas. To proceed further, we will have to alter the definition of λ slightly. We start from the observation that in any ring the congruence $a \equiv b \pmod p$ implies that $a^{p^k} \equiv b^{p^k} \pmod{p^{k+1}}$ for all k—this can be seen by expanding the binomial formula. Using this, it is not difficult to slightly change the defintion of λ in order to make it independent of σ. Namely,

$$\mu_n(a_0, \ldots, a_{n-1}) := \sum_{i=0}^{n-1} \sigma(a_i)^{p^{n-i}} p^i$$

is independent of the choice of σ modulo p^n. Hence μ_n is a better candidate for our strategy. We will thus investigate how to sum and multiply elements of this form, for $a_0, \ldots, a_{n-1} \in A^n$.

This, in turn, leads us to consider the polynomials

$$w_n(x_0, \ldots, x_n) := x_0^{p^n} + p x_1^{p^{n-1}} + \cdots + p^n x_n$$

and see if we can use them to define a ring structure on A^n, and eventually extend this to a ring structure for formal sums of elements in A.

Guided by the above remark, the point of departure for our construction is a collection of polynomials $w_n \in \mathbb{Z}[x_0, x_1, \dots]$ defined by

$$
\begin{aligned}
w_0(\underline{x}) &= x_0 \\
w_1(\underline{x}) &= x_0^p + p x_1 \\
w_2(\underline{x}) &= x_0^{p^2} + p x_1^p + p^2 x_2 \\
&\vdots \\
w_n(\underline{x}) &= x_0^{p^n} + p x_1^{p^{n-1}} + \cdots + p^n x_n \\
&\vdots
\end{aligned}
\tag{7.7.1}
$$

The definition seems mysterious at this point—the reason for the introduction of these polynomials is that one is able to find universal formulas for addition and multiplication of them. The polynomials $\{w_n\}$ are called *p-adic Witt polynomials.*

By construction, they satisfy

$$w_{n+1}(x) \equiv w_n(x^p) \pmod{p^{n+1}}, \tag{7.7.2}$$

where for simplicity of notation we denote $x = (x_0, \dots, x_n)$ and $x^p = (x_0^p, \dots, x_n^p)$. This simple observation allows us to prove the key result:

Lemma 7.7.2. *Let $f \in \mathbb{Z}[x, y, z]$. Then there are uniquely determined polynomials*

$$f_n \in \mathbb{Z}[x_0, \dots, x_n, y_0, \dots, y_n, z_0, \dots, z_n]$$

such that

$$f(w_n(\underline{x}), w_n(\underline{y}), w_n(\underline{z})) = w_n(f_0(\underline{x}, \underline{y}, \underline{z}), \dots, f_n(\underline{x}, \underline{y}, \underline{z})). \tag{7.7.3}$$

Thus, every ternary operation over the polynomials w_n is just another polynomial w_n evaluated at other polynomials. There is nothing special in choosing ternary operations here—it is just that we will not need to apply the lemma to operations of bigger arity.

Proof. One can choose $f_0(x_0, y_0, z_0) = f(x_0, y_0, z_0)$, and then use equation (7.7.3) to define f_n by induction. This can be done because f_n appears with degree 1 in (7.7.3). The only thing that needs to be proved is that the polynomials $\{f_n\}$ thus defined have coefficients in $\mathbb{Z}$, since a priori defining f_n requires a division by p^n.

This can be proved by induction using (7.7.2). For simplicity of notation, we are going to omit underlines in sets of variables. Assume all polynomials up to f_n have integer coefficients. Notice that by construction

$$w_{n+1}(x) \equiv w_n(x^p) \pmod{p^{n+1}},$$

and so

$$
\begin{aligned}
& f(w_{n+1}(x), w_{n+1}(y), w_{n+1}(z)) \\
(7.7.4) \qquad & \equiv f(w_n(x^p), w_n(y^p), w_n(z^p)) \\
& \equiv w_n(f_0(x^p, y^p, z^p), \dots, f_n(x^p, y^p, z^p)) \pmod{p^{n+1}}.
\end{aligned}
$$

At the same time we can expand

$$
(7.7.5) \qquad w_{n+1}(f_0, \dots, f_{n+1}) = f_0^{p^{n+1}} + pf_1^{p^n} + \dots + p^{n+1} f_{n+1}.
$$

If we expand the last line of (7.7.4), and compare it term by term with (7.7.5), all terms are congruent modulo p^{n+1}. There is a last summand in (7.7.5), which is $p^{n+1}f_{n+1}$, so we get that $p^{n+1}f_{n+1} \equiv 0 \pmod{p^{n+1}}$, which proves that f_{n+1} has coefficients in $\mathbb{Z}$. □

Choosing $f(x, y) = x+y$, one finds polynomials $s_n(x_0, \dots, x_n, y_0, \dots, y_n)$ that satisfy

$$
w_n(\underline{x}) + w_n(\underline{y}) = w_n(s_0, \dots, s_n).
$$

Similarly, by choosing $f(x, y) = xy$, one finds another set of polynomials $m_n(x_0, \dots, x_n, y_0, \dots, y_n)$ that satisfy

$$
w_n(\underline{x}) \cdot w_n(\underline{y}) = w_n(m_0, \dots, m_n).
$$

Definition 7.7.3. Let A be any ring, p a prime number. The ring of p-adic Witt vectors over A, denoted $W_p(A)$, is the set

$$
W_p(A) = A^{\mathbb{N}} = \{(a_0, a1, \dots) \mid a_i \in A\},
$$

endowed with the operations defined by

$$
\begin{aligned}
(a_i) + (b_i) &:= (s_0(a, b), s_1(a, b), \dots) \\
(a_i) \cdot (b_i) &:= (m_0(a, b), m_1(a, b), \dots).
\end{aligned}
$$

Of course, we have to prove that $W_p(A)$ is a ring, so the operations are associative, multiplication is distributive over the sum and so on. This is the reason why we have proved Lemma 7.7.2 with 3 variables. For instance, to prove associativity of the sum, one considers the polynomial

$$
f(x, y, z) = (x + y) + z = x + (y + z).
$$

Uniqueness in Lemma 7.7.2 implies the identity

$$
\begin{aligned}
& s_n(x_0, \dots, x_n, s_0(\underline{y}, \underline{z}), \dots, s_n(\underline{y}, \underline{z})) \\
& \quad = s_n(s_0(\underline{x}, \underline{y}), \dots, s_n(\underline{x}, \underline{y}), z_0, \dots, z_n),
\end{aligned}
$$

which amounts to the associativity of the sum in $W_p(A)$. All other ring identities can be proved in the same way. The zero element of $W_p(A)$ is just $(0, 0, \dots)$, while the multiplicative identity is $(1, 0, 0, \dots)$. One can explicitly work out the polynomials involved, but the computations rapidly become messy: for instance,

$$s_1(x_0, x_1, y_0, y_1) = \frac{x_0^p + y_0^p - (x_0 + y_0)^p}{p} + x_1 + y_1.$$

Remark 7.7.4. There are various relations between a ring and the associated Witt rings. Namely, a homomorphism $A \to B$ of rings induces a homomorphism $W_p(A) \to W_p(B)$. Moreover, by construction, the evaluation map

$$\begin{aligned} w_n\colon\ W_p(A) &\longrightarrow A \\ (\underline{a}) &\longmapsto w_n(\underline{a}) \end{aligned}$$

is a ring homomorphism such that the diagram

$$\begin{array}{ccc} W_p(A) & \xrightarrow{w_n} & A \\ \downarrow & & \downarrow \\ W_p(B) & \xrightarrow{w_n} & B \end{array} \tag{7.7.6}$$

is commutative for each map $A \to B$. Putting together all the morphisms w_n gives a morphism

$$w\colon W_p(A) \to A^{\mathbb{N}} \tag{7.7.7}$$

which in general is neither injective nor surjective.

Definition 7.7.5. Given an element $a \in W_p(A)$, the components $w_n(a)$ are called the *ghost components* of a.

The rings $W_p(A)$ are quite special: unlike most constructions that we have met until now (quotients, localizations, polynomial rings, power series, ...) they come endowed with maps $W_p(A) \to A$ which have A as target instead of source. The algebraic structure on $W_p(A)$ is more or less interesting, depending on the characteristic of A.

Remark 7.7.6. Assume that p is invertible in A. Then the homomorphism w of (7.7.7) is an isomorphism, since one can recursively solve for x_n in w_n. In this case, $W_p(A)$ is just another way to write $A^{\mathbb{N}}$ and does not carry any interesting information. This happens when A is an algebra over $\mathbb{Q}$, or when $\operatorname{char}(A)$ is a prime $q \neq p$.

This implies that the rings $W_p(A)$ are determined by the fact that

(i) the underlying set of $W_p(A)$ is $A^{\mathbb{N}}$;

(ii) the polynomials $w_n\colon W_p(A) \to A$ are homomorphisms;

(iii) the diagram in (7.7.6) commutes.

In fact, these properties determine both the addition and multiplication on $W_p(A)$ when A is a $\mathbb{Q}$-algebra. Every integral domain A of characteristic 0 embeds into its fraction field, and commutativity of (7.7.6) implies that the structure of $W_p(A)$ is uniquely determined in this case as well. Finally, every ring is a quotient of $\mathbb{Z}[\{x_i\}_{i\in I}]$ for a suitable index set I, and since these rings have characteristic 0, another application of (7.7.6) is enough to determine the operations on $W_p(A)$ for any ring A.

One can use computations with ghost components to derive many relations that hold in $W_p(A)$. We summarize some of these, but first we need a few definitions.

Definition 7.7.7. Given $a \in A$, the element $t(a) := (a, 0, 0, \dots,) \in W_p(A)$ is called the *Teichmüller representative* of a. It satisfies $w_0(t(a)) = a$.

The *Verschiebung* operation is just the shift $V_p\colon W_p(A) \to W_p(A)$ defined by

$$V_p((a_0, a_1, \dots)) = (0, a_0, a_1, \dots).$$

Proposition 7.7.8. *Let A be a ring, p a prime number.*

(i) *Given $a \in A$ and $b = (b_0, b_1, \dots) \in W_p(A)$,*

$$t(a) \cdot b = (ab_0, a^p b_1, a^{p^2} b_2, \dots).$$

In particular the Teichmüller map is multiplicative, that is, given $a, b \in A$ one has $t(ab) = t(a)t(b)$.

(ii) *The Verschiebung satisfies $w_0(V_p) = 0$ and $w_n(V_p) = p \cdot w_{n-1}$.*

(iii) *The Verschiebung is additive, namely for $a, b \in W_p(A)$*

$$w_n(V_p(a + b)) = w_n(V_p(a)) + w_n(V_p(b)).$$

(iv) *Every $a = (a_0, a_1, \dots) \in W_p$ can be written as a series*

$$a = \sum_{i=0}^{\infty} V_p^i(t(a_i)).$$

(v) *Define polynomials $p_n(\underline{x})$ by*

$$w_n(p_0, p_1, \dots, p_n) = p w_n(\underline{x}).$$

Then the $\{p_n\}$ are well defined and $p_n(\underline{x}) \equiv x_{n-1}^p \pmod{p}$.

We must be precise about the meaning of the series in iv). The ring $W_p(A)$ is endowed with ideals

$$I_k := \{(a_i) \in W_p \mid a_i = 0 \text{ for } i < k\} = V_p^k(W_p(A)).$$

These ideals define a filtration on $W_p(A)$, and the convergence in iv) has to be meant in the topology induced by this filtration (this is just a fancy way to say that the tails in that sum begin with more and more zeros).

Proof. Item i) is equivalent to stating that

$$m_n(x_0, 0, \dots, 0, y_0, \dots, y_n) = x_0^{p^n} y_n.$$

This can be proved by induction.

Item ii) is an obvious identity between the Witt polynomials. Property iii) follows from ii) and the fact that w_{n-1} is additive.

To prove iv), first notice that if $a, b \in W_p(A)$ are such that, for every i, either $a_i = 0$ or $b_i = 0$, then $w_n(a + b) = w_n(a) + w_n(b)$. With this, the result is clear since

$$V_p^i(t(a_i)) = (0, \dots, 0, a_i, 0, \dots,),$$

where a_i is in position i.

Finally, in v) the polynomials p_n are well defined as in the proof of Lemma 7.7.3, and in fact, p_n can be computed by composition of the sum polynomials $\{s_i\}$ for $i \leq n$. The equation $p_n(\underline{x}) \equiv x_{n-1}^p \pmod p$ can be proved by induction. □

At this point, we are going to assume that $\operatorname{char} A = p$. In this case, these computations simplify quite a bit.

Corollary 7.7.9. *Let A be a ring of characteristic p. Then, for $a = (a_i) \in W_p(A)$,*

$$p \cdot a = (0, a_0^p, a_1^p, \dots).$$

In particular, $\operatorname{char} W_p(A) = 0$ *and* $p = (0, 1, 0, 0, \dots)$ *in* $W_p(A)$.

Proof. This is just a restatement of item v) of Proposition 7.7.8 □

Example 7.7.10. Let $A = \mathbb{Z}/p\mathbb{Z}$. In this case, we can explicitly identify $W_p(A)$ with the ring $\mathbb{Z}_p$ of p-adic integers. In fact, by Proposition 7.7.8 iv), every element of $W_p(A)$ can be written as a series $\sum_{i=0}^{\infty} V_p^i(t(a_i))$. By the above corollary, $V_p(a) = p \cdot a$ for $a \in W_p(A)$, so every element of $W_p(A)$ admits a decomposition

$$a = \sum_{i=0}^{\infty} p^i \cdot t(a_i).$$

Moreover, t satisfies

$$t(\alpha)^p = t(\alpha^p) = t(\alpha).$$

for all $\alpha \in \mathbb{Z}/p\mathbb{Z}$.

On the other hand, using the Teichmüller character τ defined at the beginning of the section, every element $a \in \mathbb{Z}_p$ can be written as a series

$$a = \sum_{i=0}^{\infty} p^i \cdot \tau(a_i).$$

This gives a natural bijection between $W_p(A)$ and $\mathbb{Z}_p$. It is not a priori clear that this is an isomorphism, but this will follow from the next two theorems, which imply that there exists—up to isomorphism—a unique complete DVR of characteristic 0 having $\mathbb{Z}/p\mathbb{Z}$ as a residue field, and that is $W_p(\mathbb{Z}/p\mathbb{Z})$.

Corollary 7.7.9 allows us to generalize the construction of p-adic numbers to any perfect field of characteristic p. Recall that a field k of characteristic p is called *perfect* if $k^p = k$. This holds, for instance, if k is finite or if k is algebraically closed.

Theorem 7.7.11. *Let k be a perfect field of characteristic p. Then the ring $W_p(k)$ is a discrete valuation ring of characteristic 0, with maximal ideal $\mathcal{M} = I_1 = V_p(W_p(k))$. Moreover, $k \cong W_p(k)/\mathcal{M}$, $\mathcal{M}$ is generated by p, and $W_p(k)$ is complete with respect to the $\mathcal{M}$-adic topology.*

Proof. Corollary 7.7.9 tells us that $\operatorname{char}(W_p(k)) = 0$. Moreover, since k is perfect, it shows that I_1 is generated by p. The natural map $w_0 \colon W_p(k) \to k$ is surjective, and has I_1 for kernel, so I_1 is maximal. If we denote $\mathcal{M} := I_1$, the same corollary implies that $I_k = \mathcal{M}^k = (p^k)$.

By Proposition 7.7.8 iv), $W_p(A)$ is complete in the $\mathcal{M}$-adic topology. Every element $a \in W_p(k)$ with $a_0 \neq 0$ is invertible, since $\mathcal{M}$ is maximal. It follows that the only ideals of $W_p(k)$ are the powers of $\mathcal{M}$, so $W_p(k)$ is a DVR. □

Not only does the Witt construction allows us to produce a DVR out of any perfect field k of positive characteristic—the ring thus constructed enjoys a universal property among all rings having k as a quotient.

Theorem 7.7.12. *Let k be a perfect field of characteristic p, $\pi \colon A \to k$ be any surjective ring map, $M := \ker \pi$. Assume that A is complete in the M-adic topology and that $\bigcap_{i=0}^{\infty} M^i = 0$. Then there is a unique map $f \colon W_p(k) \to A$ that makes the diagram*

$$\begin{array}{ccc} W_p(k) & \xrightarrow{\;f\;} & A \\ & \searrow^{w_0} & \downarrow \pi \\ & & k \end{array}$$

commutative.

Proof. To prove the existence of f, we define maps

$$f_n \colon W_p(k)/\mathcal{M}^n \to A/M^n$$

and then try to patch them together. First notice that inside A we have $p \in M$, as $\operatorname{char}(k) = p$. By definition of the Witt polynomials, it follows that given $a_0, \dots, a_n \in M$ we have $w_n(a_0, \dots, a_n) \in M^{n+1}$. Hence, there is a commutative diagram

$$\begin{array}{ccc} W_p(A) & \xrightarrow{w_n} & A \\ \downarrow & & \downarrow \\ W_p(k) & \xrightarrow{g_n} & A/M^{n+1} \end{array}$$

where the left-hand map is induced by π. The map g_n sends $\mathcal{M}$ into M/M^{n+1}, hence it induces the desired homomorphism f_n. The construction of this diagram shows that the system of maps $\{f_n\}$ is compatible, in the sense that the diagram

$$\begin{array}{ccc} \frac{W_p(k)}{\mathcal{M}^{n+1}} & \xrightarrow{f_{n+1}} & \frac{A}{M^{n+1}} \\ \downarrow & & \downarrow \\ \frac{W_p(k)}{\mathcal{M}^{n}} & \xrightarrow{f_n} & \frac{A}{M^{n}} \end{array}$$

commutes. Hence we can define $f(a)$ as the inverse limit of the elements $f_n(a)$, which exists and is unique since A is complete and Hausdorff.

To prove uniqueness, notice that any map f as in the thesis must send $\mathcal{M}$ to M, hence $\mathcal{M}^n$ to M^n for all n. This implies that f is the inverse limit of the maps f_n defined above. □

From the two theorems together, it follows:

Corollary 7.7.13. *Let k be a perfect field of characteristic p. Then there exist a complete DVR A with maximal ideal $\mathcal{M}$ having k as residue field. Moreover, A is unique up to a unique isomorphism.*

There is a fairly straightforward generalization of p-adic Witt vectors. The ring $W_p(A)$ encodes information about characteristic p phenomena, but one can put together this information for various primes at once. To do so, we are going to introduce some notation, that is in slight conflict with the notation that we have used so far.

Definition 7.7.14. Given $n \in \mathbb{Z}_+$, we define the Witt polynomial

$$w_n(\underline{x}) = \sum_{d|n} d x_d^{\frac{n}{d}}.$$

Notice that w_n is a polynomial involving the variables x_d for $d|n$. When $n = p^m$, we recover the familiar Witt polynomials that we previously denoted w_m, save for the renaming that uses the variable x_{p^k} in place of x_k.

We can develop the theory in this more general setting, although we are going to allow ourselves to proceed a little faster.

Lemma 7.7.15. *Let $f \in \mathbb{Z}[x,y,z]$. Then there are uniquely determined polynomials*

$$f_n \in \mathbb{Z}[x_0,\dots,x_n,y_0,\dots,y_n,z_0,\dots,z_n]$$

such that

$$f(w_n(\underline{x}), w_n(\underline{y}), w_n(\underline{z})) = w_n(f_1(\underline{x},\underline{y},\underline{z}),\dots,f_n(\underline{x},\underline{y},\underline{z})).$$

Proof. As in the proof of Lemma 7.7.2, we use the equation to define f_n, and the only thing to prove is that f_n has integral coefficients. This is done by induction: for each prime p dividing n, we can use the inductive hypothesis on n/p to show that the coefficients of f_n, which a priori are in $\mathbb{Q}$, have a nonnegative p-adic valuation. Since this is true for all p that divide n, f_n is in fact integral. □

Using the lemma, one can define sum polynomials $s_n(\underline{x},\underline{y})$ that satisfy

$$w_n(\underline{x}) + w_n(\underline{y}) = w_n(s_1,\dots,s_n).$$

and multiplication polynomials $m_n(\underline{x},\underline{y})$ that satisfy

$$w_n(\underline{x}) \cdot w_n(\underline{y}) = w_n(m_1,\dots,m_n).$$

We say that a subset $S \subset \mathbb{Z}_+$ is *divisor-stable* if whenever $n \in S$ and m divides n, $m \in S$ as well. Given a divisor-stable set S and a ring A, we can define the ring of Witt vectors $W_S(A)$ as the set $A^S = \{(a_i)_{i\in S} \mid a_i \in A\}$, endowed with the operations defined by

$$(a_i) + (b_i) := (s_i(a,b))$$
$$(a_i) \cdot (b_i) := (m_i(a,b)).$$

Using Lemma 7.7.15, we get that these operations give $W_S(A)$ a ring structure. The ring $W_S(A)$ comes with homomorphisms $w_n \colon W_S \to A$ for all $n \in S$, and putting them together gives a homomorphism

$$w_S \colon W_S \to A^S.$$

Remark 7.7.16. As we have already noticed, in this more general setting we have switched to a different notation. If we take $S = \{p^k\}$ for a fixed prime p, we recover the ring that we had previously denoted $W_p(A)$, although the components are indexed $(a_1, a_p, a_{p^2}, \dots)$ in place of $(a_0, a_1, a_2, \dots)$. One can also choose the set $S = \{p^k\}_{k\le n}$—the ring $W_S(A)$ thus obtained is just $W_p(A)/p^{n+1}W_p(A)$.

When taking $S = \mathbb{Z}_+$, we get the so-called ring of *big Witt vectors.*

Remark 7.7.17. If $T \subset S$ are two divisor-stable sets, there is a surjective homomorphism $W_S(A) \to W_T(A)$ which just forgets all coordinates that are not in T. In this way, the ring of big Witt vectors has all other rings of Witt vectors as quotients.

Just as in Remark 7.7.6, the homomorphism $w_S\colon W_S \to A^S$ is bijective when A is a $\mathbb{Q}$-algebra. In this case, $W_S(A) \cong A^S$. Now if $A \to B$ is a ring map, there is an induced homomorphism $W_S(A) \to W_S(B)$, such that the diagram

$$\begin{array}{ccc} W_S(A) & \xrightarrow{w_S} & A \\ \downarrow & & \downarrow \\ W_S(B) & \xrightarrow{w_S} & B \end{array}$$

commutes. This is enough to determine the structure of $W_S(A)$ for all rings A. In fact, every integral domain A of characteristic 0 embeds into the $\mathbb{Q}$-algebra $\mathcal{F}(A)$, which implies that $W_S(A)$ is a a subring of $W_S(\mathcal{F}(A))$ formed by vectors with coefficients in A. Finally, every ring A is a quotient of an integral domain of characteristic 0, and commutativity of the diagram implies that this is enough to fix the ring structure of $W_S(A)$.

7.8. Exercises

1. Prove that a Noetherian valuation ring is either a field or a discrete valuation ring.

2. Prove that the family of arithmetic progressions is a basis for a topology on $\mathbb{Z}$ which makes $\mathbb{Z}$ into a topological ring. Prove that arithmetic progressions are both open and closed, and deduce that $\mathbb{Z}$ has infinitely many primes (otherwise $\{-1, 1\}$ would be open).

3. Give an example of a ring A, an ideal I, and an A-module M with two I-filtrations that induce the same topology on M but do not have bounded difference.

4. Show that the various p-adic norms on $\mathbb{Q}$ for different primes are not equivalent.

5. Give an alternative proof of Ostrowski's theorem 7.1.15 that does not use completions, at least for the nonarchimedean case. Namely, given a nonarchimedean absolute value $|\cdot|$ on the number field k, show that $x \to \log|x|$ is a discrete valuation, and use the results of Section 7.3 to show that $|\cdot|$ is a P-adic absolute value.

6. Given a point $x \in \mathbb{Q}_p$ and $r > 0$, define the disc

$$D(x,r) := \{y \in \mathbb{Q}_p \mid |x-y| < r\}.$$

Show that every point of a disc is a center—that is, if $y \in D(x,r)$, then $D(y,r) = D(x,r)$.

7. Prove that the p-adic expansion (7.1.3) of a number $a \in \mathbb{Z}_p$ is eventually periodic (that is, $\alpha_{i+t} = \alpha_i$ for some fixed t and i big enough) if and only if $a \in \mathbb{Q} \cap \mathbb{Z}_p = \mathbb{Z}_{(p)}$.

8. Show that in $\mathbb{Z}_5$ there is a third root of 3.

9. Let $p \in \mathbb{Z}$ be a prime and n not divisible by p. Then any element $a \in \mathbb{Z}$ which is congruent to 1 modulo p is a nth root in $\mathbb{Z}_p$.

10. Prove the following stronger version of Hensel's lemma. Let $f \in \mathbb{Z}_p[x]$ be a polynomial, $a \in \mathbb{Z}_p$ such that

$$|f'(a)|_p < |f'(a)|_p^2.$$

Then there is a unique root $\alpha \in \mathbb{Z}_p$ of f such that $|\alpha - a|_p < |f'(a)|_p$.

11. Show that there is a unique extension of the norm $|\cdot|_p$ to the algebraic closure $\overline{\mathbb{Q}_p}$ of $\mathbb{Q}_p$. On the other hand, for each finite extension K of $\mathbb{Q}$, there is an extension of $|\cdot|_p$ to K for each prime Q_i of $\mathcal{O}_K$ over p. Why this is not a contradiction?

12. In Remark 7.5.20, we state that products do not commute with intersections (even finite ones). Motivate this remark, by finding an explicit example of a ring A with ideals I_1, I_2, and I_3 such that $I_1 \cdot (I_2 \cap I_3) \neq I_1 \cdot I_2 \cap I_1 \cdot I_3$.

13. [**Car**] In the ring

$$A = \frac{\mathbb{Q}[x, z, y_1, y_2, \dots]}{(x - zy_1, x - z^2y_2, \dots)},$$

consider the ideal $I = (z)$. Show that $\bigcap_{k=1}^{\infty} I^k = (x)$ and that $z \cdot (x) \neq (x)$. This provides a counterexample to Krull intersection theorem when the ring is not Noetherian.

The following exercises (up to Exercise 20) develop the beginning of the theory of *heights* on number fields.

14. Let K be a number field. For every equivalence class of absolute values on K, choose a representative normalized as follows. If v is archimedean, take the restriction of $|\cdot|_{st}$ for a suitable embedding $K \hookrightarrow \mathbb{C}$. Otherwise, choose the normalization of Example 7.1.4e. Define $\epsilon(v) = 1$ unless v is the restriction of $|\cdot|_{st}$ for an embedding of K in $\mathbb{C}$ such that $K \not\subset \mathbb{R}$, in which case $\epsilon(v) = 2$.

Show that for any $x \in K$ we have $|x|_v = 1$ except for a finite number of v, and that

$$\prod_v |x|_v^{\epsilon(v)} = 1.$$

This is know as the *product formula* for absolute values. (*Hint*: show the result on $\mathbb{Q}$ first.)

15. Let K be a number field, and normalize absolute values on K as in Exercise 14. Given $x \in K$ define its *height*

$$H_K(x) := \prod_v \max\{|x|_v, 1\}.$$

Show that H_K is well defined, and that for a reduced rational number a/b it reduces to $H_{\mathbb{Q}}(a/b) = \max\{|a|_{st}, |b|_{st}\}$.

16. Let H_K be the height on a field K, as defined in the previous exercise. Show that if $K \subset L$ is an inclusion of number fields, we have

$$H_L(x) = H_K(x)^{[L:K]}$$

for all $x \in K$. Hence given an algebraic number x, we can define the *absolute height*

$$H(x) := H_K(x)^{\frac{1}{[K:\mathbb{Q}]}}$$

for any number field K containing x.

17. Let H be the absolute height defined in the previous exercise. Show that for an algebraic number x we have the identities $H(x^k) = H(x)^k$ for all $k \in \mathbb{N}$. Moreover, if y is another algebraic number, show the two inequalities $H(xy) \leq H(x)H(y)$ and $H(x+y) \leq 2H(x)H(y)$.

18. Given a polynomial with algebraic coefficients

$$f(x) = a_n x^n + \cdots + a_1 x + a_0,$$

define its height by

$$H(f) = \left(\prod_v \max\{|a_0|_v, \ldots, |a_n|_v\}\right)^{\frac{1}{d}},$$

where the product runs over all absolute values of the field $K{:=}\mathbb{Q}(a_0, \ldots, a_n)$, and $d = [K : \mathbb{Q}]$.

Assume that f factorizes as

$$f(x) = a_n(x - \alpha_1)\cdots(x - \alpha_n).$$

Show that

$$H(f) \leq 2^{n-1} \prod_{i=1}^{n} H(\alpha_i).$$

19. Use the previous exercise to prove the following theorem of Northcott: for any choice of constants A, B, the set

$$\{x \mid H(x) \leq A, [\mathbb{Q}(x) : \mathbb{Q}] \leq B\}$$

is finite. In particular, every number field contains a finite number of elements of bounded height. (The coefficients of the minimal polynomial over $\mathbb{Q}$ of an element of this set are bounded.)

20. Prove the following theorem of Kronecker: an algebraic number x satisfies $H(x) = 1$ if, and only if, it is a root of unity.

21. Let A be a ring, $I \subset A$ an ideal, and M an A-module. Consider the completion with respect to the I-adic topology. Show that there is a well-defined map $t : \widehat{A} \otimes_A M \to \widehat{M}$.

Assuming that M is finitely generated, show that t is surjective. (Write M as a quotient of a free module.)

22. Continuing the previous exercise, assume that M is finitely generated and A is Noetherian. Show that the map t is an isomorphism.

23. Let $A = \mathbb{R}\{\{x\}\}$ be the ring of convergent power series around 0. Show that A is not complete, but Hensel's lemma holds for A.

To better understand ideals in valuation rings, we give some definitions. Let G be an ordered group. A subset $S \subset G$ is called a *segment* if for any $g \in S$, the set $\{h \in G \mid -g \leq h \leq g\}$ is contained in S. A subgroup which is a segment is called *isolated*. Given a field k and a valuation $v \colon k^* \to G$, we can always assume that v is surjective, and then we say that G is the *value group* of v.

24. Let A be a valuation ring with value group G, and call v the valuation. Show that for any ideal I, the set

$$G_I := v(A \setminus I) \cup -v(A \setminus I)$$

is a segment of G, and that this induces a bijection between ideals of A and segments of G. Moreover, this correspondence restricts to a bijection between *prime* ideals of A and isolated subgroups of G.

25. Show that the set of isolated subgroups in an ordered group is totally ordered with respect to inclusion. Use the previous exercise to prove that the set of prime ideals in a valuation ring is totally ordered with respect to inclusion.

26. Given an ordered group G, construct a field k with a valuation v having value group G.

27. Find a valuation on $k(x_1, \ldots, x_n)$ having value group $\mathbb{Z}^n$ with the lexicographic order, and describe its valuation ring.

28. Let A be a complete DVR, $\mathcal{M}$ its maximal ideal, and let $k = A/\mathcal{M}$. If k has characteristic 0, then $A \cong k[[x]]$, where the isomorphism preserves the valuation.

29. Let $\{w_n\}$ be the p-adic Witt polynomials. Show that there is a sequence of integral polynomials $\{f_n(x_0, \ldots, x_{n+1})\}$ such that

$$w_n(f_0, \ldots, f_n) = w_{n+1}.$$

Show that for each ring A, the map $f\colon W_p(A) \to W_p(A)$ defined by

$$f(a_0, a_1, \dots) = (f_0(a_0, a_1), f_1(a_0, a_1, a_2), \dots)$$

is a homomorphism. The map f is called the *Frobenius* homomorphism of $W_p(A)$.

30. Let A be a ring of characteristic p. Show that the Frobenius homomorphism $f\colon W_p(A) \to W_p(A)$ defined in the previous exercise is given by

$$f(a_0, a_1, \dots) = (a_0^p, a_1^p, \dots),$$

justifying its name.

The next two exercises give a beautiful application of Krull's intersection theorem.

31. [**KO10**] Let A be a Noetherian integral domain, $I \subset A$ an ideal, and let $a = |A|$, $b = |A/I|$. Show that $a \le b^{\aleph_0}$. (The ring A embeds into $\prod A/I^n$.)

32. [**Bel**] One may wonder whether every Noetherian ring is the quotient of a Noetherian integral domain. Show that this is false by applying the previous exercise to the ring $K \times L$, where K is a finite field and L is a field of cardinality bigger than the continuum.

Chapter 8

Geometric Dictionary

In this chapter, we give a geometric perspective to the topics studied so far. This is useful because some notions such as dimension are much easier to grasp under the geometric point of view. We will also see that many of the notions introduced up to this point are passible of a geometric interpretation.

The first section introduces the basic definitions, like varieties in affine space, their coordinate rings, and the Zariski topology. Most of the rest of the chapter uses this new language to reinterpret old results and concepts from the geometric point of view.

In particular, in the second section we give some proofs of the celebrated Nullstellensatz and use it to illustrate the correspondence between the points of an affine variety and the maximal ideals of its coordinate ring. We also derive from it the Ax–Grothendieck theorem on affine polynomial maps. Then we introduce the local ring at a point and show that this can be obtained by localization, so the terminology finally makes sense. Next, we show that the local situation is somehow similar to the graded one, and introduce the correspondence between projective varieties and graded rings.

Toward the end of the chapter, we introduce two new geometric concepts which we will investigate from an algebraic point of view in the next chapters. The first one is the concept of dimension. Actually some different definitions are possible; the equivalence of them, in a more general setting, will be shown on the next chapter. Second, we introduce the Zariski tangent space, together with regular and singular points.

After this, one can define curves as varieties of dimension one. In the last section we show how Dedekind rings arise as coordinate rings of smooth irreducible curves, and use our knowledge about Dedekind rings to discuss morphisms of curves.

8.1. Affine varieties

We consider rings of the form $A = k[x_1, \ldots, x_n]$ where k is a field. We can interpret a polynomial $f \in A$ as an equation that defines a zero locus

$$V(f) = \{(x_1, \ldots, x_n) \in k^n | f(x_1, \ldots, x_n) = 0\}.$$

More generally, we can consider the locus defined by the vanishing of more than one equation. Given a set $S \subset A$ we define the locus

$$V(S) = \{(x_1, \ldots, x_n) \in k^n | f(x_1, \ldots, x_n) = 0 \text{ for all } f \in S\}.$$

Definition 8.1.1. A locus of the form $V(S)$ for some $S \subset A$ is called an *(affine) algebraic variety*. When S is empty we obtain the whole k^n, which we will call an *affine space* and denote by $\mathbb{A}^n(k)$, or simply $\mathbb{A}^n$ when k is fixed.

Remark 8.1.2. It is immediate to check that $V(S) = V(I)$, where I is the ideal generated by S, hence it is enough to consider the zero loci of ideals. Moreover, since ideals in $k[x_1, \ldots, x_n]$ are finitely generated, we see that a finite number of equations always suffice to define an algebraic variety.

Example 8.1.3.

(a) When $k = \mathbb{R}$, $n = 2$ and I is generated by a single quadratic polynomial, we obtain the classical example of plane conics. Notice that it can happen that the polynomial has no real zeros (such as $x^2 + y^2 + 1$)—in such case, the zero locus is just the empty set.

(b) More generally, the zero locus of a single equation $f \in A$ is called a *hypersurface*.

(c) The zero locus $V(S)$ of a set S is just the intersection of all $V(f)$ for $f \in S$.

(d) The union $V(I) \cup V(J)$ of two algebraic varieties is itself an algebraic variety, defined by the ideal $I \cdot J$. It can also be defined by the ideal $I \cap J$.

Definition 8.1.4. The last two points show that the sets of the form $V(S)$ for $S \subset A$ are the closed sets for a topology on $\mathbb{A}^n(k)$, which we will call the *Zariski topology*. Varieties $V \subset \mathbb{A}^n(k)$ inherit this topology—their closed sets are the intersections of V with other affine algebraic varieties.

Notice that this is a very coarse topology—its only open sets are complements of algebraic varieties—hence there are no small open sets such as balls in $\mathbb{R}^n$.

We can also go in the other direction: given a set $W \subset k^n$ we define

$$I(W) = \{f \in A \mid f(x_1, \ldots, x_n) = 0 \text{ for all } (x_1, \ldots, x_n) \in W\}.$$

The set $I(W)$ is always an ideal in R—in fact, it is a radical ideal.

Remark 8.1.5. The correspondences $I \to V(I)$ and $W \to I(W)$ reverse inclusions. Moreover, it is easy to check that $J \subset I(V(J))$ and $W \subset V(I(W))$. In fact, if W is itself an algebraic variety, we have the equality $W = V(I(W))$ (check it!).

The fact that J is not always the same as $I(V(J))$ can be easily seen in the case where $k = \mathbb{R}$ and J is generated by $f = x^2 + y^2 + 1$. In this case, $V(J) = \emptyset$, hence $I(V(J)) = R$. We will discuss in the next section the shape of $I(V(J))$ in the case where k is algebraically closed.

Definition 8.1.6. The ring $R(V) := A/I(V)$ is called the *coordinate ring* of the affine variety V. Notice that elements of $R(V)$ define actual polynomial functions $V \to k$.

When V is a single point $\{x\}$, $I(V)$ is just the kernel of the evaluation function

$$\begin{aligned} ev_x\colon\ A &\longrightarrow k \\ f &\longmapsto f(x). \end{aligned}$$

In particular $I(\{x\})$ is a maximal ideal.

We end this section with a property of the Zariski topology that shows that it cannot be very fine.

Definition 8.1.7. A topology is called *Noetherian* if every ascending chain of open sets (or equivalently a descending chain of closed sets) is eventually stationary.

By Hilbert's basis theorem, it is immediate that the Zariski topology is Noetherian. In fact a descending chain of closed sets

$$W_1 \supset W_2 \supset \cdots$$

gives rise to an ascending chain of ideals

$$I(W_1) \subset I(W_2) \subset \cdots$$

that must stabilize since A is Noetherian. This implies that the original chain stabilizes, since $W_i = V(I(W_i))$.

Remark 8.1.8. In particular, this implies that $\mathbb{A}^n$ is compact under the Zariski topology.

8.2. The Nullstellensatz

Starting from this section, we assume that k is algebraically closed. This simplifies the geometry of $\mathbb{A}^n(k)$ considerably.

We start with the question left open in the previous section: what can we say about an ideal of the form $I(V)$, where $V \in A^n$ is an algebraic variety? It is easy to check that $I(V)$ is a radical ideal (if f^r vanishes on V, so does f). Hence if $V = V(J)$, we see that $\sqrt{J} \subset I(V)$.

In this section, we will see that the converse holds in the algebraically closed case. We give three (equivalent) forms of the celebrated *Nullstellensatz*, also called the Hilbert zeros theorem [**Hil93**].

Theorem 8.2.1 (Nullstellensatz, strong form). *Assume that k is algebraically closed, and let $J \subset k[x_1, \dots, x_n]$ be an ideal. Then $I(V(J)) = \sqrt{J}$.*

Notice that this implies that there is a *bijective correspondence* between algebraic varieties inside $\mathbb{A}^n$ and radical ideals inside $k[x_1, \dots, x_n]$. There are other formulations of the Nullstellensatz that are apparently weaker.

Theorem 8.2.2 (Nullstellensatz, first weak form). *Assume that k is algebraically closed, and let $J \subset k[x_1, \dots, x_r]$ be an ideal. If $V(J)$ is empty, then J is the whole ring.*

Theorem 8.2.3 (Nullstellensatz, second weak form). *Assume that k is algebraically closed. The maximal ideals of $k[x_1, \dots, x_r]$ are exactly those of the form $I(\{x\})$ for some point $x \in k^n$.*

We start by proving the equivalence of the three forms, and then prove the theorem itself.

Proof of the equivalence.

(1) We first see that the strong form implies the first weak form. This is easy: if $V(J)$ is empty, $I(V(J)) = A = k[x_1, \dots, x_n]$. By the strong form, this is the radical of J, hence J itself is the whole ring.

(2) Assume that the first weak form holds, and let $\mathcal{M} \subset A$ be a maximal ideal. Then $V(\mathcal{M})$ is not empty, so that we find some point $x \in V(\mathcal{M})$. This implies that $\mathcal{M} \subset I(\{x\})$, and we must have equality since $\mathcal{M}$ is maximal; this proves the second weak form.

(3) Assume that the second weak form holds, and let $J \subset A$ be an ideal such that $V(I)$ is empty. This means that J is not contained in $I(\{x\})$ for any point x. Since this are all the maximal ideals, J is the whole ring, proving the first weak form.

(4) The trickiest part is proving that the first weak form implies the strong form. This is done by the so called Rabinowitsch trick ([**Rab29**]). Let $J \subset k[x_1, \dots, x_n]$ be an ideal, and let g be a polynomial vanishing on $V(J)$. Inside the ring $k[x_1, \dots, x_n, y]$, consider the ideal $\overline{J}$ generated by elements of J and the single polynomial $y \cdot g - 1$.

Since g is zero on $V(J)$, $V(\overline{J})$ is empty, and by the first weak form we deduce that $\overline{J}$ is the whole ring. This implies that we can find an identity of the form

$$1 = h_0(y \cdot g - 1) + h_1 f_1 + \cdots + h_r f_r$$

for some $f_1, \dots, f_r \in J$ and $h_0, \dots, h_r \in k[x_1, \dots, x_n, y]$.

We can formally substitute $y = 1/g$ in this equation and then clear the denominators by multipliying both sides by g^t for some t big enough. This leaves the identity

$$g^t = h_1' f_1 + \cdots + h_r' f_r,$$

which shows that $g \in \sqrt{J}$. □

There is another form of the Nullstellensatz that can be stated for an arbitrary field. It usually goes under the name of *Zariski's lemma.*

Theorem 8.2.4 (Zariski's lemma). *Let A be a finitely generated k-algebra that is also a field. Then A is a finite extension of k.*

Zariski's lemma implies the Nullstellensatz, as we show here.

Proof of the Nullstellensatz. We show that Zariski's lemma implies Theorem 8.2.3. Assume that k is algebraically closed, and let $\mathcal{M}$ be a maximal ideal of $k[x_1, \dots, x_r]$. Then the algebra $A := k[x_1, \dots, x_r]/\mathcal{M}$ is finitely generated over k and it is also a field, hence it is a finite extension of k. Since k it is algebraically closed, the natural inclusion $k \hookrightarrow A$ is an isomorphism.

If we denote $\lambda_i \in k$ the image of x_i under this isomorphism, it follows that $f_i := x_i - \lambda_i \in \mathcal{M}$. Since the ideal generated by the elements f_i is maximal, it follows that $\mathcal{M} = (f_1, \dots, f_r)$. □

We give two proofs of Zariski's lemma.

First proof of Zariski's lemma. Let A be a finitely generated k-algebra that is also a field. By the Noether normalization lemma 5.3.1, we can write $A = k[x_1, \dots, x_r]$, where $x_1, \dots, x_m$ are algebraically independent, and $x_{m+1}, \dots, x_r$ are integral over $k[x_1, \dots, x_m]$. The assertion that we need to prove is that $m = 0$.

If not, $1/x_1 \in A$, hence it must be integral over $k[x_1, \dots, x_m]$. This gives a nontrivial polynomial relation between $x_1, \dots, x_m$, contradicting the fact that they are algebraically independent. □

The second proof will be an immediate consequence of this more general lemma, which is useful in its own sake.

Lemma 8.2.5. *Let $A \subset B$ be integral domains and assume that B is finitely generated over A (as an A-algebra). For all nonzero $b \in B$ there exists a nonzero $a \in A$ with the following extension property: every homomorphism $f\colon A \to K$, where K is algebraically closed and $f(a) \neq 0$, can be extended to a homomorphism $g\colon B \to K$ such that $g(b) \neq 0$.*

Proof. By induction, we can assume that B is generated over A by a single element x. We now distinguish two cases, based on whether x is algebraic over (the field of fractions of) A.

If x is transcendental over A, write

$$b = a_n x^n + \dots + a_0$$

with $a_0, \dots, a_n \in A$, and choose $a := a_n$. Then, for every $f\colon A \to K$, we can extend it to B by choosing at will the image $y = g(x) \in K$. The requirement that $g(b) \neq 0$ becomes

$$f(a_n)y^n + \dots + f(a_0) \neq 0,$$

which can be guaranteed for a suitable choice of y, since K is infinite and $f(a_n) \neq 0$.

If x is algebraic over A, a little more care is needed since $g(x)$ cannot be chosen arbitrarily. In this case, x satisfies an equation

$$a_n x^n + \dots + a_0 = 0 \tag{8.2.1}$$

for some $a_0, \dots, a_n \in A$. Moreover, $1/b$ is a polynomial in x, hence it satisfies a similar equation

$$b_m b^{-m} + \dots + b_0 = 0 \tag{8.2.2}$$

for some $b_0, \dots, b_m \in A$.

In this case, we choose $a = a_n b_m$. Given $f\colon A \to K$ such that $f(a) \neq 0$, we can extend it to $f'\colon A[1/a] \to K$ by declaring $f'(1/a) = 1/f(a)$. Then we can choose a maximal extension, say $h\colon T \to K$. By Theorem 7.2.9, T is a valuation ring.

By (8.2.1), x is integral over $A[1/a]$, hence over T. But valuation rings are integrally closed by 7.2.8, hence $x \in T$, and we have the desired extension to B.

To check that $g(b) \neq 0$, it is enough to show that h is defined on $1/b$, and this follows in the same way. Namely, $1/b$ is integral over $A[1/a]$ by (8.2.2), so it belongs to T. □

Second proof of Zariski's lemma. With the notation of the above lemma, assume that A and B are fields and choose $b = 1$. Then the fact that every homomorphism from A to an algebraically closed field can be extended to B is equivalent to the fact that B is algebraic over A. □

In the final part of the section, we investigate the geometric consequences of the Nullstellensatz. By Theorem 8.2.3, points in $\mathbb{A}^n$ correspond bijectively with maximal ideals of A.

Let V be an affine variety, and consider the quotient $R(V) = A/I(V)$. Maximal ideals of $R(V)$ correspond to those maximal ideals $I(\{x\})$ of A that contain $I(V)$; this happens exactly when $x \in V$. We conclude that maximal ideals of $R(V)$ correspond bijectively to points of V. Moreover, V can be reconstructed as a topological space knowing the ring $R(V)$ alone: in fact closed sets are exactly sets of maximal ideals that contain a given ideal J. This shows that the algebraic side completely governs the geometry of algebraic varieties.

8.3. The Ax–Grothendieck theorem

In this section, we temporarily go back to our algebraic setup to give an application of the Nullstellensatz.

Theorem 8.3.1 (Ax–Grothendieck). *Let k be an algebraically closed field, $f\colon k^n \to k^n$ a polynomial map. If f is injective, then it is surjective.*

The proof by Ax in [**Ax68**] is based on model theory—we sketch it in Exercise 31. We follow a simplification of Grothendieck's argument ([**Gro66**, Theorem 10.4.11]).

To begin, we rephrase the conditions of being injective and not being surjective in more explicit terms.

Lemma 8.3.2. *Let k be an algebraically closed field, $f\colon k^n \to k^n$ an injective polynomial map. Then there exists a set of polynomials $h_1, \dots, h_n \in k[x_1, \dots, x_n, y_1, \dots, y_n]$ such that*

$$f(x) - f(y) = h_1(x,y)(x_1 - y_1) + \cdots + h_n(x,y)(x_n - y_n). \tag{8.3.1}$$

Proof. By hypothesis, $f(x) - f(y)$ only vanishes on the diagonal of $k^n \times k^n$. The radical ideal defining the diagonal is $((x_1 - y_1), \dots, (x_n - y_n))$, so this follows by the Nullstellensatz. □

Another application of the Nullstellensatz immediately translates the condition of not being surjective.

Lemma 8.3.3. *Let k be an algebraically closed field, $f\colon k^n \to k^n$ a nonsurjective polynomial map. Then there exists $y \in k^n$ and a set of polynomials $h_1, \dots, h_n \in k[x_1, \dots, x_n]$ such that*

$$(f_1(x) - y_1) \cdot h_1(x) + \cdots + (f_n(x) - y_n) \cdot h_n(x) = 1. \tag{8.3.2}$$

Notice that the Ax–Grothendieck theorem also holds when k is a finite field, trivially since any injection from a finite set to itself is also surjective. We give the proof by reducing to this case.

Proof of Ax–Grothendieck theorem. By contradiction, assume that f is injective but not surjective, hence (8.3.1) and (8.3.2) hold. Consider the ring A generated over $\mathbb{Z}$ (if k has characteristic 0) or $\mathbb{Z}/p\mathbb{Z}$ (if k has characteristic p) by all coefficients of all polynomials involved in these equations. This implies that f is still injective and not surjective on A.

Choose a maximal ideal $\mathcal{M}$ that does not contains all coefficients, so that equations (8.3.1) and (8.3.2) remain nontrivial over $k' := A/\mathcal{M}$. The map f defines a map on k'^n which is still injective and not surjective. Notice that even if k has characteristic 0, k' cannot contain $\mathbb{Q}$, hence in any case it has positive characteristic p. By Lemma 8.2.5, k' is an algebraic extension of $\mathbb{Z}/p\mathbb{Z}$, hence it is itself a finite field. But this is a contradiction, since an injective self map over k'^n, where k' is finite, is also surjective. $\square$

8.4. Morphisms

In this section we introduce morphisms between affine varieties. Let $V \subset \mathbb{A}^n(k)$ be an affine variety, and consider the ring $R(V) = k[x_1, \dots, x_n]/I(V)$. Any element $f \in k[x_1, \dots, x_n]$ defines a polynomial function $\mathbb{A}^n(k) \to \mathbb{A}^1(k)$, where we identify $\mathbb{A}^1(k)$ with k. This can be restricted to a function $V \to \mathbb{A}^1(k)$, and two polynomials f and g will define the same function on V if and only if $f - g \in I(V)$. This allows us to identify $R(V)$ with the ring of polynomial functions $V \to \mathbb{A}^1(k)$. Putting together more than one function, we give the following

Definition 8.4.1. Let $V \subset \mathbb{A}^n(k)$ be an affine variety. An *affine morphism* $V \to \mathbb{A}^m(k)$ is a function defined by the value of m elements $f_1, \dots, f_m \in R(V)$. If $W \subset \mathbb{A}^m(k)$ is another affine variety, a morphism $V \to W$ is just a morphism $V \to \mathbb{A}^m(k)$ whose image lies inside W.

Using the Nullstellensatz, we can give a precise picture of the relation between the algebraic and geometric side (assuming, as we will do from now on, that we work on an algebraically closed field k). First, we have seen that to any affine variety V, we can associate a finitely generated, reduced ring $R(V)$. The ring $R(V)$ is reduced because the ideal $I(V)$ is radical.

In the other direction, let A be any finitely generated, reduced k-algebra. Since A is finitely generated as a k-algebra, there exists a surjective map $f\colon k[x_1, \dots, x_n] \to A$. The ideal $I = \ker f$ is radical, since A is reduced. Attached to I there is the affine variety $V := V(I) \subset \mathbb{A}^n(k)$, and the Nullstellensatz implies that $I = I(V)$. It follows that in fact, $A \cong R(V)$ as k-algebras. This construction gives us a way to go from rings to varieties and vice versa.

This duality also extends to morphisms. Let $V \subset \mathbb{A}^n(k)$ and $W \subset \mathbb{A}^m(k)$ be two affine varieties. If $f\colon V \to W$ is a morphism, there is a natural composition map

$$\begin{aligned} f^*\colon\ R(W) &\longrightarrow R(V), \\ g &\longmapsto g \circ f \end{aligned}$$

and f^* is clearly a homomorphism of rings.

Vice versa, take any homomorphism $t\colon R(W) \to R(V)$. The coordinate functions $x_1, \dots, x_m$ belong to $R(W)$, and this gives us m elements $f_1, \dots, f_m \in R(V)$ defined as $f_i = t(x_i)$. Moreover, for all $g \in I(W)$ we have

$$g(f_1(x), \dots, f_m(x)) = 0$$

for all $x \in V$, just because t is well defined (do you see this?). This means that

$$(f_1(x), \dots, f_m(x)) \in W$$

for all $x \in V$, and this defines a morphism $V \to W$.

In other words, studying the algebra of reduced, finitely generated k-algebras and their homomorphisms is completely equivalent (up to inverting the direction of maps) to studying the geometry of affine varieties over k. Under this correspondence, reduced ideals of rings become closed set of varieties under the Zariski topology. For those who know the terminology, we have realized an *equivalence* between the category of reduced, finitely generated k-algebras and the opposite category of affine varieties over k.

8.5. Local rings and completions revisited

In this section we can give geometric interpretations to the operations of *localization* and *completion* of a ring.

Let $V \subset \mathbb{A}^n(k)$ be an affine variety, where k is an algebraically closed field. Section 8.2 showed that there is a bijective correspondence between points of V and maximal ideals of the ring $R(V)$—namely the point $x \in V$ corresponds to the ideal $\mathcal{M}_x$ of algebraic functions that vanish at x.

Assume that V is irreducible, so that $R(V)$ is an integral domain, and let $k(V)$ be the fraction field of $R(V)$. Elements of $k(V)$ can be seen as rational functions defined on V. An element of $k(V)$ takes the form f/g for $f, g \in R(V)$, hence it defines a function $U_g \to k$ on the Zariski open set

$$U_g = \{p \in V | g(p) \neq 0\}.$$

The localization of $R(V)$ at $\mathcal{M}_x$ is the subset of $k(V)$ consisting of those rational functions whose denominator does not vanish on x. In other words, a function $t = f/g$ belongs to the localization if, and only if, $x \in U_g$. This entails that t is actually defined on a neighborhood of x.

The ring $R(V)_{\mathcal{M}_x}$ is then ring of algebraic rational functions defined in a neighborhood of x, which gives meaning to the name *localization.*

More generally, let $P \subset R(V)$ be a prime ideal, and $V(P) \subset V$ its zero locus. The ring $R(V)_P$ is the ring of rational functions on V defined in a neighborhood of $V(P)$.

The completion of a ring gives rise to an even more local zoom around the point x. The issue is that the Zariski topology is not very fine. Zariski closed sets in V are actually subvarieties of V—or symmetrically, Zariski open subsets cannot be too small. In the case where $k = \mathbb{C}$, complex algebraic varieties are also endowed with the Euclidean topology, which contains many more open sets—for instance Euclidean balls. We want to somehow capture the notion of a function defined in a small neighborhood of $x \in V$, but the Zariski topology does not have *small* neighborhoods!

The answer comes by analogy with complex geometry. Complex differentiable functions are locally expressible as a convergent sum of a power series. In our algebraic setting, we do not have a suitable notion of convergence, but we do away with it by just considering the ring of *all* algebraic power series.

Let us first consider the case where $V = \mathbb{A}^n(k)$ is just the affine space. We might just as well choose the point $x = \underline{0}$. Then the ring of algebraic functions on V is just $k[x_1, \ldots, x_n]$, and its completion at the ideal $\mathcal{M}_x$ is the power series ring $k[[x_1, \ldots, x_n]]$, as we have seen in Example 7.5.21d.

Notice that in this case we have natural inclusions

$$k[x_1, \ldots, x_n] \subset k[x_1, \ldots, x_n]_{\mathcal{M}_x} \subset k[[x_1, \ldots, x_n]].$$

This is because a polynomial g, such that $g(\underline{0}) \neq 0$ is invertible in $k[[x_1, ..., x_n]]$, hence a rational function f/g, can be expanded as a well-defined power series.

This set of inclusions can be interpreted as taking smaller and smaller neighborhoods of x. In other words, function defined globally are a subset of functions defined in a neighborhood of x, which are themselves a subset of functions defined only formally in an infinitesimal neighborhood of x.

In the general case, Example 7.5.21(c) shows that for a ring $R(V)$, the localization at the maximal ideal $\mathcal{M}_x$ of a point sits between $R(V)$ and its completion at $\mathcal{M}_x$. Again, in this case, we interpret $\widehat{R(V)}$ as the ring of "analytic functions" defined around the point x.

8.6. Graded rings and projective varieties

In this section, we mimic what we have done for affine varieties to define varieties in projective space.

Definition 8.6.1. Let k be a field. The n-dimensional *projective space* over k is the set of lines through the origin in $\mathbb{A}^n(k)$. It can be identified with the quotient

$$\mathbb{P}^n(k) := (k^{n+1} \setminus 0)/\sim,$$

where $\sim$ is the equivalence relation given by $\underline{x} \sim \underline{y}$ if $\underline{x} = t \cdot \underline{y}$ for some $t \in k$.

We will soon give $\mathbb{P}^n(k)$ the structure of a variety. To this end, notice that if $f \in k[x_0, \ldots, x_n]$ is a polynomial and $p \in \mathbb{P}^n(k)$, the value of $f(p)$ is not well defined, since it depends on a representative for p in k^{n+1}. *But* if f is homogeneous, it is well defined whether $f(p) = 0$ or not.

Hence, we can define a zero locus

$$V(f) = \{p \in \mathbb{P}^n(k) | f(p) = 0\}.$$

More generally, we can consider the locus defined by a homogeneous ideal $I \subset k[x_1, \ldots, x_{n+1}]$ as

$$V(I) = \{p \in \mathbb{P}^n(k) | f(p) = 0 \text{ for all homogeneous } f \in I\}.$$

Definition 8.6.2. A locus of the form $V(I)$ for some homogeneous ideal I is called a *(projective) algebraic variety.* When $I = 0$, we obtain the whole $\mathbb{P}^n(k)$, which is a projective variety itself. We will sometimes denote $\mathbb{P}^n = \mathbb{P}^(k)$ when k is implied.

Remark 8.6.3. As in the affine case, every algebraic variety is defined by a finite number of equations, since $k[x_0, \ldots, x_n]$ is Noetherian. Moreover, mimicking the affine case, we can define the *Zariski topology* on $\mathbb{P}^n(k)$ whose closed sets are projective varieties.

The projective space is covered by affine charts. Namely, for every $i = 0, \ldots, n$, there is a subset $C_i \subset \mathbb{P}^k(n)$ defined by the equation $x_i \neq 0$. For each element $\overline{(x_0, \ldots, x_n)} \in C_i$, there is unique representative having $x_i = 1$, hence C_i can be naturally identified with $\mathbb{A}^n(k)$ by taking such representative and omitting the i-th coordinate.

Remark 8.6.4. The sets C_i are open in the Zariski topology and form an open covering of $\mathbb{P}^n(k)$. If $V \subset \mathbb{P}^n(k)$ is a projective variety, this restricts to an open covering of V by affine varieties.

We now give some examples of projective varieties.

Example 8.6.5.

(a) For every affine variety $V \subset \mathbb{A}^n(k)$, we can construct a projective variety by homogeneizing its equations. Namely, if $f(x_1, \ldots, x_n)$ is any polynomial of degree t vanishing on V, we can construct a homogeneous polynomial $f_h(x_0, \ldots, x_n)$ of degree t such that

$$f_h(1, x_1, \ldots, x_n) = f(x_1, \ldots, x_n).$$

It is sufficient to take any monomial in f of degree $d \leq t$ and multiply it by x_0^{t-d}.

If we do that for all $f \in I(V)$, we get a projective variety $\overline{V} \subset \mathbb{P}^n(k)$ such that $\overline{V} \cap C_0 = V$. $\overline{V}$ is called the *projective closure* of V.

(b) As in the affine case, the zero locus of a single equation f is called a *hypersurface.*

(c) The zero locus $V(I)$ of a homogeneous ideal I is just the intersection of all $V(f)$ for all f generating I.

(d) The union $V(I) \cup V(J)$ of two projective algebraic varieties is itself an algebraic variety, defined by the ideal $V(I \cdot J)$.

(e) Let $M_{m,n}(k)$ be the vector space of $m \times n$ matrices with coefficients in k. For each $r < \min\{m, n\}$, the set D_r of matrices having rank at most r is an affine variety. In fact, the condition of having rank at most r is equivalent to the vanishing of all $(r+1) \times (r+1)$ minor determinants. Since these equations are homogeneous, they also define a variety $\overline{D_r}$ in the corresponding projective space. These are called *determinantal varieties.*

Let us denote $A = k[x_0, \ldots, x_n]$. As in the affine case, attached a projective variety V there is an ideal

$$I(V) = \{f \in A \mid f(p) = 0 \text{ for all } p \in V\},$$

and a ring $R(V) = A/I(V)$. Unlike the affine case, $I(V)$ is homogeneous, hence $R(V)$ is graded. Also, unlike the affine case, elements of $R(V)$ do not correspond to polynomial functions defined on V.

You will prove in Exercise 2 that when the field k is algebraically closed there is version of the Nullstellensatz adapated to the projective case. Hence, while the algebra of finitely generated algebras over k is reflected in the geometry of affine varieties, graded finitely generated algebras are best seen as counterparts of projective varieties. Let us see what it means for morphisms.

Definition 8.6.6. Let $V \subset \mathbb{P}^n$ and $W \subset \mathbb{P}^m$ be two projective varieties. A *projective morphism* $f\colon V \to W$ is a function which locally is an affine morphism. This means that for each point $x \in V$ we can find affine charts $C_i \subset \mathbb{P}^n$ and $C'_j \subset \mathbb{P}^m$ such that $x \in C_i$, $f(x) \in C'_j$ and the affine map

$$f\,|_{V\cap C_i}\colon V \cap C_i \to W \cap C'_j$$

is a morphism of affine varieties.

As in the affine case, a morphism $V \to W$ gives a homomorphism $R(W) \to R(V)$ between the graded rings, although the elements of this rings cannot be interpreted as functions.

Remark 8.6.7. It is easy to generalize Definition 8.6.6 to speak of morphisms $V \to W$, where V is affine and W projective. Similarly, one can also consider the case where V is projective and W is affine, altough it turns out that in this case the only morphisms are constant.

Remark 8.6.8. Definition 8.6.6 is pretty cumbersome to work with. Actually, a morphism $V \to \mathbb{P}^m$, where V is projective, is given by $m+1$ elements $f_0, \dots, f_m \in R(V)$ *of the same degree*, which *do not vanish simultaneously* on V.

At least one direction is easy to see: if $f_0, \dots, f_m \in R(V)$ is such a collection of polynomials, the map $f\colon V \to \mathbb{P}^m$ given by

$$f(x) = (f_0(x), \dots, f_m(x))$$

is at least well defined. (Why?) It is also clear that f restricts to a polynomial map on each affine chart, hence it defines a projective morphism.

Example 8.6.9. Let v_n be the map $\mathbb{P}^1 \to \mathbb{P}^n$ defined by the homogeneous polynomials $x_0^n, x_0^{n-1}x_1, \dots, x_1^n$. These polynomials never vanish together on $\mathbb{P}^1$, hence the map v_n is a projective morphism, called the *Veronese map.*

The image $v_n(\mathbb{P}^1)$ is a curve inside $\mathbb{P}^n$, called a *rational normal curve.* It is easy to see that its ideal is generated by equations of the form

$$y_a y_b = y_c y_d$$

whenever $a + b = c + d$, where $y_0, \dots, y_n$ are projective coordinates on $\mathbb{P}^n$. Another way to describe these equations is by requiring that the matrix

$$\begin{pmatrix} y_0 & y_1 & \cdots & y_{n-1} \\ y_1 & y_2 & \cdots & y_n \end{pmatrix}$$

has rank 1.

8.7. A new idea: the dimension

In the previous sections, we have established a dictionary that allows us to translate algebraic concepts in a geometric framework. Under this correspondence, we have associated (reduced) k-algebras with affine varieties, graded k-algebras with projective varieties and (radical) ideals with subvarieties. Moreover, we have seen how localization and completion correspond to the geometric operations of considering smaller charts around a point.

In this section, we are going to introduce a new concept from a geometric point of view. It will be the task for the next chapter to give the full picture from the algebraic side. Actually, we will state four different definitions of dimension of an algebraic variety, and give some motivation why each one of those is at least plausible. In this chapter we deal with finitely generated reduced algebras over an algebraically closed field, while in the next one, we are going to generalize these definitions to rings more general than that. In any case, the definitions that we are going to give will rely on the intuition gained here. The next chapter will show that these definitions actually agree and give the same number whenever it makes sense.

For simplicity, we state our definitions for affine varieties, although it is easy to adapt them for projective varieties with minimal modifications. Before proceeding, we need some remark on the irreducible components of varieties.

Definition 8.7.1. Let V be a topological space. We say that V is *irreducible* if V cannot be written as a union of two proper closed subset. If V is an affine or projective variety, we call it irreducible when it is so with respect to the Zariski topology.

Remark 8.7.2. Let $V \subset \mathbb{A}^n$ be an affine variety, and assume that $I(V)$ is prime. Then V is irreducible by Proposition 1.1.20 ii).

Remark 8.7.3. Any algebraic variety can be decomposed uniquely as a finite union of irreducible varieties, which are called its *irreducible components*. For an affine variety, this is the geometric translation of the existence and uniqueness of the primary decomposition.

In fact, let V be an affine variety, and decompose the ideal $I(V)$ as the intersection $Q_1 \cap \cdots \cap Q_r$, where the ideals Q_i are primary. Letting $P_i = \sqrt{Q_i}$, we can write V as the union of the varieties $V_i := V(P_i)$, where P_i is a minimal prime of $I(V)$, and the components V_i are irreducible by the previous remark.

The same result for projective varieties follows easily by considering its intersections with the affine charts C_i.

Remark 8.7.4. In fact, by the above decomposition it follows that the affine variety V is irreducible if and only if $I(V)$ is prime.

It is easy to see that the dimension is not a well-defined concept for reducible variety. In fact, different components can have different dimensions—for instance, the equations $xy = 0$ and $xz = 0$ define the union of a line $y = z = 0$ and a plane $x = 0$ in $\mathbb{A}^3$.

There are two ways around this issue. The first one is to only define dimension for irreducible varieties, and extend it to irreducible varieties by taking the maximum across the components. The other one is to define the concept of dimension of a variety around a point, with the implication that this number can depend on the point. Given a variety V and a point $p \in V$ defined by the ideal $\mathcal{M}$, we are then lead to work with the local ring $R(V)_{\mathcal{M}}$ and to define dimension for such rings. We will follow both ways in our attempts to define dimension.

First approach. Our first attempt at defining dimension is based on the idea that if there is an inclusion of *irreducible* varieties $V \subset W$, the dimension of W should be strictly greater than that of V.

Definition 8.7.5. Let V be an irreducible algebraic variety. The *Krull dimension* of V is the maximum integer n such that there exists a chain

$$V_0 \subset V_1 \subset \cdots \subset V_n = V \tag{8.7.1}$$

of irreducible varieties.

The intution behind this definition is that the existence of a chain like (8.7.1) should imply that V has dimension at least n. Conversely, if V has dimension n, there should always be a way to obtain lower-dimensional subvarieties of any intermediate dimension, for instance by taking the intersection with a suitable linear subspace.

In the ring $R(V)$, the chain (8.7.1) corresponds to a chain of prime ideals $P_n \subset P_{n-1} \subset \cdots \subset P_0$, where P_i is the ideal defining V_i. This is the notion that we will generalize to arbitrary rings.

Second approach. Let us assume that $\mathbb{A}^n$ is an object of dimension n. If we consider a variety defined by polynomials $f_1, \dots, f_r$, we would like the dimension to drop by r, provided the equations $f_i = 0$ are in some sense independent. A measure of this is given by algebraic independence: that there exists no polynomial identity satistied by $f_1, \dots, f_r$.

To make this precise, let us take an irreducible variety V, so that the ring $R(V)$ is an integral domain. Its quotient field $k(V)$ can be considered as the field of *rational functions* on V. Each time we add an independent polynomial relation, we impose an algebraic relation between elements of $k(V)$, thereby decreasing its transcendence degree by 1.

Definition 8.7.6. Let V be an irreducible affine variety. The *transcendence dimension* of V is the transcendence degree of the field $k(V)$ over k.

This definition makes sense for algebraic varieties, but is hard to generalize over arbitrary rings, since it exploits the fact that $R(V)$ is an algebra over a field.

Third approach. This is in some sense complementary to the previous one. Transcendence dimension was an attempt to define the dimension of $V \subset \mathbb{A}^n$ as $n-r$, where r is the number of independent equations needed to define V. Dually, we can consider the minimal number of equations needed in $R(V)$ to isolate a point. Since this can, a priori, change from point to point, we work inside a localization $R(V)_{\mathcal{M}}$, where $\mathcal{M}$ is the ideal of a point $p \in V$.

Definition 8.7.7. Let V be an affine variety, $p \in V$ a point defined by the ideal $\mathcal{M}$. The *Chevalley dimension* of V at p is the minimal number of elements $x_1, \dots, x_d \in R(V)_{\mathcal{M}}$ that satisfy the equality $\sqrt{(x_1, \dots, x_d)} = \mathcal{M}$.

Intuitively, to isolate the point p in V we need as many equations in $R(V)$ as the dimension of V around p. Notice that here we do not require V to be irreducible—but we will see that, when this is the case, the Chevalley dimension does not depend on the point and agrees with the previous definitions.

Fourth approach. Our last definition of dimension is the most sophisticated and least intuitive one.

To begin, let us consider the case of $\mathbb{A}^n$, and let us take a point, say $O \in \mathbb{A}^n$. Let $\mathcal{M} \subset k[x_1, \dots, x_n]$ be the ideal of polynomials vanishing at O. The local ring around O is the ring $A = k[x_1, \dots, x_n]_{\mathcal{M}}$ of rational function whose denominator is defined at O.

A way to extract the numerical invariant n from the ring A is to consider the space A_s of functions $f \in R$ vanishing of order s at O. The quotient A_s/A_{s+1} has a basis formed by all monomials of degree s in n variables, and so its dimension is of order s^n. This allows us to define the dimension of $\mathbb{A}^n$ around O as the order of growth of this dimension.

Now let V be a variety, $p \in V$. If we had something like a local chart around p, we could try to define the dimension of V around p by following the same approach: take the ideal $\mathcal{M}$ of p and consider the dimension of $\mathcal{M}^s/\mathcal{M}^{s+1}$ and its order of growth. Unfortunately, we do not work in the differentiable setting, and this implies that we do not have this luxury.

On the other hand, we can hope to have something of the sort when we pass to the completion with respect to the $\mathcal{M}$ topology (this is not quite true, but it will turn out to be true for most points). And we know by Corollary 7.5.18 that passing to the completion should not change the dimension of $\mathcal{M}^s/\mathcal{M}^{s+1}$. Hence, we are somewhat justified to give the following definition.

Definition 8.7.8. Let V be an affine variety, $p \in V$ a point defined by the ideal $\mathcal{M}$. The *Poincaré dimension* (or *Samuel dimension*) of V at p is the order of growth of the function $n(s)$, which is the dimension of the space $\mathcal{M}^s/\mathcal{M}^{s+1}$ of polynomial functions vanishing of order s at p.

At this point, it is not even clear that this is well defined. We will make this definition precise in Section 9.2. There, we will generalize this definition to arbitrary local Noetherian rings, and show that it agrees with our previous definitions.

8.8. The Zariski tangent space

Borrowing some ideas from differential geometry, we now want to define the tangent space at point of an algebraic variety. We will use the setting of smooth manifolds (say C^k, C^∞ or analytic) as a source of inspiration—the reader that is not familiar with them can consult any standard source, such as [**Hir97**], but in any case our final definitions will be independent of this discussion.

In the differentiable (or holomorphic) setting, there are many ways to introduce the tangent space. For instance, a standard way to introduce the tangent space in a point p to a smooth manifold M is to consider classes of equivalence of smooth arcs $c\colon (-1,1) \to M$ such that $c(0) = p$, modulo the equivalence relation that identifies two such arcs if their diffentials in 0 agree, in a suitable local chart around p.

In the algebraic setting we are more constrained: for instance, given a point p on an algebraic variety V, it may very well happen that the only morphism $c\colon \mathbb{A}^1 \to V$ such that $c(0) = p$ is constant. A second complication that we face is the fact that some of the varieties we consider will not be smooth, and we would like to be able to define the tangent space even in singular points.

To work around these issues, we start from the affine case. Let $V \subset \mathbb{A}^n(k)$ be an affine variety, $p \in V$ a point, $I \subset k[x_1, \dots, x_n]$ the ideal defining V. Each polynomial $f \in k[x_0, \dots, x_n]$ has a formal differential given by the vector

$$df_p = \left(\frac{\partial f}{\partial x_1}(p), \dots, \frac{\partial f}{\partial x_n}(p) \right).$$

By analogy with the implicit function theorem, one possible definition of the tangent space would be

$$T_pV = \{v \in k^n \mid df_p \cdot v = 0 \text{ for all } f \in I\}. \tag{8.8.1}$$

This is a well-defined subspace of k^n, and using the Leibniz rule it is easy to check that if $I = (f_1, \dots, f_r)$, then

$$T_pV = \{v \in k^n \mid d(f_i)_p \cdot v = 0 \text{ for } i = 1, \dots, r\}.$$

In other words, T_pV is the kernel of the matrix of the partial derivatives

$$\begin{pmatrix} \frac{\partial f_1}{\partial x_1}(p) & \cdots & \frac{\partial f_1}{\partial x_n}(p) \\ \vdots & & \vdots \\ \frac{\partial f_r}{\partial x_1}(p) & \cdots & \frac{\partial f_r}{\partial x_n}(p) \end{pmatrix}.$$

The only disadvantage of this definition is that T_pV is not intrinsic, but is defined in terms of the ambient embedding. To recover from this, we again go by analogy with the differentiable case.

Given a smooth manifold M and a point $p \in M$, one can define the cotangent space $T_p^\star M := (T_pM)^\star$. To any smooth function f defined in a neighborhood of p, one can then attach a differential form df_p, which is an element of $T_p^\star M$. Let $\mathcal{O}_pM$ be the ring of smooth functions defined in a neighborhood of p. Then $\mathcal{O}_pM$ is a local ring with maximal ideal

$$\mathcal{M}_p = \{f \in \mathcal{O}_pM \mid f(p) = 0\}$$

(why?), and there is a natural surjective map

$$d_p\colon\ \mathcal{M}_p \longrightarrow T_p^* M$$
$$f \longmapsto df_p.$$

By the Leibniz rule, it follows that d_p is identically 0 on $\mathcal{M}_p^2$, and in fact it is a simple verification that d_p induces an isomorphism of real vector spaces

$$\frac{\mathcal{M}_p}{\mathcal{M}_p^2} \cong T_p^* M.$$

This looks like something we are able to translate to the algebraic setting. Let V be an affine variety over the field k, $p \in V$ a point defined by the maximal ideal $M_p \subset R(V)$. Let $A = R(V)_{M_p}$ be the local ring of V around p, having maximal ideal $\mathcal{M}_p = M_p \cdot A$. The set $\mathcal{M}_p/\mathcal{M}_p^2$ has the structure of vector space over the field $A/\mathcal{M}_p \cong k$.

Definition 8.8.1. The vector space $\mathcal{M}_p/\mathcal{M}_p^2$ is called the *Zariski cotangent space* of V at p, denoted by $T_p^* M$. Its dual $(\mathcal{M}_p/\mathcal{M}_p^2)^*$ is called the *Zariski tangent space* of V at p, denoted by T_pM.

Notice that $\mathcal{M}_p$ is finitely generated, hence T_pM is a finite-dimensional vector space. To reconcile this intrinsic definition with the previous one, we use the following result.

Proposition 8.8.2. *Let $V \subset \mathbb{A}^n(k)$ an affine variety, $p \in V$, and T the* extrinsic *tangent space at p, defined by* (8.8.1). *Then there is a natural duality between T and $\mathcal{M}_p/\mathcal{M}_p^2$.*

Proof. Let I be the ideal of V, $R(V) = k[x_1, \ldots, x_n]/I$ and $M_p \subset R(V)$ the ideal defining p. Given a polynomial $f \in k[x_1, \ldots, x_n]$ such that $f(p) = 0$, and a vector $v \in k^n$ we can compute the number

$$\langle f, v \rangle := df_p \cdot v.$$

By definition, $\langle f, v \rangle = 0$ for $v \in T$ and $f \in I$, hence this descends to a k-bilinear function on $M_p \times T$. By the Leibniz rule, $\langle f, \cdot \rangle$ is identically 0 if $f \in M_p^2$, hence this gives a bilinear form

$$b \colon \frac{M_p}{M_p^2} \times T \to k.$$

We claim that b is a perfect pairing. One direction is easy: if $b(\cdot, v) = 0$ identically, then v is in the kernel of $d_p(x_i - p_i)$ for each coordinate function x_i, which implies that $v = 0$.

Vice versa, take $\overline{f} \in M_p$ such that $b(\overline{f}, \cdot) = 0$ identically. We can lift $\overline{f}$ to a polynomial f, which we can expand in the form

$$f = \sum_{i=1}^{n} a_i(x_i - p_i) + g_1,$$

where g_1 vanishes of order 2 at p. The condition that $b(\overline{f}, v) = 0$ for all $v \in T$ implies that there is a polynomial $h \in I$ such that

$$h = \sum_{i=1}^{n} a_i(x_i - p_i) + g_2,$$

where g_2 vanishes of order 2 at p. It follows that

$$\overline{f} = \overline{f - h} = \overline{g_1 - g_2} \in M_p^2,$$

so b is a perfect pairing of finite-dimensional k-vector spaces.

The isomorphism between M_p/M_p^2 and $\mathcal{M}_p/\mathcal{M}_p^2$ allows us to conclude. □

Remark 8.8.3. We have defined the Zariski tangent space T_pV only when V is affine. Since T_pV is defined only in terms of the local ring $R(V)_p$, we can extend the definition to projective varieties. Namely, if V is a projective variety and $p \in V$, we can take one of the standard affine charts C_i containing p, and consider the affine variety $V \cap C_i$. The local ring of $V \cap C_i$ around p does not depend on i, up to a natural isomorphism (prove this!), hence there is a well-defined Zariski tangent space T_pV even in the projective case.

Remark 8.8.4. Using the definition of Chevalley dimension 8.7.7 and Nakayama's lemma, it follows immediately that

$$\dim T_pV \geq \dim V.$$

We will make this precise and prove this in Section 10.1.

In general, we expect by analogy with the smooth case that the two dimensions agree. We isolate this condition with a

Definition 8.8.5. Let V be an algebraic variety, p a point of V. If $\dim T_pV = \dim V$, we say that V is *regular* at p, otherwise that it is *singular*. We say that V is regular (or *smooth*) if it is regular at each of its points, singular otherwise.

It is now time to look at a few examples.

Example 8.8.6.

(a) The surface S defined by

$$f(x, y, z) = xy - z^2 - 1$$

is smooth. In fact, the differential is

$$df_{(x,y,z)} = (y, x, -2z),$$

which never vanishes on a point $p \in S$. It follows that T_pS is everywhere 2-dimensional, and $\dim S = 2$.

(b) Let $V = V(f) \subset \mathbb{A}^n$ be an affine hypersurface and take a point $p \in V$. The condition that V is regular at p amounts to saying that not all partial derivatives of f vanish at p. When $p = 0$, this means that the linear part of f is not zero. Hence, we can picture any hypersurface singularity by taking a suitable polynomial whose lowest homogeneous component has degree at least 2.

(c) Consider the variety V formed by two lines in $\mathbb{A}^2(k)$ meeting at the origin—for concreteness take the x and y axes. The ideal defining the union of the two lines is $I = (xy)$, so V is singular at the origin.

(d) With the same notation, consider now the curve defined by the equation $y^2 = x^2 + x^3$. From the analytic point of view, the singularity looks the same as the previous one: for small x, x^3 becomes negligible, and the locus looks like $y^2 = x^2$, which is the union of the two diagonals. This kind of singularity is called a *node*.

(e) Consider the curve given by the equation $y^2 = x^3$. In this case, from the analytic point of view the locus has a double tangent line given by $y^2 = 0$. This kind of singularity is called a *cusp*.

The following figure depicts the planar singularities that we have described above.

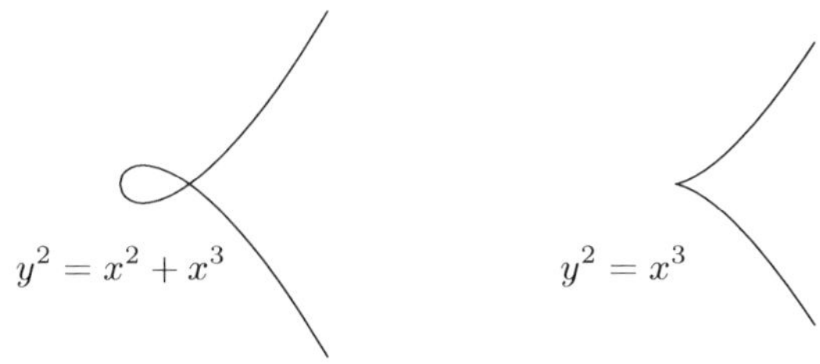

8.9. Curves and Dedekind rings

Having gained some intuition about dimension and smoothness, we now turn to the simplest varieties, apart from points: those of dimension 1, which we will call *curves*. Let $C \subset \mathbb{A}^n$ be a regular, irreducible variety of Krull dimension 1. This implies that its only Zariski-closed subsets are finite collections of points.

The ring $R(C)$ is an integral domain, because C is irreducible. Moreover, it is Noetherian, and the fact that C is a curve says that the nonzero prime ideals of $R(C)$ are maximal. As we will see in Chapter 10, the condition that C is regular ensures that $R(C)$ is integrally closed. By Theorem 5.1.19, we conclude that $R(C)$ is a Dedekind ring.

This means that in our dictionary, Dedekind rings become the algebraic counterpart to smooth, irreducible curves. Let $x \in C$ be any point, corresponding to the maximal ideal $\mathcal{M} \subset R(C)$. By our definition of Dedekind rings, the localization $R_x = R(C)_{\mathcal{M}}$ is a discrete valuation ring. In fact, let $f \in \mathcal{M} \setminus \mathcal{M}^2$, that is, a function vanishing at x of order 1—f is called a *uniformizer* at x.

Since the only ideals of R_x are powers of $\mathcal{M}$, every other function $g \in R_x$ can be written as $g = u \cdot f^k$ for some u invertible in R_x—that is—not vanishing at x. This shows that elements of R_x are determined, up to invertible elements, by their order of vanishing at x.

In the infinitesimal case, the same geometric picture holds as in the local case. Namely, let $\widehat{R_x}$ be the completion of R_x at $\mathcal{M}$. As we have discussed before, elements of $\widehat{R_x}$ can be interpreted as germs of functions in infinitesimal neighborhoods of x. By Corollary 7.5.18, the only ideals of $\widehat{R_x}$ are powers of $\widehat{\mathcal{M}}$, hence every element $g \in \widehat{R_x}$ can be written as $g = u \cdot f^k$, where $u \in \widehat{R_x}$ is unit and $f \in \widehat{\mathcal{M}} \setminus \widehat{\mathcal{M}}^2$ is a uniformizer.

We now turn to morphisms. Let $f \colon C \to D$ be a nonconstant morphism of affine smooth irreducible curves. There is a pullback map on the rings $f^* \colon R(D) \to R(C)$. We claim that f^* is injective. In fact, let $s \in R(D)$

such that $s \circ f = 0$ identically. This means that s vanishes on $f(C)$, and if s is not 0, this implies that $f(C)$ is a finite set of points. By irreducibility, it follows that f is constant on C.

The inclusion $f^* \colon R(D) \hookrightarrow R(C)$ gives $R(C)$ the structure of a $R(D)$ algebra, and clearly $R(C)$ is finitely generated as algebra. We will consider what happens when we assume in addition that $R(C)$ is an *integral* extension of $R(D)$. This happens at least in two interesting cases.

Example 8.9.1.

(a) Let C be any smooth irreducible curve. By the Noether normalization lemma 5.3.1, we can write $R(C)$ as integral extension of $k[x_1, \ldots, x_d]$ for some d—and in fact, $d = 1$ for dimensional reasons. This gives us a morphism $C \to \mathbb{A}^1$ with the desired properties.

(b) Another case where one can prove that f^* is an integral extension is when f is the restriction of a morphism of projective curves. This is a nontrivial result, and follows for instance by the Stein factorization; see [**Har77**, Corollary III.11.5].

In any case, assume that $R(C)$ is integral over $R(D)$, and let $k(C)$ be the fraction field of $R(C)$. Since we assumed that $R(C)$ is integrally closed, $R(C)$ is the integral closure of $R(D)$ in $k(C)$. In this situation, we can apply the results of Section 6.3. Let $y \in D$ be a point, corresponding to a maximal ideal P of $R(D)$. We can then factorize

$$P \cdot R(C) = Q_1^{e_1} \cdots Q_r^{e_r},$$

where $Q_1, \ldots, Q_r$ are the primes over P. Since we assumed that k is algebraically closed, the quotient fields $R(C)/Q_i$ and $R(D)/P$ are all isomorphic to k, hence $f(Q_i|P) = 1$ for all i.

Let $x_i \in C$ be the point defined by the ideal Q_i. These are exactly the preimages of y. The ramification index $e_i = e(Q_i|P)$ expresses in a precise way the multiplicity of x_i in $f^{-1}(y)$. In fact, let $g \in P \setminus P^2$ be a uniformizer at y. We can pull back g to a function $f^*(g) = g \circ f \in R(C)$, and it is easy to verify that $f^*(g) \in Q_i^{e_i} \setminus Q_i^{e_i+1}$, so $f^*(g)$ vanishes of order e_i at x_i.

In this setting, Theorem 6.3.2 simplifies to

$$\sum_{i=1}^{r} e(Q_i|P) = n,$$

where n is the degree of the field extension $[k(C) : k(D)]$. We can interpret this as saying that the number of preimages of y is constant, when they are counted with multiplicities.

The map f can be seen as the algebraic analogue of a branched covering in topology or complex analysis. Unramified primes P (if any) correspond to points which have exactly n preimages—these are the points where the covering is unramified. Over the other points, we can have multiple branches of the covering converging into one, hence we need a local notion of multiplicity to keep the number of preimages constant.

8.10. Exercises

1. Prove the following geometric version of Noether normalization lemma: any affine variety V over a field k admits a surjective map onto an affine space $\mathbb{A}^n(k)$ such that $R(V)$ is an integral extension of $k[x_1, \ldots, x_n]$. If moreover the field is infinite, a generic projection to an affine subspace of suitable dimension will do.

2. Prove the projective version of the Nullstellensatz: given a projective variety V over an algebraically closed field, points of V are in bijective correspondence with maximal homogeneous ideals of $R(V)$ *different from the irrelevant ideal* $R(V)_+$, by the correspondence

$$p \to \mathcal{M}_p = \{f \in R(V) \mid f(p) = 0\}.$$

3. Prove the claim of Remark 8.6.8.

4. Let $f \colon \mathbb{P}^1(k) \to \mathbb{P}^2(k)$ be the map given by $x_0^3, x_0x_1^2, x_1^3$ (a projection of the rational normal curve). Show that the image of f is given by the equation $y_0y_2^2 = y_1^3$. In the affine chart C_0, this is simply $y^2 = z^3$—draw it when $k = \mathbb{R}$, and check that it has a cusp at the origin.

The following exercises, up to Exercise 8, describe a result called the *combinatorial Nullstellensatz*, by Noga Alon ([**Alo99**]), and some applications of it. This is a variant of the standard Nullstellensatz: on the one hand, it works on any field, and does not require taking radicals; on the other hand it only concerns ideals generated by polynomials of a certain special form.

5. Let k be a field, $p \in k[x_1, \ldots, x_n]$ a polynomial of degree t_i in the variable x_i. Let $S_i \subset k$ be a subset of size $t_i + 1$. If p vanishes on $S_1 \times \cdots \times S_n$, then $p = 0$.

6. Use the previous exercise to prove thc *combinatorial Nullstellensatz.* Let k be a field, $p \in k[x_1, \ldots, x_n]$ a polynomial and $S_1, \ldots, S_n \subset k$ be finite, nonempty subsets. Define polynomials

$$g_i(x_i) = \prod_{s \in S_i} (x_i - s).$$

If p vanishes on $S_1 \times \cdots \times S_n$, then there are polynomials $h_i \in k[x_1, \ldots, x_n]$, satisfying $\deg h_i \leq \deg p - \deg g_i$, such that

$$p = g_1h_1 + \cdots + g_nh_n.$$

7. Use the previous exercise to prove the following result. Let k be a field, $p \in k[x_1, \ldots, x_n]$ a polynomial of total degree t, say by a monomial $x_1^{t_1} \cdots x_n^{t_n}$. Let $S_i \subset k$ be a subset of size at least $t_i + 1$. Then there exists $s \in S_1 \times \cdots \times S_n$ such that $p(s) \neq 0$.

8. Use the previous exercise to prove the *Cauchy–Davenport theorem*: let A, B be nonempty subsets of $\mathbb{Z}/p\mathbb{Z}$, for some prime p. Then

$$|A + B| \geq \min\{p, |A| + |B| - 1\}.$$

9. Let A be the coordinate ring of an affine variety, M a finitely generated module over A. For concreteness, think of the case where M is the module of sections of a vector bundle over a subvariety. Give a geometric interpretation of the associated primes of M, of its support, and a plausible explanation why the minimal primes of the two sets are the same.

10. Let k be an uncountable, algebraically closed field. Give an alternative proof of the Nullstellensatz in the form of Theorem 8.2.3, as follows. Let $\mathcal{M}$ be a maximal ideal of $k[x_1, \ldots, x_n]$ and consider the field $k' := k[x_1, \ldots, x_n]/\mathcal{M}$. Assume that k' contains an element t transcendental over k and show that the elements $1/(t - x)$ for $x \in k$ are linearly independent. Derive a contradiction and conclude that $k' = k$, from which Theorem 8.2.3 follows.

11. To generalize the result of the previous exercise to any algebraically closed field, argue as in [**CL**]. Namely, given a field k, consider the quotient of $k^{\mathbb{N}}$ by a maximal ideal $\mathcal{M}$ that contains the ideal

$$I := \left\{ (x_i) \in k^{\mathbb{N}} \mid x_i = 0 \text{ except for a finite number of } i \right\}.$$

The field k' thus obtained is called an *ultrapower* of k. Show that k' is uncountable and algebraically closed, hence the previous exercise applies. Deduce the Nullstellensatz for k.

12. Generalize the Veronese map to a map

$$v_d \colon \mathbb{P}^n \to \mathbb{P}^N,$$

where $N+1$ is the number of monomial of degree d in $n+1$ variables. Show that the image of v_d is a projective variety.

13. Show that there is well-defined map

$$s_{m,n} \colon \mathbb{P}^m \times \mathbb{P}^n \to \mathbb{P}^{(m+1)(n+1)-1}$$

which sends a pair $((x_i), (y_j))$ to the point having as coordinates all possible pairwise products $(x_i y_j)$. Moreover, $s_{m,n}$ is injective. This is called the *Segre map*.

14. Show that the image of the Segre map defined in Exercise 13 is a subvariety of $\mathbb{P}^{(m+1)(n+1)-1}$. Since $s_{m,n}$ is injective, this allows us to give the structure of a projective variety to a product of projective spaces, and by extension to a product of projective varieties.

15. Show that the restriction of the Segre map $s_{n,n}$ defined in Exercise 13 to the diagonal $\Delta \subset \mathbb{P}^n \times \mathbb{P}^n$ agrees with the Veronese map v_2 defined in Exercise 12, after having identified Δ with $\mathbb{P}^n$ in the natural way.

16. Give a coordinate-free description of the Segre map defined in Exercise 13, as a map $\mathbb{P}(V) \times \mathbb{P}(W) \to \mathbb{P}(V \otimes W)$.

17. Let $p \in \mathbb{P}^n$, and $H \subset \mathbb{P}^n$ a hyperplane not meeting p. Define the *projection* from p as the map $\pi_p \colon \mathbb{P}^n \setminus \{p\} \to H$ which sends a point q to the intersection of the line pq and H:

$$\pi_p(q) = \overline{pq} \cap H.$$

Show that in suitable coordinates, this is just the map sending $(x_0, \ldots, x_n)$ to $(x_0, \ldots, x_{n-1})$. Deduce that if $V \subset \mathbb{P}^n$ is a subvariety not meeting p, the map $\pi_p \colon V \to H$ is an algebraic map.

18. Let $V \subset P^n$ be a projective variety, $p \notin V$ a point and consider the projection map $\pi_p \colon V \to H$ of the previous exercise. Show that the image $\pi_p(V)$ is a projective subvariety. (For a point $q \in H$ show, using coordinates, that the line $l = \overline{pq}$ meets V if and only if every pair of polynomials $f, g \in I(V)$ have a common zero on l, then use resultants.)

19. A *quasiprojective variety* is a Zariski open set in a projective variety. Show that affine and projective varieties are quasiprojective, and that a quasiprojective is covered by open sets that are affine varieties. Use the last fact to give a definition of morphism between quasiprojective varieties, by reducing to the affine case.

20. Let V, W be quasiprojective varieties, as in the previous exercise, and assume that V is irreducible. A *rational map* $f \colon V \dashrightarrow W$ is an equivalence class of maps $U \to W$, where $U \subset V$ is open and dense, and two maps are equivalent if they agree on the intersection of their domains. Show that this notion is well defined and that every rational map is defined on a maximal open set. Can you give examples of rational maps that are *not* morphisms?

21. Let V be an irreducible affine variety. Show that the set of rational maps $V \dashrightarrow \mathbb{A}^1$ (see Exercise 20) can be identified with the fraction field of the coordinate ring $R(V)$.

The following exercises, up to Exercise 26, define the *Grassmann varieties*, or *Grassmannians*, which are one of the most important family of projective varieties, and study their basic properties.

22. Let V be an n-dimensional vector space over a field k. For $m \leq n$, define the set

$$G(m, V) := \{H \subset V \text{ subspace of dimension } m\}.$$

Given a subspace $H \in G(m, V)$, choose a basis $v_1, \dots, v_m$ of H. Show that the tensor

$$v_1 \wedge \cdots \wedge v_m \in \Lambda^m V$$

is determined, up to scalar multiples, by H only, hence it is a well-defined map

$$\psi \colon G(m, V) \to \mathbb{P}(\Lambda^m V),$$

which is called the *Plücker map*. Show that the Plücker map is injective. Finally, identify $G(1, V)$ with $\mathbb{P}(V)$.

23. Let V be an n-dimensional vector space, $m \leq n$ and let $\phi \in \Lambda^m V$. Show that ϕ is the product of elements of V if and only if the rank of the linear map

$$\begin{aligned} m_\phi \colon \quad V &\longrightarrow \Lambda^{m+1} V \\ v &\longmapsto \phi \wedge v \end{aligned}$$

is at most $n - m$. Deduce that the image of $G(m, V)$ via the Plücker map is a closed subvariety of $\mathbb{P}(\Lambda^m V)$, in fact a determinantal variety. By the Plücker embedding, the Grassmannians take the structure of a projective variety.

24. Let k be a field not of characteristic 2, V a finite-dimensional vector space over k. Let $\phi \in \Lambda^2 V$. Show that ϕ is the product of two vectors of V if and only if $\phi \wedge \phi = 0$. Deduce that the Grassmannian $G(2, V)$ is defined by quadratic equations.

25. Let V be a finite-dimensional vector space and consider the Grassmannian $G(m, V)$—this can be identified with the set of $m-1$-dimensional projective subspaces of $\mathbb{P}(V)$. Define the *universal family*

$$U(m, V) := \{(H, p) \mid p \in H\} \subset G(m, V) \times \mathbb{P}(V).$$

Show that a point $(H, p) \in U(m, V)$ if and only if

$$v_1 \wedge \cdots \wedge v_m \wedge w = 0,$$

where $v_1, \dots, v_m$ span H and w spans p. Deduce that $U(m, V)$ is a closed subvariety of $G(m, V) \times \mathbb{P}(V)$ (the latter inherits the structure of variety via the Plücker and Segre embeddings).

26. The universal family $U(m, V)$ described in the previous exercise has two natural morphisms given by projections, $U(m, V) \to G(m, V)$ and $U(m, V) \to \mathbb{P}(V)$. Describe the fibers of these morphisms. We have not studied the dimension formally yet, but what do you think is the dimension of $U(m, V)$? And the dimension of $G(m, V)$?

27. Consider the subset

$$B_0(\mathbb{A}^n) := \{(x, y) \mid x_i y_j = x_j y_i\} \subset \mathbb{A}^n \times \mathbb{P}^{n-1}.$$

The product $\mathbb{A}^n \times \mathbb{P}^{n-1}$ is a quasiprojective variety (Exercise 19) via the Segre embedding (Exercise 13). Show that $B_0(\mathbb{A}^n)$ is a closed subvariety of $\mathbb{A}^n \times \mathbb{P}^{n-1}$, hence it is itself a quasiprojective variety. $B_0(\mathbb{A}^n)$ is called the *blow-up*[1] of $\mathbb{A}^n$ in the origin.

28. Let $B_0(\mathbb{A}^n)$ be the blow-up of $\mathbb{A}^n$ in the origin, defined in the previous exercise, and consider the morphisms given by projection $\pi_1 \colon B_0(\mathbb{A}^n) \to \mathbb{A}^n$ and $\pi_2 \colon B_0(\mathbb{A}^n) \to \mathbb{P}^{n-1}$. Show that π_1 is bijective outside the origin, hence it admits an inverse *rational* map (Exercise 20) $\mathbb{A}^n \dashrightarrow B_0(\mathbb{A}^n)$. What is the fiber of π_1 over the origin? Show that the fibers of π_2 are all lines—can you link this to the universal family of Exercise 25?

For the next exercises, we need some definitions. Let V be an affine (resp., projective) variety of dimension k. Say $V \subset \mathbb{A}^n$ (resp., $V \subset \mathbb{P}^n$) so that its codimension is $n - k$. We say that V is a *complete intersection* if the ideal $I(V)$ is generated by $n - k$ elements (resp., $n - k$ homogeneous elements). We say that V is a *set-theoretic complete intersection* if there exists an ideal (resp., a homogeneous ideal) I generated by $n - k$ elements such that $V(I) = V$. Clearly a complete intersection is also a set-theoretic complete intersection, in general the second condition is weaker. Even if we do not have a formal definition of dimension, for the exercises assume that the rational normal curve has dimension 1, while points have dimension 0.

29. Let $V \subset \mathbb{A}^n$ be a finite set of points. Show that V is a set-theoretic complete intersection. Find an example of a finite set in $\mathbb{A}^2$ that is not a complete intersection.

30. Let $c \colon \mathbb{P}^1 \to \mathbb{P}^3$ be the rational normal curve; its image C is called the *twisted cubic*. Find explicit equations for C and show that C is not a complete intersection, but it is a set-theoretic complete intersection. Also, the intersection of C with each affine chart of $\mathbb{P}^3$ is an affine curve that *is* a complete intersection. (To prove that C is a set-theoretic complete intersection, start from 3 equations for C and combine two of them with a suitable linear combination with coefficients in $k[x_0, \ldots, x_3]$.)

[1] Actually, the term "blow up" is kind of a mistranslation. The original Italian term is *scoppiamento*, which can be translated as blow up, but in its original form is more akin to "decoupling".

31. This exercise requires some model theory. Give an alternative proof of the Ax–Grothendieck theorem as follows. First, prove it for finite fields, and deduce that the result holds for the algebraic closure of a finite field. To get the proof in general, consider the axioms for an algebraically closed field of characteristic p on the language $\{0, 1, +, \cdot\}$ of rings. Up to isomorphism, there is at most one algebraically closed field of given characteristic and cardinality. From the Löwenheim–Skolem theorem, it follows that the theory of algebraically closed fields of characteristic p is complete. Since the Ax–Grothendieck theorem is a first order statement, use Gödel's completeness theorem to prove that it holds for all algebraically closed fields of characteristic p. By a similar reasoning, derive the result in characteristic 0.

Chapter 9

Dimension Theory

In this chapter, we revisit the notion of dimension, which was introduced from a geometric point of view in Section 8.7. Here, our focus becomes more algebraic, since we want to extend these ideas to rings that do not necessarily appear as coordinate rings of algebraic varieties. In particular, in the earlier chapter we were concerned with finitely generated algebras over an algebraically closed field, while many rings of interest do not even contain a field.

We will start with some basic definitions of the dimension of rings, which we soon generalize to modules, since it is easier to develop the dimension theory of rings and modules at the same time. We then go on to prove the main results about Hilbert functions, which will be the essential technical tool to prove that all definitions we give are equivalent. Once we have proved this important result, we introduce the notion of *height* of ideals, which is a slight generalization of the Krull dimension, and use the theory developed so far to prove the famous Krull Hauptidealsatz. In the rest of the chapter, we study the nonlocal case. We derive the main properties of Krull dimension, and its behavior under quotients, polynomial extensions, and integral extensions, as well as investigate the dimension of graded rings. A good reference for the topics of this chapter and the next one is [**Ser00**].

9.1. Dimension of rings and modules

Definition 9.1.1. Let A be a ring. The *Krull dimension* of A, denoted $\dim A$, is the maximum length of a chain of prime ideals

$$P_0 \subset P_1 \subset \cdots \subset P_n$$

contained in A.

This is of course a direct generalization of Definition 8.7.5.

Remark 9.1.2. Since every prime ideal contains the nilradical, it is clear that $\dim A = \dim A/\mathcal{N}(A)$.

We have already met some examples of rings of small Krull dimension.

Example 9.1.3.

(a) A ring A has dimension 0 if and only if every prime ideal is maximal. This holds for Artinian rings by Corollary 2.4.2. The converse holds assuming that A is Noetherian, by Theorem 2.4.11

(b) Let A be an integral domain, so that 0 is prime. Then $\dim A = 1$ if and only if every prime other than 0 is maximal. This holds for Dedekind rings by Theorem 5.1.19.

The next definition of dimension will only make sense in certain cases. We introduce the class of rings where the theory works nicely.

Definition 9.1.4. Let A be a ring. We say that a is *semilocal* if A has only finitely many maximal ideals.

This class includes local rings, as well as Artinian rings, by Proposition 2.4.3.

Definition 9.1.5. Let A be a semilocal ring, $\mathcal{M}_1, \dots, \mathcal{M}_n$ its maximal ideals, and let

$$\mathcal{M} = \mathcal{M}_1 \cap \dots \cap \mathcal{M}_n$$

be its Jacobson radical. The minimum number d (if any) for which there exist $a_1, \dots, a_d \in \mathcal{M}$ such that the length

$$\ell\left(\frac{A}{(a_1, \dots, a_d)}\right)$$

is finite is called the *Chevalley dimension* of A, and denoted (temporarily) by $\delta(A)$.

This definition is in fact temporary, since we will see that this number agrees with $\dim A$.

Remark 9.1.6. Let $J = (a_1, \dots, a_d)$. Then $\ell(A/J) < \infty$ implies that the sequence $\frac{\mathcal{M}^k + J}{J}$ must be stationary, hence $\mathcal{M}^k \subset J \subset \mathcal{M}$ for some $k \gg 0$.

If A is in fact local, this implies that J is $\mathcal{M}$-primary by Proposition 3.2.6. Conversely, if A is Noetherian and local with $\mathcal{M}^k \subset J \subset \mathcal{M}$, A/J has finite length by Corollary 2.5.17.

It follows that for a local Noetherian ring A, δA is the minimum number of elements $a_1, \dots, a_d \in \mathcal{M}$ such that $\mathcal{M}^k \subset (a_1, \dots, a_d)$ for some $k \geq 1$, or in other words, the minimum number of generators of an $\mathcal{M}$-primary ideal.

The above remark should make clear the connection with Definition 8.7.7. We extrapolate the condition, since it is useful terminology.

Definition 9.1.7. Let A be a local ring with maximal ideal $\mathcal{M}$. Elements $a_1, \ldots, a_d \in \mathcal{M}$ are called a *system of parameters* for A if the ideal $(a_1, \ldots, a_d)$ is $\mathcal{M}$-primary, and moreover d is minimal, that is, $d = \delta(A)$.

We can generalize the above definitions to modules, in such a way that the theorem on dimension will be valid in this more general setting.

Definition 9.1.8. Let A be a ring, M an A-module. The *Krull dimension* of M is defined as $\dim M := \dim A/\operatorname{Ann}(M)$.

Remark 9.1.9. When talking about a module of the form $M = A/I$, there may be an ambiguity in the definition of $\dim(M)$, since it can be considered as a module over A or as a ring. In fact, there is no ambiguity, as the two definitions agree. This follows with a moment's thought from the fact that $\operatorname{Ann}(A/I) = I$.

Similarly we can generalize the Chevalley dimension.

Definition 9.1.10. Let A be a semilocal ring, and let

$$\mathcal{M} = \mathcal{M}_1 \cap \cdots \cap \mathcal{M}_n$$

be its Jacobson radical. If M is an A-module, we define the *Chevalley dimension* of M—denoted δM—as the least number d such that there exist $a_1, \ldots, a_d \in \mathcal{M}$ such that the length

$$\ell\left(\frac{M}{a_1 M + \cdots + a_d M}\right)$$

is finite.

Remark 9.1.11. If A is Noetherian and M is finitely generated, then δM is finite. In fact, if $a_1, \ldots, a_d$ is any system of parameters of A and $I = (a_1, \ldots, a_d)$, A/I is Artinian, hence $M/I \cdot M$ has finite length.

9.2. Hilbert functions

In order to make Definition 8.7.8 precise, and generalize it to a wider class of rings, we need to understand the growth of the size of the homogeneous components of graded rings and modules.

Remark 9.2.1. Let $A = \bigoplus_{i=0}^{\infty} A_n$ be a graded Noetherian ring, and $M = \bigoplus_{i=0}^{\infty} M_n$ a finitely generated graded A-module. Then each M_n is finitely generated as an A_0-module.

In fact, let $x_1, \dots, x_s$ be homogeneous generators of A as an A_0-algebra (these exist by Proposition 1.7.15). Let $m_1, \dots, m_t$ be homogeneous generators of M as an A-module, of degrees $d_1, \dots, d_t$. Then every $m \in M_n$ can be written as

$$m = y_1 m_1 + \cdots + y_t m_t,$$

where $y_i \in A_{n-d_i}$. Each y_i is a polynomial in the x_j, hence M_n is generated by the products $x_1^{a_1} \cdots x_s^{a_s} m_i$ having total degree n.

Assume now that A_0 is in fact Artinian. Then each component M_n has finite length $\ell(M_n)$ and we can put these lengths together in a generating function.

Definition 9.2.2. Let $A = \bigoplus_{i=0}^{\infty} A_n$ be a graded ring with A_0 Artinian, and let $M = \bigoplus_{i=0}^{\infty} M_n$ be a finitely generated graded A-module. The *Hilbert–Poincaré series* of M is the formal series

$$P(M,t) = \sum_{n=0}^{\infty} \ell(M_n) t^n \in \mathbb{Z}[[t]].$$

The starting point is the following observation, that was proved by Hilbert in the polynomial setting, and later extended by Serre in this generality.

Proposition 9.2.3 (Hilbert, Serre). *Let A be a graded ring, and assume that A_0 is Artinian, and let M be a finitely generated graded A-module. The Hilbert–Poincaré series $P(M,t)$ is rational, more precisely*

$$P(M,t) = \frac{f(t)}{\prod_{i=1}^{s}(1 - t^{k_i})},$$

where $f \in \mathbb{Z}[t]$ is a polynomial, and $A = A_0[x_1, \dots, x_s]$ with x_i homogeneous of degree k_i.

Proof. By induction on s. When $s = 0$, we have $A_n = 0$ for $n \geq 1$. Since M is finitely generated over A, we must have $M_n = 0$ for all large n, hence $P(M,t)$ is in fact a polynomial.

For the inductive step, consider the multiplication map

$$f_s \colon M \longrightarrow M, \qquad m \longmapsto x_s \cdot m.$$

which gives rise to an exact sequence

$$0 \longrightarrow K_n \longrightarrow M_n \longrightarrow M_{n+k_s} \longrightarrow L_{n+k_s} \longrightarrow 0,$$

where K and L are the kernel and cokernel respectively.

Both K and L are finitely generated graded A-modules, and they are annihilated by x_s, hence they are finitely generated over $A' = A[x_1, \dots, x_{s-1}]$. The additivity of the length gives

$$\ell(K_n) - \ell(M_n) + \ell(M_{n+k_s}) - \ell(L_{n+k_s}) = 0,$$

and multiplying by t^{n+k_s} and summing over n gives

$$t^{k_s} P(K,t) - t^{k_s} P(M,t) + P(M,t) - P(L,t) = g(t),$$

where $g(t) \in \mathbb{Z}[t]$ is a polynomial. We can obtain

$$P(M,t) = \frac{1}{(1-t^{k_s})}\left(-t^{k_s} P(K,t) + P(L,t) + g(t)\right)$$

and we conclude by the inductive hypothesis. □

Remark 9.2.4. The above proof goes unchanged—without assuming that A_0 is Artinian—for every integer-valued function $\lambda(M)$ over finitely generated A_0-modules having the property of being additive over exact sequences.

Definition 9.2.5. Under the hypothesis of Proposition 9.2.3, the order of pole of $P(M,t)$ at $t = 1$ is called the *Poincaré dimension* (or *Samuel dimension*) of M and denoted $d(M)$.

Corollary 9.2.6. *Let A, M as in Proposition* 9.2.3 *and assume that each x_i is homogeneous of degree* 1*. Then $\ell(M_n) = p(n)$ for all n large enough, where $p(n) \in \mathbb{Q}[n]$ is a polynomial function of degree $d-1$, and d is the Poincaré dimension of M. Moreover, the first coefficient of p can be written as $e/(d-1)!$, where e is an integer.*

Proof. Under the hypothesis we have

$$P(M,t) = \frac{f(t)}{(1-t)^s} = \frac{g(t)}{(1-t)^d}$$

after simplification, so that $g(1) \neq 0$. The generalized binomial theorem allows us to write

$$\frac{1}{(1-t)^d} = \sum_{n=0}^{\infty} \binom{d+n-1}{d-1} t^n,$$

with the convention that $\binom{-1}{-1} = 1$ and $\binom{m}{-1} = 0$ for $m \geq 0$.

Writing $g(t) = a_0 + a_1 t + \cdots + a_r t^r$, we get

$$\ell(M_n) = \sum_{k=0}^{r} a_k \binom{d+n-k-1}{d-1}$$

for all n such that $d + n - r - 1 \geq 0$, and this is a polynomial in n of degree at most d. To check that the degree is exactly d, we compute the first coefficient, which is

$$\sum_{k=0}^{r} a_k \frac{1}{(d-1)!} = \frac{g(1)}{(d-1)!} \neq 0. \qquad \square$$

Example 9.2.7. Let $A = A_0[x_1, \dots, x_s]$, where the x_i are indeterminates. Then A_n is generated by all monomials $x_1^{a_1} \cdots x_s^{a_s}$ such that $a_1 + \cdots + a_s = n$. Letting $\ell_0 = \ell(A_0)$, we have

$$A_n \cong A_0^{\binom{n+s-1}{s-1}},$$

hence $\ell(A_n) = \ell_0\binom{n+s-1}{s-1}$.

Let now A be a semilocal ring, with Jacobson radical $\mathcal{M}$, and let M be a finitely generated A-module. Any ideal I such that $\mathcal{M}^k \subset I \subset \mathcal{M}$ is called an *ideal of definition* for A. In this case, we can consider the graded ring $A' = \mathrm{Gr}_I(A)$ and the graded A'-module $M' = \mathrm{Gr}_I(M)$.

Since $A'_0 = A/I$ is Artinian, the above results apply. In particular, if $I = (x_1, \dots, x_s)$, we can write $A' = A'_0[\xi_1, \dots, \xi_s]$, where ξ_i is the class of x_i in A'.

Definition 9.2.8. The *Hilbert function* of M at I is

$$\chi_M^I(n) := \sum_{i=0}^{n} \ell(M'_n) = \ell(M/I^n M).$$

By Corollary 9.2.6, $\chi_M^I(n)$ agrees for all large n with a rational polynomial of degree d, where d is the Poincaré dimension of M. This polynomial is called the *Hilbert polynomial* of M at I.

Remark 9.2.9. As in Remark 9.1.9, one may wonder whether the definition of Poincaré dimension is ambiguous for a module of the form $M = A/I$, which can be seen as a module over A or as a ring. Since $d(M)$ is the degree of the Hilbert polynomial, which can be computed as $\ell(M/I^n M)$ for n large, and since this length does not depend on whether we see M as a module over A or over itself, the definition of Poincaré dimension agrees in the two cases.

Proposition 9.2.10. *The degree of $\chi_M^I(n)$ does not depend on the choice of I.*

Proof. If J is another ideal of definition, we have $I^a \subset J$ and $J^b \subset I$ for suitable integers a, b. It follows that

$$\chi_M^I(a(n+1)-1) \geq \chi_M^J(n),$$

and conversely

$$\chi_M^J(b(n+1)-1) \geq \chi_M^I(n).$$

This implies that the two polynomials χ_M^I and χ_M^J have the same order of growth. □

The following result has a similar proof, but in fact it is much more subtle due to the use of the Artin–Rees lemma.

Proposition 9.2.11. *Let A be a semilocal Noetherian ring, I an ideal of definition for A, and let*

$$0 \longrightarrow M' \longrightarrow M \longrightarrow M'' \longrightarrow 0$$

be a short exact sequence of finitely generated A-modules. Then $\chi_M^I - \chi_{M''}^I$ and $\chi_{M'}^I$ have the same degree and first coefficient.

Proof. It is easy to check that

$$\ell\left(\frac{M}{I^n M}\right) = \ell\left(\frac{M''}{I^n M''}\right) + \ell\left(\frac{M'}{I^n M \cap M'}\right),$$

so $\chi_M^I = \chi_{M''}^I + \psi(n)$ for

$$\psi(n) := \ell\left(\frac{M'}{I^n M \cap M'}\right).$$

By Lemma 7.5.15, we know that $I^n M \cap M'$ and $I^n M'$ have bounded difference, hence ψ is a polynomial of the same degree and first coefficient as $\chi_{M'}^I$. □

9.3. The main theorem on dimension

In this section, we are going to prove that all definitions that we have given for dimension actually agree with each other.

Theorem 9.3.1 (Dimension theorem)**.** *Let A be a semilocal Noetherian ring, M a finitely generated A-module. Then*

$$\dim(M) = \delta(M) = d(M).$$

Hence, after this section, we will not need to distinguish between Krull, Chevalley, and Poincaré dimension, and we will just talk about the dimension of a module. We will prove this theorem as a consequence of three inequalities. The first one, we prove separately in the ring case.

Proof that $d(A) \geq \dim(A)$. By induction on $d = d(A)$. Let $\mathcal{M}$ be the Jacobson radical of A. When $d = 0$, $\chi_A^{\mathcal{M}}(n) = \ell(A/\mathcal{M}^n)$ is eventually constant, hence $\mathcal{M}^n = \mathcal{M}^{n+1}$ for n big enough. By Nakayama's lemma 1.3.19, $\mathcal{M}^n = 0$. If $P \subset A$ is a prime ideal, since $0 \subset P$, it follows that $\mathcal{M} \subset P$. Writing $\mathcal{M} = \mathcal{M}_1 \cap \cdots \cap \mathcal{M}_k$, where the $\mathcal{M}_i$ are maximal, we get that $P \supset \mathcal{M}_i$ for some i by Proposition 1.1.20, so $\dim(A) = 0$.

For the induction step, we can assume that $\dim(A) > 0$. Let

$$P_0 \subsetneq P_1 \subsetneq \cdots \subsetneq P_r$$

be a chain of primes. Choose $x \in P_1 \setminus P_0$ and consider the exact sequence

$$0 \longrightarrow \frac{A}{P_0} \xrightarrow{\cdot x} \frac{A}{P_0} \longrightarrow \frac{A}{(x)+P_0} \longrightarrow 0.$$

Denote $B = A/((x) + P_0)$. Proposition 9.2.11 implies that $d(B) < d(A)$, hence we can apply the inductive hypothesis to get

$$\dim(B) \leq d(B) \leq d(A) - 1.$$

However, in B we have the chain

$$\overline{P_1} \subsetneq \overline{P_2} \subsetneq \cdots \subsetneq \overline{P_r},$$

which implies that $r - 1 \leq d(A) - 1$, that is, $r \leq d(A)$. □

Since $d(A)$ is finite, we can already state a corollary of the first part of the proof, which we will use in the sequel.

Corollary 9.3.2. *Let A be a semilocal Noetherian ring. Then* $\dim(A)$ *is finite—in particular, every* descending *chain of* prime *ideals is stationary.*

Remark 9.3.3. The Krull dimension is defined for arbitrary rings, and the above Corollary is not valid for a general ring, not even a Noetherian one. The first example of a Noetherian ring of *infinite* Krull dimension was given in [**Nag62**, Appendix, example E1], as follows.

Example 9.3.4 (Nagata)**.** Take integers m_i such that the differences $m_{i+1} - m_i$ are increasing. In the ring $A := k[x_1, x_2, \ldots]$ take the prime ideals $P_i := (x_{m_i+i}, \ldots, x_{m_{i+1}})$. Let S be the complement of the union of the P_i. Then the localization $S^{-1}A$ is Noetherian and has infinite dimension. You will prove both claims in Exercises 18-22.

Having proved the first inequality for rings, we consider the module case.

Proof that $d(M) \geq \dim(M)$**.** By Theorem 2.5.16, we find a chain

$$0 \subsetneq M_1 \subsetneq M_2 \subsetneq \cdots \subsetneq M_q = M, \tag{9.3.1}$$

where $M_{i+1}/M_i \cong A/P_i$, and $P_i \in \operatorname{Ass}(M)$ is an associated prime of M. We will prove the inequality by linking the dimension of M and that of the rings A/P_i.

To get there, let

$$0 \longrightarrow M' \longrightarrow M \longrightarrow M'' \longrightarrow 0$$

be an exact sequence of A-modules. By Proposition 3.3.3, if $P \supset \operatorname{Ann}(M)$ is a prime ideal of A, then either $P \supset \operatorname{Ann}(M')$ or $P \supset \operatorname{Ann}(M'')$. Hence if we have a chain of primes,

$$\operatorname{Ann}(M) \subset P_0 \subsetneq P_1 \subsetneq \cdots \subsetneq P_r,$$

we have either $\operatorname{Ann}(M') \subset P_0$ or $\operatorname{Ann}(M'') \subset P_0$. In conclusion,

$$\dim(M) = \max\{\dim(M'), \dim(M'')\}.$$

Similar reasoning can be done for the Poincaré dimension. By Proposition 9.2.11, $\chi_M - \chi_{M''}$ has the same leading coefficient as $\chi_{M'}$, hence either $\chi_{M'}$ or $\chi_{M''}$ has the same degree as χ_M, which implies

$$d(M) = \max\{d(M'), d(M'')\}.$$

Using (9.3.1), we obtain

$$\begin{aligned} \dim(M) &= \max\{\dim(A/P_i)\} \\ d(M) &= \max\{d(A/P_i)\}, \end{aligned}$$

and we can conclude since the A/P_i are rings. □

This is probably the hardest inequality. We do not to treat the ring case separately for the next ones.

Proof that $\dim(M) \geq \delta(M)$**.** Again, we use induction on $\dim(M)$. When $\dim(M) = 0$, we have $\dim(A/\operatorname{Ann}(M)) = 0$, which means that every prime in that ring is in fact maximal. This implies that $A/\operatorname{Ann}(M)$ is Artinian, and since M is finitely generated over $A/\operatorname{Ann}(M)$, it is an Artinian and Noetherian module. Hence M has finite length, and $\delta(M) = 0$.

For the inductive step, consider a maximal chain of prime ideals containing $\operatorname{Ann}(M)$. That is, the smallest prime ideal in the chain is a minimal prime of M. By primary decomposition for modules, M has finitely many minimal primes, say $P_1, \ldots, P_t$. Since $\dim(M) > 0$, these are not maximal.

We claim that this allows us to find some elements $x \in \mathcal{M}$ such that $x \notin P_j$ for any j, where $\mathcal{M}$ is the Jacobson radical. In fact, write $\mathcal{M} = \mathcal{M}_1 \cap \cdots \cap \mathcal{M}_k$. By Proposition 1.1.20,

$$\mathcal{M}_i \not\subset P_1 \cup \cdots \cup P_t,$$

so we can find $y_i \in \mathcal{M}_i$ such that $y_i \notin P_j$ for any j. Then, taking $x := y_1 \cdots y_k$ does the job.

If we take $M_1 = M/xM$, then $\operatorname{Ann}(M_1) \supset (x) + \operatorname{Ann}(M)$. In particular, no minimal prime P_j contains $\operatorname{Ann}(M_1)$, and a chain for M_1 must be shorter than one for M. This implies that $\dim(M_1) < \dim(M)$. Also, by construction $\delta(M) \le \delta(M_1) + 1$, since M_1 is just the quotient of M by an element of $\mathcal{M}$. By induction,

$$\delta(M) \le \delta(M_1) + 1 \le \dim(M_1) + 1 \le \dim(M). \qquad \square$$

We finally turn to the last inequality in the theorem.

Proof that $\delta(M) \ge d(M)$. Again, we do induction on $\delta(M)$. When $\delta(M) = 0$, M has finite length. In this case, the polynomial

$$\chi_M(n) = \ell\left(\frac{M}{\mathcal{M}^n M}\right) \le \ell(M)$$

is bounded, hence it must have degree 0, so $d(M) = 0$.

For the inductive step, assume $\delta(M) = s$, and take $x_1, \dots, x_s \in \mathcal{M}$ such that

$$\ell\left(\frac{M}{x_1 M + \cdots + x_s M}\right) < \infty.$$

For $i = 1, \dots, s$, call $M_i = M/(x_1 M + \cdots + x_i M)$. It is a simple verification that $\delta(M_i) = s - i$.

We are going to relate χ_M and χ_{M_1}. To do this, observe that

$$\frac{M_1}{\mathcal{M}^n M_1} \cong \frac{M}{\mathcal{M}^n M + x_1 M},$$

so

$$\ell\left(\frac{M_1}{\mathcal{M}^n M_1}\right) = \ell\left(\frac{M}{\mathcal{M}^n M + x_1 M}\right) = \ell\left(\frac{M}{\mathcal{M}^n M}\right) - \ell\left(\frac{x_1 M}{\mathcal{M}^n M \cap x_1 M}\right).$$

On the other hand, we have a surjective homomorphism

$$M \xrightarrow{\cdot x_1} x_1 M \longrightarrow \tfrac{x_1 M}{\mathcal{M}^n M \cap x_1 M},$$

whose kernel is $(\mathcal{M}^n : x_1)$. Hence

$$\frac{x_1 M}{\mathcal{M}^n M \cap x_1 M} = \frac{M}{(\mathcal{M}^n : x_1)},$$

and

$$\ell\left(\frac{M_1}{\mathcal{M}^n M_1}\right) = \ell\left(\frac{M}{\mathcal{M}^n M}\right) - \ell\left(\frac{M}{(\mathcal{M}^n : x_1)}\right).$$

Finally, the inclusion $\mathcal{M}^{n-1} \subset (\mathcal{M}^n : x_1)$ allows us to simplify this to

$$\ell\left(\frac{M_1}{\mathcal{M}^n M_1}\right) \ge \ell\left(\frac{M}{\mathcal{M}^n M}\right) - \ell\left(\frac{M}{\mathcal{M}^{n-1} M}\right).$$

This just reads $\chi_{M_1}(n) \ge \chi_M(n) - \chi_M(n-1)$, hence χ_{M_1} has smaller degree than χ_M. By induction,

$$d(M) \le d(M_1) + 1 \le \delta(M_1) + 1 = \delta(M). \qquad \square$$

9.4. Height

Let A be a Noetherian ring, $P \subset A$ a prime ideal. By Corollary 9.3.2, every descending chain of prime ideals

$$P = P_n \supset P_{n-1} \supset \cdots \supset P_0 \tag{9.4.1}$$

must terminate, and in fact has length bounded by $\dim A_P$, which is finite.

Definition 9.4.1. The maximum length of a chain as in (9.4.1) is called the *height* of the prime ideal P, denoted $\operatorname{ht}(P)$. By Proposition 1.6.6, we have $\operatorname{ht}(P) = \dim A_P$.

The height of a prime ideal is a more refined invariant than the dimension, in the sense that

$$\dim A = \sup_{\substack{P \subset A \\ \text{prime}}} \operatorname{ht}(P).$$

It turns out that this is not only finite, but we can give a precise bound in terms of the number of generators.

Theorem 9.4.2. *Let A be a Noetherian ring, $I \subset A$ an ideal and P a minimal prime of I. If $I = (a_1, \ldots, a_r)$, then $\operatorname{ht}(P) \leq r$.*

Proof. First, notice that the only prime ideal containing IA_P is PA_P. In fact, let Q be a prime of A such that $IA_P \subset QA_P$. This means that $I \subset Q \subset P$, and since P is minimal, we must have $Q = P$.

In particular, $\sqrt{IA_P} = PA_P$, which is maximal, so IA_P is PA_P-primary. This means that IA_P is an ideal of definition for A_P, so $\dim A_P = dA_P \leq r$. □

The theorem was first discovered in the following restricted form as Krull's Hauptidealsatz or Krull's principal ideal theorem.

Corollary 9.4.3 (Krull). *Let A be a Noetherian ring, $I \subset A$ a principal ideal, say $I = (x)$ where x is not a unit or a divisor of zero. Let P be a minimal prime of I. Then* $\operatorname{ht}(P) = 1$.

Proof. The fact that $\operatorname{ht}(P) \leq 1$ is a special case of the theorem. Assume that $\operatorname{ht}(P) = 0$. If $I \subsetneq A$, then P is a minimal prime of 0, and by Corollary 3.2.23, the union of the primes belonging to 0 is the set of zero divisors. □

Remark 9.4.4. The assumption that the ring is Noetherian is fundamental for Krull's Hauptidealsatz; see Exercise 11 for a counterexample in the non-Noetherian case.

We can also state a partial converse. Taken together, these results demonstrate a strong relationship between the height of ideals and the number of generators.

Proposition 9.4.5. *Let P be a prime of height r in the Noetherian ring A.*

(i) *P is the minimal prime of an ideal $I = (a_1, \dots, a_r)$.*

(ii) *If $b_1, \dots, b_s \in P$, then $\operatorname{ht}(P/(b_1, \dots, b_s)) \geq r - s$.*

(iii) *If $a_1, \dots, a_r$ are as in* i) *and $s \leq r$, then $\operatorname{ht}(P/(a_1, \dots, a_s)) = r - s$.*

Proof.

(i) A_P is a local ring of dimension r, hence it has a PA_P-primary ideal generated by r elements. Call it $IA_P = (\alpha_1, \dots, \alpha_r)$. Write $\alpha_i = \frac{a_i}{s_i}$, where $a_i \in P$. Then P is a minimal prime of $I = (a_1, \dots, a_r)$.

(ii) Let $\overline{A} := A/(b_1, \dots, b_s)$, $\overline{P} := P/(b_1, \dots, b_s)$, and denote $t = \operatorname{ht}(\overline{P})$. By the first point, $\overline{P}$ is the minimal prime of an ideal $\overline{I} = (\overline{c_1}, \dots, \overline{c_t})$. Then P is a minimal prime of $I := (b_1, \dots, b_s, c_1, \dots, c_t)$, and by Theorem 9.4.2 $\operatorname{ht}(P) \leq s + t$.

(iii) One inequality is the previous point. The converse comes from the fact that $P/(a_1, \dots, a_s)$ is a minimal prime of the ideal $(\overline{a_{s+1}}, \dots, \overline{a_r})$ in $A/(a_1, \dots, a_s)$. □

The form in which we state Krull's principal ideal theorem may look a little awkward. One may wonder whether a minimal prime of a principal ideal is itself principal, allowing to simplify the statement. In fact, this is true only in a very special case.

Proposition 9.4.6. *Let A be a ring. The following are equivalent:*

(i) *for every principal ideal $I \subset A$ and minimal prime P of I, P is also principal*

(ii) *A is a unique factorization domain.*

Proof. Assume that A is a unique factorization domain, take $a \in A$ and let $I = (a)$. Factorize $a = p_1^{e_1} \cdots p_r^{e_r}$. If P is a minimal prime of I, P must contain one of the p_i, hence $P = (p_i)$ by minimality.

For the converse, assume i). If $Q \subset A$ is a prime ideal, take any $a \in Q$ and let P be a minimal prime of A. By the hypothesis, P is principal, and we conclude the proof by Kaplansky's theorem 3.1.19. □

9.5. Properties of dimension

In this section, we take a closer look to the dimension of rings, and in particular we study how the dimension changes under various operations.

Remark 9.5.1. In a sense, we have already discussed the case of localization by a prime ideal. By definition, $\dim A_P = \operatorname{ht}(P)$, and we have already reviewed the Hauptidealsatz and its converse that allow us to control the height of ideals.

The first thing we look at is the behavior of dimension under quotients. It is not always well-behaved, but there are simple cases where we can control the drop in dimension.

Proposition 9.5.2. *Let A be a local Noetherian ring with maximal ideal $\mathcal{M}$, $x \in \mathcal{M}$, and assume that x is not a zero divisor. Then* $\dim(A/(x)) = \dim A - 1$.

Proof. We have the exact sequence

$$0 \longrightarrow (x) \longrightarrow A \longrightarrow A/(x) \longrightarrow 0$$

and we know by Proposition 9.2.11 that $\chi_A - \chi_{A/(x)}$ and $\chi_{(x)}$ have the same first coefficient.

On the other hand, since x is not a zero divisor, A and (x) are isomorphic as A-modules, which implies that $\deg \chi_{A/(x)} < \deg \chi_A$, so $dA/(x) < dA$.

The reverse inequality is clear by using the Chevalley dimension: if $a_1, \ldots, a_d \in A$ and $(\overline{a_1}, \ldots, \overline{a_d})$ is $\mathcal{M}/(x)$-primary in $A/(x)$, then $(x, a_1, \ldots, a_d)$ is $\mathcal{M}$-primary in A, hence $\delta(A) \leq \delta(A/(x)) + 1$. □

Another useful property is that the dimension does not change under completion, as the geometric intuition suggests.

Proposition 9.5.3. *Let $A, \mathcal{M}$ be a local Noetherian ring, $\widehat{A}$ its completion in the $\mathcal{M}$-adic topology. Then* $\dim A = \dim \widehat{A}$.

Proof. This follows at once from Corollary 7.5.18, since A and $\widehat{A}$ have the same associated graded ring, hence the same Hilbert polynomial. □

Next, we take on studying the dimension of polynomial rings. For that, we need a lemma, which is a useful result of its own.

Lemma 9.5.4. *Let $f \colon A \to B$ be a homomorphism of Noetherian rings. Take a prime $Q \subset B$ and let $P = f^{-1}Q$ be the contraction. Then*

$$\operatorname{ht}(Q) \leq \operatorname{ht}(P) + \dim \frac{B_Q}{PB_Q}.$$

Proof. By localization, we can assume that A is local with maximal ideal P and B is local with maximal ideal Q, in which case the thesis reads

$$\dim B \leq \dim A + \dim \frac{B}{PB}.$$

Let $r = \dim A$, and $s = \dim B/(PB)$. Then by Proposition 9.4.5, P is minimal over an ideal $I = (a_1, \ldots, a_r)$ and $Q/(PB)$ is minimal over an ideal $(\overline{b_1}, \ldots, \overline{b_s})$. The quotient $Q/(PB + (\overline{b_1}, \ldots, \overline{b_s}))$ is both maximal and minimal inside $B/(PB + ((\overline{b_1}, \ldots, \overline{b_s}))$, hence it is nilpotent. (Why?) It follows that

$$Q^n \subset PB + (b_1, \ldots, b_s)$$

for n big enough. By a similar reasoning,

$$P^m \subset (a_1, \ldots, a_r)$$

for m big enough, and so

$$Q^{m+n} \subset J := (f(a_1), \ldots, f(a_r), b_1, \ldots, b_s).$$

This implies that Q is a minimal prime of J, and the thesis follows by Theorem 9.4.2. □

This allows us to study the relation between a ring and its polynomial extension. For Noetherian rings, we have a very neat formula.

Proposition 9.5.5. *If A is a Noetherian ring,* $\dim A[x] = \dim A + 1$.

Proof. The inequality $\dim A[x] \geq \dim A + 1$ is easy. If

$$P_0 \subsetneq P_1 \subsetneq \cdots \subsetneq P_d$$

is a chain of primes in A, we get ideals $P_i[x] \subset A[x]$. These are primes, since $A/P_i[x]$ is a domain, hence $\dim A[x] \geq \dim A$. Moreover, we have a strict inequality, since $P_d[x]$ is not maximal, as $A/P_d[x]$ is not a field.

For the converse, let Q be a maximal ideal of $A[x]$, and $P = Q \cap A$. By Lemma 9.5.4,

$$\operatorname{ht}(Q) \leq \operatorname{ht}(P) + \dim \frac{A[x]_Q}{PA[x]_Q},$$

hence it is enough to prove that the last term is at most 1. Now

$$\left(\frac{A}{P}[x]\right)_Q = k[x]_Q,$$

where k is the fraction field of A/P. But $\dim k[x] = 1$, and localization at Q can only lower the dimension. □

The same proof also gives the following consequence (Exercise 7).

Proposition 9.5.6. *If A is Noetherian ring,* $\dim A[[x]] = \dim A + 1$.

Corollary 9.5.7. *If k is a field,* $\dim k[x_1, \ldots, x_n] = n$.

Proposition 9.5.5 does not hold for rings that are not Noetherian, but still we can state a partial result.

Proposition 9.5.8. *Let A be a ring. Then*

$$\dim A + 1 \leq \dim A[x] \leq 2 \dim A + 1.$$

Proof. The first inequality follows as in Proposition 9.5.5, as its proof does not use the Noetherian hypothesis. The second one follows once we prove that there are no three prime ideals $Q_1 \subsetneq Q_2 \subsetneq Q_3 \subset A[x]$ having the same restriction to A. In fact, any chain of prime ideals in $A[x]$ can have at most twice the length of a chain in A.

To prove the claim, assume

$$Q_1 \cap A = Q_2 \cap A = Q_3 \cap A = P.$$

By taking the quotient at P, we can assume that A is domain and $P = 0$. Localizing at Q_3 we get a chain of three prime ideals inside $A[x]_{Q_3}$. This ring is a further localization of $k[x]$, where $k = \mathcal{F}(A)$ is the fraction field of A. This is a contradiction, since $\dim k[x] = 1$. □

In fact, this is not a limitation of our techniques: all intermediate dimensions can actually appear, as shown in [**Sei53**]. We only give a simple example.

Example 9.5.9. [**Sla**] Let t, y be two indeterminates, k a field, and take the ring

$$A := \{f(y) \in k(t)[[y]] \mid f(0) \in k\}.$$

The ideal

$$P := \{f(y) \in k(t)[[y]] \mid f(0) = 0\} \subset A$$

is prime, and $A/P \cong k$ by evaluation, so in fact P is maximal. In fact, P is the only nontrivial prime ideal of A (why?), so $\dim A = 1$.

On the other hand, $\dim A[x] = 3$. To see this, consider the map

$$\begin{aligned} ev_t\colon\ A[x] &\longrightarrow k(t)[[y]] \\ p(x) &\longmapsto p(t). \end{aligned}$$

Since, $k(t)[[y]]$ is a domain, $Q = \ker ev_t$ is a prime ideal, and clearly $yx - yt \in Q$, so $Q \neq 0$. On the other hand, $Q \subset P[x]$, and the inclusion is strict since $y \in P$, but $y \notin Q$. This show that we have the chain of 4 prime ideals

$$0 \subsetneq Q \subsetneq P[x] \subsetneq \mathcal{M},$$

where $\mathcal{M}$ is any maximal ideal containing $P[x]$. Hence, $\dim A[x] \geq 3$, and the reverse inequality is Proposition 9.5.8.

Now that the picture for polynomial extensions is clear, we turn to the opposite case: the dimension of integral extensions. Here things are much simpler. In fact, the going-up theorem 5.2.6 immediately gives:

Proposition 9.5.10. *Let $A \subset B$ be rings, with B integral over A. Then* $\dim A = \dim B$.

Putting the two results together, we are finally able to conclude that our definition of transcendence dimension 8.7.6 also agrees with the other definitions we have given. The precise statement is the following.

Theorem 9.5.11. *Let k be a field and A a finitely generated k-algebra which is an integral domain. Let $K = \mathcal{F}(A)$ be the fraction field of A. Then the transcendence degree of K over k equals* $\dim A$.

Proof. Let s be the transcendence degree of K over k. By Noether normalization lemma 5.3.1, we can find algebraically independent $a_1, \dots, a_s \in A$ such that A is integral over $k[a_1, \dots, a_s]$. But $\dim k[a_1, \dots, a_s] = s$, and integral extensions preserve the dimension. □

9.6. Dimension of graded rings

If we track our definition of Poincaré dimension, it starts by considering the lengths of homogeneous components in a graded ring A, where A_0 is Artinian. In turn, we were led to consider the special case of the graded ring associated to a local Noetherian ring.

Now that the picture for local Noetherian rings is complete, thanks to Theorem 9.3.1, it makes sense to go back to investigate graded rings. In particular, one may wonder whether the dimension of graded rings can be expressed in terms of its Hilbert polynomial, and whether the dimension of a local Noetherian ring and its associated graded ring are related. As it turns out, both things are true.

First, we establish a few results that allow us to only consider homogeneous ideals when dealing with the Krull dimension of graded rings.

Proposition 9.6.1. *Let A be a graded ring, $I \subset A$ a homogeneous ideal, and P a minimal prime of I. Then P is homogeneous.*

Proof. We can write $P = \operatorname{Ann}(x)$ for some $x \in A/I$. Since A/I is a graded ring, we can decompose $x = x_0 + x_1 + \cdots + x_d$ into its homogeneous components. Choose any $a \in P$, and again decompose $a = a_0 + a_1 + \cdots + a_r$. The condition $ax = 0$ becomes

$$\begin{aligned} a_0 x_0 &= 0 \\ a_1 x_0 + a_0 x_1 &= 0 \\ a_2 x_0 + a_1 x_1 + a_0 x_2 &= 0 \\ &\vdots \end{aligned}$$

By multiplying each of these in turn by a_0, we recursively get $a_0^{k+1} x_k = 0$. In particular, $a_0^{d+1} \in \operatorname{Ann}(x) = P$, so in fact $a_1 + \cdots + a_r \in P$. We can then repeat the reasoning for all a_i to find that all of them lie in P. □

Proposition 9.6.2. *Let A be a graded ring, and $P \subset A$ a homogeneous prime of height r. Then, there exists a chain*

$$P_0 \subsetneq P_1 \subsetneq \cdots \subsetneq P_r = P$$

composed by homogeneous *primes.*

Proof. Let P_0 be the smallest element in a maximal chain of primes ending at P. Then P_0 is a minimal prime of 0, and by Proposition 9.6.1 it is homogeneous. By taking the quotient with respect to P_0, we can prove the result assuming that A is an integral domain.

Take any homogeneous $a \in P$, $a \neq 0$. By Proposition 9.4.5, $\mathrm{ht}(P/a) \geq r-1$. Since A is an integral domain, any maximal chain ending in P has to start from 0, so in fact $\mathrm{ht}(P/a) = r-1$. This gives us a chain

$$P_1 \subsetneq \cdots \subsetneq P_r = P,$$

where P_1 is a minimal prime over a. By Proposition 9.6.1, P_1 is homogeneous and we can then finish the proof by induction. $\square$

Having established these basic results, we turn to the relationship between a local ring and its associated graded ring.

Theorem 9.6.3. *Let k be a field, and consider a graded ring of the form $A = k[x_1, \ldots, x_r]$, where the x_i have degree 1. Let $M = (x_1, \ldots, x_r)$, and consider the localization A_M, with maximal ideal $\mathcal{M} = MA_M$.*

(i) *Let*

$$\phi(n) = \ell(A_n) \text{ (for } n \text{ large enough)}$$

be the Hilbert polynomial of A and

$$\chi(n) = \ell\left(\frac{A_M}{\mathcal{M}^n}\right) \text{ (for } n \text{ large enough)}$$

be the Hilbert polynomial of A_M. Then $\phi(n) = \chi(n) - \chi(n-1)$.

(ii) *The dimensions are equal, in fact*

$$\dim A = \dim A_M = \deg \phi + 1.$$

(iii) *As graded rings,*

$$A \cong \mathrm{Gr}_{\mathcal{M}}(A_M).$$

Proof. (i) There is an isomorphism

$$A_n \cong \frac{M^n}{M^{n+1}} \cong \frac{\mathcal{M}^n}{\mathcal{M}^{n+1}},$$

hence

$$\phi(n) = \ell(A_n) = \ell\left(\frac{\mathcal{M}^n}{\mathcal{M}^{n+1}}\right) = \chi(n) - \chi(n-1).$$

(ii) By the previous point,

$$\deg \phi + 1 = \deg \chi = \dim A_M.$$

On the other hand, since A_M is local, $\dim A_M = \text{ht}(\mathcal{M}) = \text{ht}(M)$. The problem is, a priori A may have other maximal ideals of greater height.

First, assume that A is an integral domain, and let t be its transcendence degree over k. We know that $t = \dim A$ by Theorem 9.5.11. It is not restrictive to assume that in fact $x_1, \dots, x_t$ are algebraically independent over k. Then, the monomials of a given degree n in $x_1, \dots, x_t$ are linearly independent over k. This implies

$$\ell(A_n) \geq \binom{n+t-1}{t-1},$$

hence $\deg \phi \geq t - 1 = \dim A - 1$, which is what we need.

In the general case, take any maximal chain of primes in A, terminating at P_0. Then P_0 is a minimal prime of A, and by Proposition 9.6.1 it is homogeneous. Then the inequalities

$$\dim A = \dim \frac{A}{P_0} = \text{ht}\, \frac{M}{P_0} \leq \text{ht}\, M \leq \dim A$$

show that $\dim A = \text{ht}\, M = \dim A_M$.

(iii) This follows from by verifying that the group isomorphism

$$A \cong \bigoplus_n A_n \cong \bigoplus \tfrac{\mathcal{M}^n}{\mathcal{M}^{n+1}} \cong \text{Gr}_{\mathcal{M}}(A_M)$$

is in fact a ring homomorphism. □

Corollary 9.6.4. *Let A be a local Noetherian ring with maximal ideal $\mathcal{M}$, $k = A/\mathcal{M}$. Then*

$$\dim A = \dim \text{Gr}_{\mathcal{M}}(A).$$

Proof. Just apply the previous theorem by taking as x_i the images in $\mathcal{M}/\mathcal{M}^2$ of a finite set of generators of $\mathcal{M}$. □

9.7. Exercises

1. Let k be a field and $A = k[x_1, \dots, x_s]$. Let $f \in A$ be a polynomial of degree d. Prove that for n large enough we have

$$\ell(A_n) = \dim A_n = \binom{n+s}{s} - \binom{n-d+s}{s},$$

and that this is a polynomial of degree $s - 1$, with leading coefficient $\frac{d}{(s-1)!}$.

2. Find an explicit expression for the Hilbert–Poincaré series of the ring $k[x_1, x_2^2, \dots, x_n^n]$ (as a module over itself) as a rational function.

The following exercises, up to Exercise 6, follow [**Car80**] to give an alternative proof of Krull's principal ideal theorem that does not rely on the theory developed in this chapter, and especially avoids the usage of the Artin–Rees lemma.

3. Let A be an integral domain, $a, b \in A$ nonzero. Show that there is an isomorphism $((a) : (b)) \cong ((b) : (a))$, and that this induces an isomorphism

$$\frac{((a) : (b))}{(a)} \cong \frac{((b) : (a))}{(b)}.$$

4. Let A be an integral domain, $a, b \in A$ nonzero. Show that there is an isomorphism

$$\frac{A}{((a) : (b))} \cong \frac{(a, b)}{(a)}.$$

5. Let A be a Noetherian local domain with maximal ideal $\mathcal{M}$, and assume that there exists $x \in A$ such that (x) is $\mathcal{M}$-primary. Prove that $\operatorname{ht}(\mathcal{M}) = 1$. (Assume $0 \subsetneq P \subsetneq \mathcal{M}$ with P prime and take a nonzero $z \in P$. The chain $((z) : (x^n))$ is eventually stationary. Use the previous two exercises to deduce that $\ell(A/(x^n, z))$ is eventually constant, and so (x^n, z) is also stationary. From this, find the contradiction that $x \in P$.)

6. Reduce Krull's principal ideal theorem to the local version proved in the previous exercise.

7. Prove in detail Proposition 9.5.6—in particular you will need to prove that $\dim k[[x]] = 1$.

8. Give an alternative proof of Proposition 9.5.6 by reducing to the local case and then showing that for a Noetherian local ring A, the ring $A[[x]]$ is also local and the Chevalley dimension satisfies $\delta(A[[x]]) = \delta(A) + 1$.

9. Show that $\mathbb{R}$ is not the field of fractions of a Noetherian ring other than itself. (By localization at a suitable prime and the Krull–Akizuki theorem, you can assume that this ring is integrally closed.)

10. Let A_n be the valuation ring with value group $(\mathbb{Z}^n, \mathrm{LEX})$ constructed in Exercise 27 of Chapter 7. Show that $\dim A_n = n$, and in fact all prime ideals of A_n fit into a chain

$$P_0 \subsetneq P_1 \subsetneq \cdots \subsetneq P_n.$$

11. With the notation of the previous exercise, find a principal ideal $I \subset A_n$ and a minimal prime P of I having $\operatorname{ht}(P) = n$. This shows a counterexample for Krull's Hauptidealsatz 9.4.3 when the ring is not Noetherian. Why this does not contradict Proposition 2.2.12?

12. Let A be Noetherian local ring of dimension 2. Show that A admits infinitely many primes of height 1. Deduce the more general case: if A is a Noetherian ring of finite dimension d, and there exist finitely many primes of height k, then either $k = 0$ or $k = d$.

13. Let A be a valuation ring. Using the previous exercise and Exercise 25 in Chapter 7, show that if $\dim A \geq 2$, then A is not Noetherian

14. Lest the previous exercise seduces you into thinking that valuation rings of dimension 1 are Noetherian, here is a counterexample from [**Knab**]. Consider the rings

$$k[x] \subset k[x^{1/2}] \subset \cdots \subset k[x^{1/2^k}] \subset \cdots.$$

Abstractly, each of these rings is isomorphic to $k[x]$, hence if we let $P_k = (x^{1/2^k}) \subset k[x^{1/2^k}]$, the localization

$$A_k := k[x^{1/2^k}]_{P_k}$$

is a DVR. Show that there are natural inclusions $A_k \subset A_{k+1}$ and that the union $A := \bigcup_{k=0}^{\infty} A_k$ is a valuation ring having a value group isomorphic to the dyadic rationals

$$\left\{ \frac{a}{2^k} \in \mathbb{Q} \mid a \in \mathbb{Z}, b \in \mathbb{N} \right\}.$$

Conclude that A is a non-Noetherian valuation domain of dimension 1.

15. Let A be a Noetherian ring, and consider a chain of prime ideals

$$P_0 \subsetneq P_1 \subsetneq \cdots \subsetneq P_n.$$

Given any $a \in P_n$, show that we can find a different chain

$$P_0' \subsetneq P_1' \subsetneq \cdots \subsetneq P_{n-1}' \subsetneq P_n$$

ending at P_n, such that $a \in P_1$.

16. Let S be the multiplicative set generated by x and the cyclotomic polynomials. Prove that $S^{-1}\mathbb{Z}[x]$ is a principal ideal domain.

17. Let A be a principal ideal domain, so $\dim A = 1$. Show that a maximal prime ideal of $A[x]$ can have height at most 2 by Krull's Hauptidealsatz, and conclude that $\dim A[x] = 2$. This gives a simple proof of Proposition 9.5.5 for the case of a PID.

The next Exercises, up to 22, go in detail in Nagata's example of a Noetherian ring of infinite dimension. We use the notation of Example 9.3.4. Our presentation is taken from [**gne**]. Recall the notation: $A = k[x_1, x_2, \ldots]$, m_i is an increasing sequence of integers such that $m_{i+1} - m_i$ is increasing, $P_i = (x_{m_i+1}, \ldots, x_{m_{i+1}})$ and $S = A \setminus \bigcup_i P_i$.

18. Prove that $\mathrm{ht}(P_i) = m_{i+1} - m_i$, and deduce that $S^{-1}A$ has infinite dimension.

19. Prove that the ideals of the form $S^{-1}P_i$ are maximal in $S^{-1}A$, and that every nonzero $a \in S^{-1}A$ is contained in finitely many ideals of the form $S^{-1}P_i$.

20. Let $I \subset A$ an ideal. Show that if $I \subset \bigcup_i P_i$, then $I \subset P_i$ for some i. Deduce that the ideals of the form $S^{-1}P_i$ are the only maximal ideals in $S^{-1}A$. (Find a way to avoid cancellation of monomials)

21. Prove that the localizations $(S^{-1}A)_{S^{-1}P_i}$ are Noetherian.

22. Let A be a ring, and assume that for each maximal ideal $\mathcal{M} \subset A$, the localization $A_{\mathcal{M}}$ is Noetherian. Assume further that each nonzero $a \in A$ is contained in finitely many maximal ideals. Show that A is Noetherian. Use this criterion, together with the previous exercises, to show that the ring $S^{-1}A$ in Nagata's example is Noetherian.

23. Use the converse Hauptidealsatz Proposition 9.4.5 to give an alternative proof of Proposition 3.4.9.

24. Let A be a Noetherian ring, $B = A[x_1, \ldots, x_r]$, Q a prime of B and $P = Q \cap A$. Prove that

$$\operatorname{ht}(Q) = \operatorname{ht}(P) + \dim \frac{B_Q}{PB_Q}.$$

25. Prove *Nagata's height formula*: Let $A \subset B$ be Noetherian integral domains, with B finitely generated as an A-algebra, Q a prime of B and $P = Q \cap A$. Let $k(P) = A_P/PA_P$ be the residue field at P and $k(Q) = B_Q/QB_Q$ that at Q. Then

$$\operatorname{ht}(Q) \leq \operatorname{ht}(P) + \operatorname{trdeg}_{\mathcal{F}(A)} \mathcal{F}(B) - \operatorname{trdeg}_{k(P)} k(Q),$$

where trdeg denotes the transcendence degree of a field extension. (By induction, one can assume that B is generated over A by a single element b. Distinguish the cases where b is transcendental or algebraic over A, and use the previous exercise.)

Chapter 10

Local Structure

In this chapter, we investigate in more detail the local structure of rings—in particular, we study the condition of regularity and the related notion of multiplicity. Regular rings are the algebraic counterpart to smooth algebraic varieties. Smoothness for an algebraic variety is measured by looking at the dimension of its Zariski tangent space—this is at least the dimension of the variety, and equality happens in the smooth case. By Nakayama's lemma, the dimension of the Zariski tangent space at a point $x \in V$ is the same as the minimal number of generators of the maximal ideal of the local ring $R(V)_x$. This notion can readily be generalized to a Noetherian local ring, giving rise to the concept of regular ring.

In the first section, we study the elementary properties of regular rings—we show that from an analytic point of view they are all very similar, and that a regular ring is necessarily integral, which translates to the fact that the union of two algebraic varieties is singular along the intersection. Many more results are known for regular local rings—in particular, they are unique factorization domains—but we are only able to prove the simplest of them, since the most important results require homological techniques and will be proved in the sequel of this book.

Next, we define the notion of multiplicity of a local ring. This is, in some sense, a measure of how much the ring fails to be regular, or how complex a singularity is. In fact, regular rings have multiplicity 1, and the converse is true under some additional hypotheses. The latter implication is not easy, though, and a good part of the chapter builds enough theory to prove at least a special case.

We can also define the multiplicity of a finitely generated module, although the ring case is the most interesting one. The multiplicity of a local ring is related to the degree of the associated graded ring via the *tangent cone construction.*

In the next sections, we develop various results around multiplicity, in particular, the very useful *additivity formula.* We also study the behavior of the multiplicity of rings of the form $A/(a)$, where A is a fixed local ring and $a \in A$ varies. This is a way to express the notion of order of vanishing of the element a, and, in some cases, it can be used to define a valuation on the ring A.

We end the chapter with the famous structure theorem of Cohen, which gives a precise description of complete local Noetherian rings.

10.1. Regular rings

Let A be a local Noetherian ring with maximal ideal $\mathcal{M}$. From the previous chapter we know that A has finite dimension d, and there is an $\mathcal{M}$-primary ideal Q generated by elements $a_1, \ldots, a_d$. In general, though, we cannot just take $Q = \mathcal{M}$.

Definition 10.1.1. Let $A, \mathcal{M}$ be a local Noetherian ring, $a_1, \ldots, a_d$ a system of parameters. We say that $a_1, \ldots, a_d$ is a *regular* system of parameters if $(a_1, \ldots, a_d) = \mathcal{M}$. If $\mathcal{M}$ admits a regular system of parameters, then we say that A is *regular.*

Definition 10.1.2. Let A be a Noetherian ring. We say that A is *regular* if A_P is a regular local ring for all primes $P \subset A$.

Remark 10.1.3. In the definition we require that A is Noetherian, so that by Corollary 9.3.2 we know that A_P has finite dimension for all primes P.

Remark 10.1.4. Let $A, \mathcal{M}$ be a local Noetherian ring. The minimum number of generators of $\mathcal{M}$ is called the *embedding dimension* of A, denoted $\operatorname{embdim} A$. By definition, we have

$$\operatorname{embdim} A \geq \dim A,$$

and A is regular exactly when the above is an equality.

Before giving examples, we can look at the definition from a different angle.

Remark 10.1.5. Let $a_1, \ldots, a_d$ be any system of parameters for $A, \mathcal{M}$. The quotient module $\mathcal{M}/\mathcal{M}^2$ is a vector space over $k := A/\mathcal{M}$. By Nakayama's lemma 1.3.19, $a_1, \ldots, a_d$ is regular if and only if the images $\overline{a_1}, \ldots, \overline{a_d}$ are linearly independent over k.

The above remark makes the geometric meaning of regularity more apparent. As discussed in Section 8.8, in the geometric case this notion corresponds to a nonsingular variety. To be more specific, take a variety V and a point $p \in V$, corresponding to a maximal ideal $\mathcal{M} \subset R(V)$. Assume that V has dimension d in p—then elements $a_1, \dots, a_d \in R(V)_{\mathcal{M}}$ form a regular system of parameters if and only if their linear components are independent over k. This can be guaranteed exactly when the Zariski tangent space $T_p V$ has dimension d. In general, the dimension of $T_p V$ is the embedding dimension of the ring $R(V)_{\mathcal{M}}$.

The name embedding dimension comes from the remark that V cannot be embedded in $\mathbb{A}^n$ for $n < \operatorname{embdim} R(V)_{\mathcal{M}}$, since its tangent space is a subspace of k^n, so the embedding dimension is a lower bound for the dimension of an affine space in which V can be embedded. Notice that in any case this bound is not at all sharp, as the example of a regular variety should immediately show.

Example 10.1.6.

(a) Let A be a local ring of dimension 0. Then A is regular if and only if it is a field.

(b) More generally, let k be a field, $A = k[[x_1, \dots, x_d]]$. Then A is local with maximal ideal $(x_1, \dots, x_d)$. Since $\dim A = d$, A is regular of dimension d. This is our prototypical example: we will prove with Cohen's theorem that the completion of a regular local rings has this form whenever it has the same characteristic as its residue field.

(c) Let $A, \mathcal{M}$ be a regular local ring of dimension d. Then the completion $\widehat{A}$ has dimension d, and generators for $\mathcal{M}$ map to generators of $\widehat{\mathcal{M}}$. This implies that $\widehat{A}$ is regular as well.

(d) Let $A, \mathcal{M}$ be a regular local ring of dimension 1. Then $\mathcal{M}$ has a single generator a. It turns out (we will not prove this here) that a regular local ring is a UFD, and in particular integrally closed. By Proposition 7.3.2, it follows that A is a DVR. You will prove this directly in Exercise 1.

(e) Vice versa, every DVR has dimension 1, and its maximal ideal is principal by Proposition 7.3.1. Hence local regular rings of dimension 1 are the same as discrete valuation rings. This gives a lot of examples of regular rings, such as the p-adic integers $\mathbb{Z}_p$.

(f) For a nongeometric example, take $A = \mathbb{Z}[x]$ and the maximal ideal $\mathcal{M} = (p, x)$ for a prime p. Then $A_{\mathcal{M}}$ has dimension 2, and its maximal ideal is generated by two elements, so $A_{\mathcal{M}}$ is a regular local ring.

Remark 10.1.5 helps understanding when a quotient of a regular ring remains regular.

Proposition 10.1.7. *Let A be a regular local ring, with maximal ideal $\mathcal{M}$. Elements $a_1, \ldots, a_i$ are a subset of a regular system of parameters if and only if $A/(a_1, \ldots, a_i)$ is regular.*

Proof. Let $I = (a_1, \ldots, a_i)$. Notice that $\dim A/I = d - i$ by Proposition 9.4.5. Assume that $a_1, \ldots, a_d$ is a regular system of parameters. Then $\overline{a_{i+1}}, \ldots, \overline{a_d}$ are a regular system of parameters for the quotient A/I, because their images remain linearly independent over $k = A/\mathcal{M}$. Conversely, any regular system of parameters for A/I lifts to a regular system of parameters for A together with $a_1, \ldots, a_i$. □

From an analytic point of view, regular local rings have a very simple structure. This is expected, since they correspond to nonsingular points on a variety. In differential geometry, the inverse function theorem ensures that the nonsingular points of the zero locus of finitely many C^∞ functions admits a local chart, hence the local structure around all nonsingular points looks the same. We have an analogous statement in the algebraic setting.

Proposition 10.1.8. *Let $A, \mathcal{M}$ be a regular local ring of dimension d. Then, as graded rings,*

$$\mathrm{Gr}_{\mathcal{M}}(A) \cong k[x_1, \ldots, x_d],$$

where $k = A/\mathcal{M}$.

Proof. Let $a_1, \ldots, a_d$ be a regular system of parameters. There is a surjective homomorphism of graded rings $\phi\colon k[x_1, \ldots, x_d] \to \mathrm{Gr}_{\mathcal{M}}(A)$, so

$$\mathrm{Gr}_{\mathcal{M}}(A) \cong \frac{k[x_1, \ldots, x_d]}{I}$$

for some homogeneous ideal I. Assuming $I \neq 0$, take a homogeneous $f \in I$, say of degree r. Then every multiple of f lies in I, so we can bound the length

$$\ell(\mathrm{Gr}_{\mathcal{M}}(A)_n) \leq \binom{n+d-1}{d-1} - \binom{n+d-r-1}{d-1}$$

for n big enough. But the right-hand side is a polynomial of degree $d-1$, while Theorem 9.6.3 guarantees that $\dim \mathrm{Gr}_{\mathcal{M}}(A) = d$. □

This result has an important corollary.

Theorem 10.1.9. *A regular local ring is an integral domain.*

Proof. Let $A, \mathcal{M}$ be a regular local ring, and take some nonzero $a, b \in A$. By Krull's intersection theorem 7.5.24, $\bigcap_{n \in \mathbb{N}} \mathcal{M}^n = 0$, hence $a \in \mathcal{M}^r \setminus \mathcal{M}^{r+1}$ and $b \in \mathcal{M}^s \setminus \mathcal{M}^{s+1}$ for some $r, s \in \mathbb{N}$. This implies that $\overline{a} \in \mathrm{Gr}_{\mathcal{M}}(A)_r$

and $\overline{b} \in \mathrm{Gr}_{\mathcal{M}}(A)_s$ are not zero. By the above Proposition, $\mathrm{Gr}_{\mathcal{M}}(A)$ is a domain—in particular $\overline{ab} \neq 0$, which implies that $ab \neq 0$. □

In the nonlocal case, we cannot hope to have such a result: a variety made by multiple smooth, not intersecting components has a coordinate ring that is regular but not integral. This is essentially the only thing that can go wrong:

Proposition 10.1.10. *Let A be a regular ring. Then A is a finite direct sum of integral domains.*

Proof. Let $P_1, \ldots, P_r$ be the minimal primes of A, so that

$$\mathcal{N}(A) = \sqrt{0} = P_1 \cap \cdots \cap P_r.$$

Since A is Noetherian, $\mathcal{N}(A)$ is finitely generated, so $\mathcal{N}(A)^n = 0$ for n big enough.

Let $\mathcal{M}$ be a maximal ideal of A. Then $A_{\mathcal{M}}$ is a regular local ring, in particular it is a domain. The only minimal prime of $A_{\mathcal{M}}$ is 0, hence $\mathcal{M}$ contains exactly one of the P_i by Corollary 1.6.9. It follows that the primes $P_1, \ldots, P_r$ are all coprime. By the Chinese remainder theorem,

$$A \cong \frac{A}{P_1^n} \oplus \cdots \oplus \frac{A}{P_r^n}.$$

Let $A_i := A/P_i^n$, and take a divisor of zero $a \in A_i$. By construction, a is nilpotent in A_i. For every maximal ideal $\mathcal{M} \subset A$, the image of a inside $A_{\mathcal{M}}$ is 0, since $A_{\mathcal{M}}$ is a domain. But then $\mathrm{Ann}(a)$ is not contained in any maximal ideal, so $a = 0$. It follows that each A_i is an integral domain, and A is a finite sum of integral domains. □

We have another way to express the fact that the analytic structure of a regular ring is especially simple.

Lemma 10.1.11. *Let A be a local Noetherian ring of dimension d, with a system of parameters $a_1, \ldots, a_d$, and let $Q = (a_1, \ldots, a_d)$. Let $f \in A[x_1, \ldots, x_d]$ be a homogeneous polynomial of degree s, and assume that $f(a_1, \ldots, a_d) \in Q^{s+1}$. Then $f \in \mathcal{M}[x_1, \ldots, x_d]$.*

Proof. Evaluation at $a_1, \ldots, a_d$ gives a surjective homomorphism

$$\tfrac{A}{Q}[x_1, \ldots, x_d] \xrightarrow{ev_a} G_Q(A) = \bigoplus \tfrac{Q^n}{Q^{n+1}}.$$

By hypothesis, $ev_a(\overline{f}) = 0$, where $\overline{f}$ is the class of f modulo Q. If $f \notin \mathcal{M}[x_1, \ldots, x_d]$, $\overline{f}$ has an invertible coefficient, hence $\overline{f}$ is not a zero divisor.

By Proposition 9.5.2,

$$\dim G_Q(A) \leq \dim \frac{A/Q[\underline{x}]}{(\overline{f})} = d - 1,$$

which is a contradiction since $\dim G_Q(A) = d$. □

The following theorem expresses the fact that a system of local coordinates for a regular ring is given by analytically independent parameters. It is just a special case of the above lemma.

Theorem 10.1.12. *Let A be a regular local ring of dimension d, with a* regular *system of parameters $a_1, \dots, a_d$. Then the a_i are* analytically independent, *that is, if $f \in A[x_1, \dots, x_d]$ is a homogeneous polynomial such that $f(a_1, \dots, a_d) = 0$, then $f \in \mathcal{M}[x_1, \dots, x_d]$.*

When the ring contains a field, this result has a simpler statement.

Corollary 10.1.13. *Let $A, \mathcal{M}$ be a regular local ring of dimension d, with a* regular *system of parameters $a_1, \dots, a_d$. Assume that there is a field $k \subset A$ that maps isomorphically onto $A/\mathcal{M}$. Then the a_i are algebraically independent over k.*

Proof. Take a polynomial $f \in k[x_1, \dots, x_d]$ such that $f(a_1, \dots, a_d) = 0$. Let s be the minimal degree of a monomial of f, so that $f = f_s + g$, where each monomial of g has degree at least $s + 1$. Then $f_s(a_1, \dots, a_d) \in \mathcal{M}^{s+1}$, and Lemma 10.1.11 guarantees that $f_s \in \mathcal{M}[x_1, \dots, x_d]$. Since $f_s \in k[x_1, \dots, x_d]$, this means that $f_s = 0$, hence $f = 0$. □

There are a few properties about regular rings that are too important to omit. However, we cannot prove them with the techniques we have at hand. The proof of these theorems marks the beginning of the usage of homological methods in commutative algebra. We will expand on this circle of ideas in the following volume [**Fer**].

The most important result is that Theorem 10.1.9 can be strengthened considerably:

Theorem 10.1.14 (Auslander–Buchsbaum). *A regular local ring is a unique factorization domain.*

Proof. See [**Eis95**, Theorem 19.19] or [**Fer**, Theorem 8.4.1]. □

Remark 10.1.15. In particular, this means that regular local rings are integrally closed. A singular point of variety having an integrally closed local ring is called *normal*—this is considered a mild form of singularity.

Example 10.1.16. The planar singularity $y^2 = x^2 + x^3$ (a node) is not normal. In fact, this is a rephrasing of the content of Exercise 5 in Chapter 5.

The other important result is

Theorem 10.1.17. *Let A be regular local ring, $P \subset A$ a prime. Then the local ring A_P is regular as well.*

Proof. See [**Eis95**, Corollary 19.14] or [**Fer**, Corollary 8.2.5]. □

A consequence of this fact is that in order for a ring to be regular it is sufficient that all localizations $A_{\mathcal{M}}$ for a maximal ideal $\mathcal{M}$ are regular local rings (that is, it is redundant to ask this for all prime ideals). Some texts even adopt this as the definition of regular ring.

10.2. Multiplicity and degree

In this section, we study the leading coefficient of the Hilbert polynomial in more detail, both in the local and in the graded case. In the local case, this term is an algebraic expression of the multiplicity of a singularity. In particular, it gives an interesting invariant for local Noetherian rings that are not regular. In the graded case, it computes the degree of a projective variety. We also show how the two notions are related via the *tangent cone* construction.

We start with a semilocal Noetherian ring A of dimension d. Let Q be any ideal of definition for A, so that $\sqrt{Q} = \mathcal{J}(A)$, and fix a finitely generated A-module M. Then we know by Corollary 9.2.6 that for n big enough we can express the length of $M/Q^{n+1}M$ with a polynomial

$$\chi_M^Q(n) = \ell\left(\frac{M}{Q^{n+1}M}\right).$$

We can bound the degree of $\chi_M^Q(n)$ since

$$\dim M = \dim A/\operatorname{Ann}(M) \leq d.$$

Hence, we can write

$$\chi_M^Q(n) = \frac{e}{d!}n^d + a_{d-1}n^{d-1} + \cdots + a_0 \tag{10.2.1}$$

for some $e \in \mathbb{N}$.

Definition 10.2.1. Let A be a semilocal Noetherian ring with $\dim A = d$, M a finitely generated A-module. The natural number e in (10.2.1) is called the *Hilbert–Samuel multiplicity*, or simply *multiplicity*, of M at Q, denoted $e(Q, M)$.

As a notation, we will simply write $e(Q) = e(Q, A)$. When A is local with maximal ideal $\mathcal{M}$, we let $e(A) = e(\mathcal{M}) = e(\mathcal{M}, A)$.

In the geometric case, let V be an affine variety, $\mathcal{M} \subset R(V)$ the maximal ideal of a point $p \in V$. We will call $e(R(V)_{\mathcal{M}})$ the multiplicity of V at the point p.

Remark 10.2.2. Some authors introduce a notation for all coefficients of the Hilbert polynomial, and so denote by $e_0(Q, M)$ what we denote by $e(Q, M)$.

Remark 10.2.3. Let A be a regular local ring. Then Proposition 10.1.8 gives the expression

$$\chi_A^{\mathcal{M}}(n) = \binom{n+d}{d} = \frac{1}{d!}n^d + \cdots,$$

hence $e(A) = 1$. It follows that the multiplicity is an interesting invariant of the ring only in the singular case.

The converse is not true without additional assumptions.

Example 10.2.4. Consider the ring $A = k[x, y]/(x^2, xy)$ and the maximal ideal $M = (x, y) \subset A$. Let $B = A_M$, which is a local ring with maximal ideal $\mathcal{M} = M \cdot B$.

For $n \geq 2$ we have

$$\ell\left(\frac{A}{\mathcal{M}^n}\right) = \dim_k \frac{A}{\mathcal{M}^n} = n + 1,$$

which shows that $\dim A = 1$ and $e(A) = 1$. On the other hand, A is not regular, because it is not even an integral domain (in fact x is nilpotent in A).

We can also consider the $\mathcal{M}$-primary ideal $Q = (y)$. Then, by the same computation,

$$\ell\left(\frac{A}{Q^n}\right) = n + 1,$$

so $e(Q, A) = 1$ as well.

Remark 10.2.5. In the above example, we may be tempted to describe A as the local ring in 0 of the variety defined by the ideal $I = (x^2, xy)$. This is not correct, since I is not radical, and in fact $V(I)$ is just the line $x = 0$. Talking properly of such singularities requires the language of schemes, which are geometric objects that are not fully described by their set of geometric points. In this language, the zero locus of I would be a line with an embedded double point.

Our counterexample is not integral, but this is not the only restriction. For an example of an integral domain of multiplicity 1 which is not regular, see [**Nag62**, Appendix A1, Example 2]. In general, the appendix of Nagata's book is a fantastic source of counterexamples in commutative algebra.

On the other hand, the condition having multiplicity 1 is not too far from being regular. To state this precisely, we need the following definition.

Definition 10.2.6. Let A be a local Noetherian ring, $\widehat{A}$ its completion with respect to the topology defined by its maximal ideal. We say that A is *unmixed* if for every associated prime P of 0 in $\widehat{A}$ we have

$$\dim \frac{\widehat{A}}{P} = \dim \widehat{A} = \dim A.$$

With this definition, we can state the following *multiplicity* 1 *criterion.*

Theorem 10.2.7 (Nagata). *Let A be a local Noetherian ring. Then A is regular if and only if $e(A) = 1$ and A is unmixed.*

The proof of this result is surprisingly subtle, see [**Nag62**, Theorem 40.6]. Following [**Now97**], we will give the proof in the case where the residue field of A has characteristic 0, as a consequence of Cohen's theorem.

We now pass to the graded case, where we can give similar definitions. Let A be a graded ring of dimension d, and assume that A_0 is Artinian. For a finitely generated graded A-module M we have the bound

$$\dim M = \dim A/\operatorname{Ann}(M) \leq d.$$

Hence, for n big enough we can write

$$\ell(M_n) = \frac{e}{d!}n^d + a_{d-1}n^{d-1} + \cdots + a_0 \tag{10.2.2}$$

for some $e \in \mathbb{N}$.

Definition 10.2.8. Let A be a graded Noetherian ring of dimension d, M a finitely generated graded A-module. The natural number e in (10.2.2) is called the (Hilbert–Samuel) *multiplicity* of M, denoted $e(M)$.

In particular, we are interested in the multiplicity of a graded ring as a module over itself. In this case, we will also call $e(A)$ the *degree* of A, denoted $\deg A$.

Remark 10.2.9. The notions of multiplicity and degree are strictly related. In fact, let $A, \mathcal{M}$ be a local Noetherian ring of dimension d. Then by construction the Hilbert polynomials of $A, \mathcal{M}$ and $\operatorname{Gr}_{\mathcal{M}}(A)$ are the same, since

$$\chi_A^{\mathcal{M}}(n) = \sum_{i=0}^{n} \ell\left(\frac{\mathcal{M}^i}{\mathcal{M}^{i+1}}\right) = \chi_{\operatorname{Gr}_{\mathcal{M}}(A)}(n).$$

In particular

$$e(A) = \deg \mathrm{Gr}_{\mathcal{M}}(A).$$

To better understand this relation, we consider the geometric case. Let $V \subset \mathbb{A}^n(k)$ be an affine variety, p a point of V defined by the maximal ideal $M \subset R(V)$. To this we associate the local ring $A = R(V)_M$, with maximal ideal $\mathcal{M} = M \cdot A$. If we translate the variety so that $p = 0$, each polynomial $f \in I(V)$ has zero constant term, hence we can write

$$f = f_d + f_{d+1} + \cdots,$$

where f_k is homogeneous of degree k and $d > 0$. In particular, we can consider the homogeneous component of lowest degree, f_d. For the purpose of this section, we denote $H(f) := f_d$.

Definition 10.2.10. Let $V \subset \mathbb{A}^n(k)$ be an affine variety with $0 \in V$. Let $H(I(V))$ be the homogeneous ideal generated by all polynomials $H(f)$ for $f \in I(V)$. The affine variety C_0V defined by $H(I(V))$ is called the *tangent cone* of V in 0. Since $H(I(V))$ is homogeneous, it also defines a projective variety in $\mathbb{P}^{n-1}(k)$ called the *projective tangent cone* of V in 0, and denoted $\mathbb{P}C_0V$.

Remark 10.2.11. Assume that $I(V) = (f_1, \dots, f_r)$. Then

$$H(I(V)) = (H(f_1), \dots, H(f_r)).$$

This is especially useful when V is regular in 0. In this case, we can choose a regular system of parameters $f_1, \dots, f_r$. Then, each f_i has a nonzero linear term, and the tangent cone is exactly the Zariski tangent space to V in 0. However, when V is singular, the tangent cone contains strictly more information.

By construction, the ring associated to the projective tangent cone $\mathbb{P}C_0V$ is exactly the associated graded ring to the local ring $A, \mathcal{M}$. In particular, we can translate Remark 10.2.9 as follows:

Proposition 10.2.12. *Let $V \subset \mathbb{A}^n(k)$ be an affine variety with $0 \in V$. Then the multiplicity of V in 0 is equal to the degree of the projective variety $\mathbb{P}C_0V$.*

Notice that we have defined the notion of tangent cone in the point 0, but using a suitable translation this concept readily generalizes to other points in V.

Example 10.2.13. Let C be the node defined by the equation $y^2 = x^2 + x^3$. The tangent cone is defined by the equation $y^2 = x^2$, so it is the union of two lines. This expresses the fact that—while C itself is irreducible—locally there are two different branches.

The notion of multiplicity has a much simpler interpretation for singularities of a hypersurface, as the degree of vanishing of a singular polynomial. The following Proposition makes this precise.

Proposition 10.2.14. *Let k be a field and $V \subset \mathbb{A}^n(k)$ an affine hypersurface, given by the equation $f = 0$. Assume that $0 \in V$, which means that $f(0) = 0$, and write f as a sum of homogeneous components*

$$f = f_d + f_{d+1} + \cdots ,$$

where $\deg f_k = k$ and $f_d \neq 0$. Let A be the local ring of V in 0. Then $e(A) = d$.

Proof. Let $\mathcal{M} = I(0) \subset k[x_1, \ldots, x_n]$ be the ideal of the point 0, and let $I = (f)$, $\mathcal{M}' = \mathcal{M}/I$, so that $A = R(V)_{\mathcal{M}'}$. To compute the multiplicity, we have to evaluate the length of $A/\mathcal{M}'^k$.

Let $B_k := k[x_1, \ldots, x_n]/(f, \mathcal{M}^k)$. First, we claim that

$$A/\mathcal{M}'^k \cong B_k.$$

In fact, since quotients and localizations commute, $A/\mathcal{M}'^k$ is a localization of B_k at the multiplicative set S constisting of images of polynomials with nonzero constant term. But such polynomials are already invertible in B_k. In fact, take a polynomial $g \in k[x_1, \ldots, x_n]$ with $g(0) \neq 0$. Multiplication by $\overline{g}$ is an injective map $B_k \to B_k$, and since B_k is a finite-dimensional k-vector space, it must be surjective as well. This means that $\overline{g}$ has an inverse in B_k, proving the claim.

To compute $\dim B_k$, for $k > d$, we look at the map

$$\mu_f \colon \frac{k[x_1, \ldots, x_n]}{\mathcal{M}^{k-d}} \to \frac{k[x_1, \ldots, x_n]}{\mathcal{M}^k}$$

given by multiplication by f. Since $f_d \neq 0$, μ_f is injective, giving an exact sequence

$$0 \longrightarrow \tfrac{k[x_1,\ldots,x_n]}{\mathcal{M}^{k-d}} \longrightarrow \tfrac{k[x_1,\ldots,x_n]}{\mathcal{M}^k} \longrightarrow B_k \longrightarrow 0 .$$

This allows us to compute

$$\begin{aligned} \dim \frac{A}{\mathcal{M}'^k} &= \dim B_k = \binom{n+k}{n} - \binom{n+k-d}{n} \\ &= \frac{d}{(n-1)!} k^{n-1} + \text{lower order terms}, \end{aligned} \tag{10.2.3}$$

which implies that $e(A) = d$. □

With exactly the same proof we get an analogous result for the graded case:

Proposition 10.2.15. *Let k be a field and $V \subset \mathbb{P}^n(k)$ a projective hypersurface, given by the equation $f = 0$. Let A be the ring associated to V. Then* $\deg A = \deg f$.

In fact, the notions of multiplicity and degree were originally understood in simple cases such as this, and the definition with the Hilbert polynomial was introduced later [**Sam49**].

Remark 10.2.16. The concepts of multiplicity and degree are both fundamental in intersection theory. To understand what this is about, we recall some notions on the topology of manifolds.

Let M be a compact, oriented smooth manifold of real dimension d. The cup product gives a ring structure on the graded sum of the cohomology groups $H^*(M, A)$, for all rings A. To a compact oriented submanifold $S \subset M$ one can associate a fundamental class $c(S) \in H^*(M, \mathbb{Z})$ using Poincaré duality. If $S, T \subset M$ are two such submanifolds, one can then compute the product $c(S) \cdot c(T) \in H^*(M, \mathbb{Z})$.

Let $s = \dim S$, $t = \dim T$. Assuming $s + t = d$, and that S and T are transverse, the product

$$c(S) \cdot c(T) \in H^d(M, \mathbb{Z}) \cong \mathbb{Z}$$

computes the number of points of intersection between S and T, counted with sign. For a point $p \in S \cap T$, the sign is positive when the natural isomorphism

$$T_pS \oplus T_pT \cong T_pM$$

preserves the orientation on the tangent space, negative otherwise. In the case where S and T are not transverse, one can deform S in its tubular neighborhood to a submanifold S' having $c(S') = c(S)$, in such a way that S' and T *are* transverse, and then the product $c(S') \cdot c(T)$ has this geometric interpretation as the number of signed intersections.

One would like to be able to obtain a similar theory for projective algebraic varieties, but there are some subtleties. First, for fields other than $\mathbb{R}$ or $\mathbb{C}$, there is not an obvious replacement for singular cohomology. In any case, one would like to be able to compute products even inside singular varieties, where something like Poincaré duality cannot be expected to hold. Third, since subvarieties are defined algebraically, there is no obvious way to deform them to obtain tranversality, as one can do in the differentiable case.

It turns out that one can develop intersection theory in this algebraic setting, but some care has to be taken. In particular, the notion of multiplicity is fundamental in computing products in cases where the intersections cannot be made transverse. The degree of a projective variety $V \subset \mathbb{P}^k$, instead, measures the number of intersections (counted with multiplicities) with a generic linear space of dimension $k - \dim V$. For much more about intersection theory, the standard reference is [**Ful84**].

10.3. Formulas for multiplicity

In this section, we are going to investigate some properties of multiplicity. In particular we will derive some convenient formulas to compute the multiplicity in a local ring in terms of multiplicity in smaller rings such as quotients and localizations. We start by mentioning some elementary facts, which are immediate.

Proposition 10.3.1. *Let A be a semilocal Noetherian ring of dimension d, Q and Q' two ideals of definition for A and M a finitely generated A-module.*

(i) $e(Q, M) = \lim_{n\to\infty} \frac{d!}{n^d}\ell\left(\frac{M}{Q^n M}\right)$;
(ii) $e(Q^r, M) = e(Q, M) \cdot r^d$
(iii) *if* $Q \subset Q'$, $e(Q', M) \le e(Q, M)$.

Also, we can rephrase Proposition 9.2.11 as follows:

Proposition 10.3.2. *Let A be a semilocal Noetherian ring, $Q \subset A$ an ideal of definition. If we have an exact sequence of finitely generated A-modules*

$$0 \longrightarrow M' \longrightarrow M \longrightarrow M'' \longrightarrow 0\ ,$$

then $e(Q, M) = e(Q, M') + e(Q, M'')$.

Although we have defined the multiplicity for an A-module M, in many cases of interest only the multiplicity of A carries some new information. In particular, we can reduce the computation of the multiplicity of an A-module to that of a ring in many cases. The following result, known as the *additivity formula*, achieves this, and at the same time reduces the computation of multiplicity to the case of integral domains. Some authors call this the associativity formula (see for instance [**Eis95**, Ex. 12.11]), but following [**Lec57**] we reserve this name for Theorem 10.3.8.

Proposition 10.3.3 (Additivity formula). *Let A be a local Noetherian ring, $Q \subset A$ an ideal of definition, M a finitely generated A-module. Let $d = \dim A$ and let $P_1, \ldots, P_r$ be the minimal primes of A for which $\dim A/P_i = d$. Then*

$$e(Q, M) = \sum_{i=1}^{r} e(\overline{Q_i}, A/P_i)\ell(M_{P_i}),$$

where $\overline{Q_i} := \frac{Q+P_i}{P_i}$.

Proof. We use an induction over $s(M) := \sum \ell(M_{P_i})$.

When $s(M) = 0$, we must have $M_{P_i} = 0$ for all i, that is, $P_i \notin \operatorname{Supp}(M)$ for all i. But this means that $\operatorname{Ann}(M)$ is not contained in any minimal prime P_i of A such that $\dim A/P_i = d$. It follows that $\dim(M) = \dim(A/\operatorname{Ann}(M)) \leq d-1$, and in this case $e(Q, M) = 0$.

For the inductive step, choose a minimal prime $P \in \operatorname{Supp}(M)$ that is one of $P_1, \dots, P_r$, say $P = P_1$. By Corollary 3.3.15, the minimal primes of $\operatorname{Supp}(M)$ and $\operatorname{Ass}(M)$ are the same, hence P is associated to M. This means that we can find $N \subset M$ such that $N \cong A/P$. By Proposition 10.3.2, we have

$$e(Q, M) = e(Q, N) + e(Q, M/N). \tag{10.3.1}$$

Remark that $N_P \cong \frac{A_P}{PA_P}$, so $\ell(N_P) = 1$, while $N_{P_i} = 0$ for $i \geq 2$, since these are different minimal primes. This allows us to compute

$$s(M) = \sum_{i=1}^{r} \ell(N_{P_i}) + \sum_{i=1}^{r} \ell((M/N)_{P_i}) = 1 + s(M/N).$$

By inductive hypothesis,

$$e(Q, M/N) = \sum_{i=1}^{r} e(\overline{Q_i}, A/P_i)\ell((M/N)_{P_i}).$$

Moreover, by our choice of N we have $e(Q, N) = e(Q, A/P) = e(\overline{Q_1}, A/P)$. Equation (10.3.1) becomes

$$e(Q, M) = e(\overline{Q_1}, A/P) + \sum_{i=1}^{r} e(\overline{Q_i}, A/P_i)\ell((M/N)_{P_i}).$$

For all $i \geq 2$, we have $(M/N)_{P_i} \cong M_{P_i}$, while for $i = 1$ we have $\ell((M/N)_{P_1}) = \ell(M_{P_1}) - 1$, and the thesis follows. □

Corollary 10.3.4. *Under the hypothesis of Proposition* 10.3.3, *assume that A is an integral domain. Then $e(Q, M) = e(Q)\operatorname{rk}(M)$.*

Proof. In this case, the only minimal prime is 0, hence $e(Q, M) = e(Q)\ell(M_0)$ and $\ell(M_0) = \dim_k M \otimes k$, where k is the fraction field of A. □

In particular, if A is a regular local ring, the multiplicity just measures the rank of an A-module, and gives no new information.

Remark 10.3.5. Proposition 10.3.3 can also be used taking $M = A$. In this case, it reduces the computation of $e(Q, A)$ to that of $e(Q, A/P)$ for various primes P.

Consider an affine variety V and a point $p \in V$. From a geometric point of view, the additivity formula allows us to compute the multiplicity of V at

p in term of the multiplicity of the components of V of maximal dimension passing through p.

Remark 10.3.6. The previous remark does not add anything in the case where V is irreducible. But even then, the additivity formula can be useful. In fact, let A be a local Noetherian ring with maximal ideal $\mathcal{M}$, and $\widehat{A}$ its completion in the $\mathcal{M}$-adic topology. By Corollary 7.5.18, A and $\widehat{A}$ have the same associated graded ring, hence the same Hilbert polynomial. It follows that $e(A) = e(\widehat{A})$, and it can happen that A is an integral domain while $\widehat{A}$ is not. Geometrically, this can happen if an irreducible variety has two branches at a point that are analytically separable.

Example 10.3.7. Let k be a ring of characteristic different from 2. The curve C defined by $y^2 = x^2 + x^3$ in $\mathbb{A}^2(k)$ is irreducible. But in the ring $k[[x, y]]$ one has the factorization $y^2 = (x\sqrt{1+x})^2$, where

$$\sqrt{1+x} = 1 + \frac{1}{2}x - \frac{1}{8}x^2 + \frac{1}{16}x^3 + \cdots,$$

so the completion of the local ring of C at 0 is not an integral domain, and Proposition 10.3.3 applies. Again, we see that while C is irreducible, around 0 we can consider C as composed by two different components.

Related to the additivity formula, there is the following *associativity formula* of Lech from [**Lec57**], generalizing a previous result from [**Che45**].

Theorem 10.3.8 (Lech). *Let $A, \mathcal{M}$ be a local ring, $a_1, \ldots, a_d$ a system of parameters for A and $Q = (a_1, \ldots, a_d)$. Fix a natural number $m \leq d$ and let $Q_1 = (a_1, \ldots, a_m)$ and $Q_2 = (a_{m+1}, \ldots, a_d)$. Then*

$$e(Q) = \sum_P e\left(\frac{Q_1 + P}{P}\right) e\left(Q_2 \cdot A_P\right),$$

where the sum ranges over the primes $P \subset A$ that are minimal over Q_2 and that satisfy $\dim A/P + \operatorname{ht} P = \dim A$.

Notice that the elements of the formula are well defined: for such a prime P, $\overline{a_1}, \ldots, \overline{a_m}$ is a system of parameters for A/P, so $(Q_1 + P)/P$ is $\mathcal{M}/P$-primary, and $a_{m+1}/1, \ldots, a_d/1$ is a system of parameters for A_P, so $Q_2 \cdot A_P$ is $P \cdot A_P$-primary.

Let us set up some terminology. Just for the purpose of this proof, we will call a prime $P \subset A$ *balanced* if $\dim A/P + \operatorname{ht} P = \dim A$, and a chain

$$P_d \subsetneq P_1 \subsetneq \cdots \subsetneq P_0$$

compatible with $a_1, \ldots, a_d$ if for each $k = 0, \ldots, d$ we have

$$(a_{k+1}, \ldots, a_d) \subset P_k.$$

This allows us to state some simple lemmas.

Lemma 10.3.9. *Each prime P appearing in a chain compatible with $a_1, \dots, a_d$ is balanced.*

Proof. The chain can be split into a chain for A/P and one for A_P. $\square$

Lemma 10.3.10. *The set of chains compatible with $a_1, \dots, a_d$ is finite.*

Proof. The hypothesis implies that P_k is a minimal prime of the ideal $(a_{k+1}, \dots, a_d)$, and minimal primes of an ideal are finite in a Noetherian ring. $\square$

Lemma 10.3.11. *The set of chains compatible with $a_1, \dots, a_d$ is not empty.*

Proof. There exists a minimal prime of 0 which is balanced, call it P_d. Recursively choose P_k as a minimal prime of $P_{k+1}+(a_{k+1})$ which is balanced. $\square$

Lemma 10.3.12. *Each balanced prime P_k such that $(a_{k+1}, \dots, a_d) \subset P_k$ and $\dim A/P_k = k$ appears in a chain compatible with $a_1, \dots, a_d$.*

Proof. Apply the previous lemma to find a chain for A/P_k and one for A_{P_k}, and join them. $\square$

We are now ready to prove the associativity formula.

Proof of Theorem 10.3.8. Let Σ be the set of chains compatible with $a_1, \dots, a_d$, which is finite and nonempty by the previous lemmas. A repeated application of the additivity formula (Proposition 10.3.3) gives

$$e(Q, A) = \sum_{\underline{P} \in \Sigma} \ell\left((A/P_1)_{P_0}\right) \cdots \ell\left((A/P_d)_{P_{d-1}}\right) \ell(A_{P_d}). \tag{10.3.2}$$

A similar repeated application, just stopped earlier, gives

$$e(Q, A) = \sum_{\underline{P} \in \Sigma} e\left(\frac{Q + P_m}{P_m}\right) \ell\left((A/P_{m+1})_{P_m}\right) \cdots \ell\left((A/P_d)_{P_{d-1}}\right) \ell(A_{P_d}).$$

By applying (10.3.2) to A_{P_m}, we simplify this to

$$e(Q, A) = \sum_{P_m} e\left(\frac{Q + P_m}{P_m}\right) e(Q_{P_m}, A_{P_m}),$$

where the sum is over all balanced primes that contain $(a_{m+1}, \dots, a_d)$ appearing in a member of Σ. By the previous lemmas, these are exactly the balanced primes minimal over $(a_{m+1}, \dots, a_d)$, and we get the conclusion. $\square$

We conclude this section with two classical result by Samuel. The first one ([**Sam53**]) relates length and multiplicity.

Proposition 10.3.13 (Samuel). *Let $A, \mathcal{M}$ be a local Noetherian ring of dimension d, Q an ideal of definition generated by the system of parameters $a_1, \ldots, a_d$. Then*

$$e(Q, A) \leq \ell\left(\frac{A}{Q}\right).$$

Proof. Evaluation at $a_1, \ldots, a_d$ gives a surjective homomorphism of graded rings

$$\frac{A}{Q}[x_1, \ldots, x_d] \to \mathrm{Gr}_Q(A).$$

This gives a corresponding inequality between their Hilbert polynomials, and since both rings have dimension d, an inequality between their first coefficient

$$e\left(\frac{A}{Q}[x_1, \ldots, x_d]\right) \geq e(\mathrm{Gr}_Q(A)) = e(Q, A).$$

To conclude, we just note that the Hilbert polynomial of the first ring is

$$\ell\left(\frac{A}{Q}\right)\binom{n+d}{d}. \qquad \square$$

The next theorem gives a way to relate multiplicities in integral extensions. It appears in [**ZS76b**, Theorem 24, Chapter 10], but we give a slightly simplified statement.

Theorem 10.3.14. *Let $A \subset B$ be Noetherian integral domains, and assume that A is local with maximal ideal $\mathcal{M}$. Let $Q \subset A$ be an ideal of definition, and assume that B is integral over A. Take a primary decomposition*

$$Q \cdot B = \bigcap_{i=1}^{r} Q_i,$$

where Q_i is P_i-primary in B. Then the polynomials

$$[\mathcal{F}(B) : \mathcal{F}(A)]\chi_A^Q(n) \tag{10.3.3}$$

and

$$\sum_{i=1}^{r} [B/P_i : A/\mathcal{M}]\chi_{B_{P_i}}^{Q_i}(n) \tag{10.3.4}$$

have the same degree and leading term.

The statement can be simplified when all localizations B_{P_i} have the same dimension as A. In this case, all summands contribute to the leading term, and we get

Corollary 10.3.15. *In the above theorem, assume that* $\operatorname{ht} P_i = \operatorname{ht} \mathcal{M}$ *for all primes* P_i. *Then*

$$[\mathcal{F}(B) : \mathcal{F}(A)]e(Q, A) = \sum_{i=1}^{r} [B/P_i : A/\mathcal{M}]e(Q_i, B_{P_i}).$$

Remark 10.3.16. If we assume that A is integrally closed, then we can apply the above corollary. In fact, for every prime P_i as in Theorem 10.3.14, we have $Q \subset P_i \cap A$, which implies that $P_i \cap A = \mathcal{M}$. In this case, we can apply the going down Theorem 5.2.13 to conclude that $\operatorname{ht} P_i = \operatorname{ht} \mathcal{M}$.

Remark 10.3.17. If B is local, there is only one prime P_i that appears in the sum, and the corollary applies again.

Proof of Theorem 10.3.14.

For a given M, we denote $\ell_A(M)$ the length of M as an A-module and $\ell_B(M)$ the length of M as a B-module. Take n big enough, and let $d = \chi_{B_{P_i}}^{Q_i}(n)$. By Theorem 2.5.16, there is a chain of B_{P_i}-modules

$$Q_i^n = M_d \subsetneq M_{d-1} \subsetneq \cdots \subsetneq M_1 = P_i \cdot B_{P_i}$$

such that $\ell_{B_{P_i}}(M_k/M_{k+1}) = 1$. This is also a chain of B-modules, and their length as B-modules is the same—and in fact it is the same as $\dim_{B/P_i}(M_k/M_{k+1})$. As $A/\mathcal{M}$-vector spaces, though, each term M_k/M_{k+1} has dimension $[B/P_i : A/\mathcal{M}]$. This implies that

$$[B/P_i : A/\mathcal{M}]\chi_{B_{P_i}}^{Q_i}(n) = [B/P_i : A/\mathcal{M}]d = \ell_A(B/Q_i^n).$$

Moreover, $Q^n \cdot B = \bigcap_{i=1}^r Q_i^n$, and by the Chinese remainder theorem we have

$$\sum_{i=1}^{r} \ell_A(B/Q_i^n) = \ell_A(B/Q^n \cdot B).$$

Putting the two equations together, we recognize that the sum in (10.3.4) is just $\ell_A(B/Q^n \cdot B)$. On the other hand, the term in (10.3.3) amounts to $[\mathcal{F}(B) : \mathcal{F}(A)]\ell_A(A/Q^n)$, so we need to compare these two lengths.

Let $k = [\mathcal{F}(B) : \mathcal{F}(A)]$. If B is a free module over A, its rank must be k, in which case the equality

$$\ell_A(B/Q^n \cdot B) = k \cdot \ell_A(A/Q^n)$$

holds trivially. In general we cannot assume that, but we can find $b_1, \ldots, b_k \in B$ that span the field $\mathcal{F}(B)$ over $\mathcal{F}(A)$, and then one can fit B in

$$C \subset B \subset \frac{1}{d}C,$$

where $C = \langle b_1, \ldots, b_n \rangle_A$ and $d \in A$. We use this to compare $\ell_A(B/Q^n \cdot B)$ with $\ell_A(C/Q^n \cdot C)$.

Namely, there is a surjection

$$\frac{C}{Q^n \cdot C} \to \frac{C + Q^n \cdot B}{Q^n \cdot B},$$

which implies the inequality

$$\begin{aligned}\ell_A\left(\frac{C}{Q^n \cdot C}\right) &\geq \ell_A\left(\frac{C + Q^n \cdot B}{Q^n \cdot B}\right) \geq \ell_A\left(\frac{d \cdot B + Q^n \cdot B}{Q^n \cdot B}\right) \\ &= \ell_A\left(\frac{B}{Q^n \cdot B}\right) - \ell_A\left(\frac{B}{d \cdot B + Q^n \cdot B}\right).\end{aligned}$$

By a symmetric reasoning,

$$\ell_A\left(\frac{C}{Q^n \cdot C}\right) \leq \ell_A\left(\frac{B}{Q^n \cdot B}\right) + \ell_A\left(\frac{C}{d \cdot C + Q^n \cdot C}\right).$$

Hence, to understand the difference between $\ell_A(B/Q^n \cdot B)$ and $\ell_A(C/Q^n \cdot C)$, we look at the terms

$$\ell_A\left(\frac{B}{d \cdot B + Q^n \cdot B}\right)$$

and

$$\ell_A\left(\frac{C}{d \cdot C + Q^n \cdot B}\right).$$

In both cases, these are Hilbert functions for modules over the ring A/dA, which has dimension strictly less than $\dim A$, so $\ell_A(B/Q^n \cdot B)$ and $\ell_A(C/Q^n \cdot C)$ have the same degree and leading coefficient. Since for C we have

$$\ell_A(C/Q^n \cdot C) = k \cdot \ell_A(A/Q^n),$$

the theorem is proved. □

10.4. Multiplicity and valuations

In this section, we investigate the meaning of multiplicity as the order of vanishing of a suitable element. The prototypical result in this spirit is Proposition 10.2.14, which links the order of vanishing of a polynomial f in 0 to the multiplicity of the hypersurface defined by f in 0. We are going to greatly generalize this result, and in doing so, we state some conditions under which the order of vanishing of an element can be interpreted as a valuation.

Let A be a Noetherian ring, I an ideal. Then by Krull intersection theorem 7.5.24 we have

$$\bigcap_{n=0}^{\infty} I^n = 0,$$

hence for each element $a \in A \setminus \{0\}$ we can find a unique n such that $a \in I^n \setminus I^{n+1}$.

Definition 10.4.1. Let A be a Noetherian ring, I an ideal. If $a \in I^n \setminus I^{n+1}$, we denote $v_I(a) = n$. The function $v_I \colon A \setminus \{0\} \to \mathbb{N}$ is called the *order function* of I. When A is local with maximal ideal $\mathcal{M}$, we denote $v_A = v_{\mathcal{M}}$.

Now assume that A is local with maximal ideal $\mathcal{M}$. By construction, we have the inequality

$$v_A(a+b) \geq \min\{v_A(a), v_A(b)\}, \tag{10.4.1}$$

which makes v_A something similar to a discrete valuation, but this is not always the case. For one thing, valuations are defined on a field, and A need not be an integral domain, so it may not have a fraction field. But even if A is a domain, v_A can fail to be multiplicative, hence it is not always a valuation.

Remark 10.4.2. In fact, we always have the inequality

$$v_A(ab) \geq v_A(a) + v_A(b),$$

but it can be strict. The condition that we always have equality is equivalent to saying that the associated graded ring $A' = \mathrm{Gr}_{\mathcal{M}}(A)$ is an integral domain. In fact let $a \in \mathcal{M}^r \setminus \mathcal{M}^{r+1}$ and $b \in \mathcal{M}^s \setminus \mathcal{M}^{s+1}$. Then $\overline{a}$ is a nonzero element of A'_r and $\overline{b}$ is nonzero in A'_s, so their product is nonzero in A'_{r+s} if and only if $v_A(ab) = v_A(a) + v_A(b)$.

In this section, following [**Hor76**], we want to relate the order function of A to the function that measures multiplicity of an element $a \in A$.

Definition 10.4.3. Let $A, \mathcal{M}$ be a Noetherian local ring of dimension d. Given $a \in \mathcal{M} \setminus \{0\}$, we define

$$\mu_A(a) = e\left(\frac{A}{(a)}\right)$$

if $\dim A/(a) = d-1$, and $\mu_A(a) = \infty$ otherwise. We define $\mu(a) = 0$ for $a \in A \setminus \mathcal{M}$ (this is consistent, since in this case $A/(a)$ is trivial). The function

$$\mu_A \colon A \setminus \{0\} \to \mathbb{N} \cup \{\infty\}$$

is called the *multiplicity function* of A.

Remark 10.4.4. By Proposition 9.4.5, $\dim A/(a) \geq d-1$, so the condition $\mu_A(a) = \infty$ means that $\dim A/(a) = d$. This can only happen if a is a divisor of 0 by Proposition 9.5.2. Hence, on an integral domain, the multiplicity function takes values in $\mathbb{N}$.

It is no wonder that the functions v_A and μ_A are related, since both measure some kind of order of vanishing of an element. In some cases, we have already established a relation between the functions v_A and μ_A.

Example 10.4.5.

(a) Let A be a DVR. In this case, v_A is exactly the discrete valuation on A. Moreover, a quotient $A/(a)$ has dimension 0, and if $v_A(a) = n$, then $\ell(A/(a)) = n$ as well, since all ideals are powers of the maximal ideal. Hence $\mu_A = v_A$ in this case.

(b) Let k be a field, let $\mathcal{M} \subset k[x_1, \ldots, x_n]$ be the ideal of 0, and let $A = k[x_1, \ldots, x_n]_{\mathcal{M}}$. Then we can rephrase Proposition 10.2.14 by saying that $\mu_A = v_A$. Moreover, it is immediate that v_A is multiplicative by looking at the monomials of lowest total degre. Hence, we can extend v_A to a valuation on $k(x_1, \ldots, x_n)$.

In both examples, things behave as well as we can hope: the functions v_A and μ_A agree and both are valuations. We will prove in this section that these two phenomena are strictly related. In fact, while v_A satisfies (10.4.1), the multiplicity function μ_A is almost always multiplicative.

Proposition 10.4.6. *Let $A, \mathcal{M}$ be a local Noetherian ring. If $\mu_A(a) = \infty$ or a is not a divisor of zero, then*

$$\mu_A(ab) = \mu_A(a) + \mu_A(b).$$

The hypothesis that a is not a divisor of zero is necessary; see Exercise 8.

Proof. In the case where $\mu_A(a) = \infty$, there is a minimal prime $P \subset A$ such that $a \in P$ and $\dim A/P = \dim A$. Hence, $ab \in P$ as well, and $\mu_A(ab) = \infty$.

Thus, we can assume that $\mu_A(a)$ and $\mu_A(b)$ are both finite. Since a is not a divisor of 0, multiplication by a induces an isomorphism

$$\frac{A}{(b)} \cong \frac{(a)}{(ab)}.$$

Using the exact sequence

$$0 \longrightarrow \frac{(a)}{(ab)} \longrightarrow \frac{A}{(ab)} \longrightarrow \frac{A}{(a)} \longrightarrow 0\,,$$

we deduce the equality

$$\ell\left(\frac{A}{(ab)}\right) = \ell\left(\frac{A}{(a)}\right) + \ell\left(\frac{A}{(b)}\right). \tag{10.4.2}$$

Let $d = \dim A$. We can specialize the additivity formula in Proposition 10.3.3 to get

$$e\left(\frac{A}{(a)}\right) = \sum_P e\left(\frac{A}{P}\right) \ell\left(\frac{A_P}{a \cdot A_P}\right),$$

where the sum ranges over all primes $P \ni a$ such that $\dim A/P = d-1$. We can also include primes P that do not contain a, since in that case $A_P/(a \cdot A_P)$ is 0. A similar formula holds for b and ab.

Using additivity of the lengths (10.4.2) in the rings of the form A_P, we obtain the conclusion. □

Corollary 10.4.7. *Let $A, \mathcal{M}$ be a local Noetherian integral domain. Then the multiplicity function μ_A is multiplicative.*

We now investigate under what conditions v_A is multiplicative as well. A simple case is the following.

Proposition 10.4.8. *Let A be a regular local ring. Then the order function v_A is a discrete valuation on A.*

Remark 10.4.9. Notice that A need not be a discrete valuation ring—in fact this cannot happen unless $\dim A = 1$. What we mean is just that v_A extends to a valuation with values in $\mathbb{Z}$ on the fraction field $\mathcal{F}(A)$. Here we are implicitly using the fact that A is an integral domain (Theorem 10.1.9).

Proof. By Remark 10.4.2, it is enough to prove that $\mathrm{Gr}_{\mathcal{M}}(A)$ is a domain, and Proposition 10.1.8 tells us that $\mathrm{Gr}_{\mathcal{M}}(A)$ is isomorphic to $k[x_1, \ldots, x_d]$, where $d = \dim A$ and $k = A/\mathcal{M}$. □

In order to connect the functions v_A and μ_A, we start with a lemma that simplifies the computation of μ_A.

Lemma 10.4.10. *Let A be a semilocal Noetherian ring, Q an ideal of definition, $a \in Q$. Then*

$$\chi_A^Q(n) - \chi_{A/(a)}^{Q/(a)}(n) = \ell\left(\frac{A}{(Q^n : a)}\right).$$

Proof. This is just a computation:

$$\begin{aligned}
\chi_A^Q(n) - \chi_{A/(a)}^{Q/(a)}(n) &= \ell\left(\frac{A}{Q^n}\right) - \ell\left(\frac{A}{Q^n + (a)}\right) \\
&= \ell\left(\frac{Q^n + (a)}{Q^n}\right) = \ell\left(\frac{(a)}{Q^n \cap (a)}\right) \\
&= \ell\left(\frac{(a)}{(a) \cdot (Q^n : a)}\right) = \ell\left(\frac{A}{(Q^n : a)}\right).
\end{aligned}$$

The last equality uses the fact that, while multiplication by a is not necessarily injective, its kernel is contained in $(Q^n : a)$ anyway. □

Using this lemma, we see that to control the difference between $e(Q, A)$ and $e(Q/(a), A/(a))$ we need to understand the ideal $(Q^n : a)$. Assume that

$a \in Q^s$. Then we have the inclusion $Q^{n-s} \subset (Q^n : a)$. Samuel introduced in [**Sam53**] the following definition to capture the case where we are able to control the behavior of $(Q^n : a)$.

Definition 10.4.11. Let A be a semilocal Noetherian ring, Q an ideal of definition. We say that the element $a \in Q^s$ is *superficial* (of order s) for Q if there exist an integer c such that

$$(Q^n : a) \cap Q^c = Q^{n-s}$$

for all n large enough.

When $A, \mathcal{M}$ is a local ring, we call the element superficial of order s if it is so for $\mathcal{M}$. Notice that this implies that $a \notin \mathcal{M}^{s+1}$, so in fact the order $v_A(a) = s$. This gives a hint that superficial elements can be used to connect v_A to μ_A. We can specialize the previous lemma to the case of superficial elements:

Proposition 10.4.12. *Let A be a semilocal Noetherian ring, Q an ideal of definition, a superficial of order s for Q. Then there exists an integer c such that*

$$\chi_A^Q(n) - \chi_A^Q(n-s) \leq \chi_{A/(a)}^{Q/(a)}(n) \leq \chi_A^Q(n) - \chi_A^Q(n-s) + \chi_A^Q(c)$$

for n large enough.

Proof. By definition of superficial element, we get

$$\begin{aligned} \ell\left(\frac{(Q^n : a)}{Q^{n-s}}\right) &= \ell\left(\frac{(Q^n : a)}{(Q^n : a) \cap Q^c}\right) \\ &= \ell\left(\frac{Q^c + (Q^n : a)}{Q^c}\right) \leq \ell\left(\frac{A}{Q^c}\right), \end{aligned}$$

which we can translate to

$$0 \leq \chi_A^Q(n-s) - \ell\left(\frac{A}{(Q^n : A)}\right) \leq \chi_A^Q(c).$$

We conclude by Lemma 10.4.10. □

By comparing just the first coefficients in the Proposition, we get

Corollary 10.4.13. *Let A be a semilocal Noetherian ring of dimension $d > 1$, $a \in Q^s$ superficial of order s for the ideal of definition Q. Then $e(Q/(a), A/(a)) = e(Q, A) \cdot s$.*

Proof. We need to compute the first coefficient of the polynomial $\chi_A^Q(n) - \chi_A^Q(n-s)$. The coefficients in the same degree cancel each other, so the first nonzero monomial is in degree $d-1$ and is given by

$$\frac{e(Q, A)}{d!}(n^d - (n-s)^d) = \frac{e(Q, A)}{d!} \cdot dsn^{d-1} + \cdots,$$

giving $e(Q/(a), A/(a)) = e(Q, A) \cdot s$. □

Corollary 10.4.14. *Let $A, \mathcal{M}$ be a Noetherian local ring. Then $A[[x]]$ is local as well, and $e(A[[x]]) = e(A)$.*

Proof. The ring $A[[x]]$ is clearly local with maximal ideal $(\mathcal{M}, x)$. Moreover, it is easy to check that x is superficial of order 1. If $\dim A > 0$, then Corollary 10.4.13 applies and gives the conclusion. If $\dim A = 0$, then $e(A) = \ell(A)$ and the Hilbert polynomial of $A[[x]]$ is $\chi_{A[[x]]}(n) = \ell(A) \cdot n$. □

We can finally prove the result that links v_A to μ_A.

Theorem 10.4.15. *Let $A, \mathcal{M}$ be a local Noetherian integral domain. Then v_A is a valuation if and only if $v_A = k \cdot \mu_A$, in which case $k = e(A)$.*

Proof. If v_A is a multiple of μ_A, then it is multiplicative by Corollary 10.4.7, hence it is a valuation.

Conversely, assume that v_A is multiplicative. Then the ring $G_{\mathcal{M}}(A)$ is an integral domain, and this implies that every nonzero element $a \in A$ is superficial (this should be clear, but see Lemma 10.5.2). Assuming $\dim A > 1$, by Corollary 10.4.13, we get

$$\mu_A(a) = e\left(\frac{A}{(a)}\right) = v_A(a) \cdot e(A).$$

For the case where $\dim A = 1$, apply the result to $A[[x]]$. It is a simple check that the order function $v_{A[[x]]}$ is also a valuation, hence $v_{A[[x]]} = k \cdot \mu_{A[[x]]}$. For $a \in A$, we have $v_{A[[x]]}(a) = v_A(a)$, and by Corollary 10.4.14 also $\mu_{A[[x]]}(a) = \mu_A(a)$, so we get the conclusion. □

Putting this together with Proposition 10.4.8, we get

Corollary 10.4.16. *Let A be a regular local integral domain. Then $v_A = \mu_A$, and both functions are valuations.*

In the case where v_A and μ_A disagree, we cannot expect that μ_A is a valuation, but still we have the following result that we quote without proof (see [**Hor76**, Theorem 4]).

Theorem 10.4.17. *Let A be a local Noetherian integral domain. Then there exist r discrete valuations $v_1, \ldots, v_r$ on $\mathcal{F}(A)$ and corresponding integers $n_1, \ldots, n_r$ such that*

$$\mu_A = n_1 v_1 + \cdots + n_r v_r.$$

10.5. Superficial elements

Let A be a semilocal ring, $Q \subset A$ an ideal of definition. In the previous section, we defined an element $a \in Q^s$ to be *superficial* of order s if it has the property that

$$(Q^n : a) \cap Q^c = Q^{n-s}$$

for a fixed $c \in \mathbb{N}$ and for all n large enough. We used this to prove Corollary 10.4.13 that states that in this case we have $e(Q/(a), A/(a)) = e(Q, A) \cdot s$, provided $\dim A > 1$. This allows us to prove properties of multiplicity by induction on the dimension.

In this section, we investigate this technical tool in more detail, starting from some existence results for superficial elements, and use it to prove some deeper properties of multiplicity. Most of this material is taken from [**ZS76b**]. The main existence result is

Theorem 10.5.1. *Let A be a semilocal Noetherian ring, $Q \subset A$ an ideal of definition. Then there exists a superficial element of order s for some $s \geq 1$.*

In order to prove it, we rephrase the condition of being superficial.

Lemma 10.5.2. *Let A be a semilocal Noetherian ring, $Q \subset A$ an ideal of definition, $A' = G_Q(A)$ the associated graded ring. Given an element $a \in Q^s$, let $\overline{a} \in A'_s$ be its image. Then a is superficial of order s if and only if $\operatorname{Ann}_{A'}(\overline{a}) \subset A'_{<c}$ for some $c \in \mathbb{N}$.*

Proof. This is just a matter of spelling out the definitions.

To say that a is *not* superficial (with constant c) means that we can find $b \in Q^c \setminus Q^{n-s}$ such that $ab \in Q^n$. Rephrasing in terms of the order function v_Q of Q, this means that $c \leq v_Q(b) < n - s$ and $v_Q(ab) \geq n$.

To say that $\operatorname{Ann}_{A'}(a) \not\subset A'_{<c}$ means that we can find b with $v_Q(b) \geq c$ such that $\overline{b} \in A'_{v_Q(b)}$ satisfies $\overline{a}\overline{b} = 0$, which is the same as $ab \in Q^{s+v_Q(b)+1}$.

The two conditions are now seen to be equivalent. □

Proof of Theorem 10.5.1.

Let $A' = \operatorname{Gr}_Q(A)$ be the associated graded ring. By the lemma, we need to understand divisors of 0 in A', so we investigate the associated primes of 0.

Let A'_+ be the irrelevant ideal, and partition the associated primes $P_1, \dots, P_r$ of 0 in A', so that the last $r - h$ contain A_+ and the first h do not. We claim that we can find a homogeneous element $\overline{a} \in A'_+$ such that $\overline{a} \notin P_i$ for $i = 1, \dots, h$.

To prove this, choose homogeneous elements $b_i \in A'_+ \setminus P_i$ and $c_{ij} \in P_j \setminus P_i$ for $i, j = 1, \dots, h$, and then take

$$a_i = b_i \prod_{j \neq i} c_{ij},$$

so that $a_i \in A'_+$, $a_i \notin P_i$ but at the same time $a_i \in P_j$ for all $j \neq i$. Finally, letting

$$\overline{a} = \sum_{i=1}^{h} a_i^{d_i}$$

for suitable d_i such that $\overline{a}$ is homogeneous, we have the desired element.

Say $\overline{a}$ is homogeneous of order s, and let $a \in Q^s$ be a representative of $\overline{a}$. We want to show that a is superficial of order s. Using Lemma 10.5.2, we look at an element $b \in \operatorname{Ann}(\overline{a})$. We can assume that b is homogeneous.

Consider a primary decomposition of 0

$$0 = \bigcap_{i=1}^{r} Q_i,$$

where Q_i is P_i-primary. For a suitable exponent c we have

$$(A'_+)^c \subset \bigcap_{i=h+1}^{r} Q_i. \tag{10.5.1}$$

By our choice of $\overline{a}$ we have $b \in P_i$ for $i = 1, \ldots, h$. If we assume $b \in A'_{\geq c}$, we find from (10.5.1) that $b \in P_i$ for $i = h+1, \ldots, r$, hence $b = 0$. This proves that $\operatorname{Ann}(\overline{a}) \subset A'_{<c}$, and the theorem. □

Unfortunately, we are not free to choose the order s at will in the above theorem. This is not a limitation of our approach.

Example 10.5.3. Let $k = \mathbb{Z}/2\mathbb{Z}$ and consider the ring

$$A := \frac{k[[x,y]]}{(xy(x+y))}.$$

Then A does not have superficial elements of order 1 for its maximal ideal $\mathcal{M}$. For if $a \in A$ is such an element, then $a \in \mathcal{M} \setminus \mathcal{M}^2$. But there are only 3 such elements: x, y and $x+y$, and each of them is annihilated by a homogeneous element of arbitrarily large order in $\operatorname{Gr}_{\mathcal{M}}(A)$.

We can however guarantee the existence of superficial elements of any order, assuming the residue field is infinite.

Theorem 10.5.4. *Let $A, \mathcal{M}$ a Noetherian local ring of positive dimension, and assume that $k = A/\mathcal{M}$ is infinite. Let Q be a $\mathcal{M}$-primary ideal. Then for any $s \geq 1$ there exists a superficial element of order s for Q. Moreover, we can take such an element outside all minimal primes of 0 in A.*

Proof. As in Theorem 10.5.1, we consider the graded ring $A' := G_Q(A)$ and take a primary decomposition of 0 in A':

$$0 = \bigcap_{i=1}^{r} Q_i,$$

where Q_i is P_i-primary, and the indices have been chosen so that $A'_+ \subset P_i$ exactly for $i = h+1, \dots, r$, for some $h \le r$.

Let $D_1, \dots, D_t$ be the minimal primes of 0 in A. Given a nonzero element $a \in A$ with $v_Q(a) = s$, denote $\overline{a} \in A'_s$ its class, and given an ideal $I \subset A$ denote

$$\overline{I} := (\overline{a} \mid a \in I, a \neq 0).$$

We claim that $A'_s \not\subset \overline{D_i}$ for any $i = 1, \dots, t$ and any $s \in \mathbb{N}$. Otherwise, $Q^s \subset Q^{s+1} + D_i$, and by Nakayama's lemma $Q^s \subset D_i$. Since D_i is prime, this implies that $\mathcal{M} \subset D_i$, which can only happen if $\dim A = 0$.

Now choose any $s > 0$. Then A_s is not contained in any of the $\overline{D_i}$, as well as in P_j for $j \le h$. Let $M = A'_s/A'_{s+1}$, which we regard as a module over $A/Q \cong A'/A'_+$. Every ideal J of A' defines a submodule $J \cap A'_s/A'_{s+1}$, and in particular the ideals $\overline{D_i}$ and P_j for $j \le h$ define a collection of submodules N_i, $i = 1, \dots, h+t$. By what we have just said, we have strict inclusions $N_i \subsetneq M$.

If we prove that $M \neq \bigcup_{i=1}^{h+t} N_i$, then we can find $\overline{a} \in A'_s$ such that $a \notin D_i$ for $i = 1, \dots, t$ and $\overline{a} \notin P_j$ for $j = 1, \dots, h$. As in the proof of Theorem 10.5.1, this means that a is superficial of order s, and we are done.

It remains to prove that $M \neq \bigcup N_i$. This is clear when $Q = \mathcal{M}$, for in this case M is a vector space over the infinite field k, hence it cannot be the union of finitely many subspaces. In general, we reduce to this case as follows. Choose g such that $\mathcal{M}^g \subset Q$. Given a submodule $N \subsetneq M$, we have

$$N + \frac{\mathcal{M}}{Q} M \subsetneq M,$$

otherwise, we would have the chain

$$\begin{aligned} M = N + \frac{\mathcal{M}}{Q} M &= N + \frac{\mathcal{M}}{Q}\left(N + \frac{\mathcal{M}}{Q} M\right) = N + \frac{\mathcal{M}^2}{Q} M \\ &= \cdots = N + \frac{\mathcal{M}^g}{Q} M = N. \end{aligned}$$

Therefore, as vector spaces,

$$\frac{N + \mathcal{M}M}{\mathcal{M}M} \subsetneq \frac{M}{\mathcal{M}M}.$$

In particular, $(N_i + \mathcal{M}M)/\mathcal{M}M$ is a strict subspace of $M/\mathcal{M}M$, and the union over all i cannot be the whole space. $\square$

Superficial elements are the basis for many results on multiplicities, including the proof of the multiplicity 1 criterion (Theorem 10.2.7). We will just give an application in this section.

Proposition 10.5.5. *Let $A, \mathcal{M}$ be a Noetherian local ring, $Q \subset A$ an ideal of definition, and assume that $k = A/\mathcal{M}$ is infinite. Then there exists an ideal of definition $Q' \subset Q$ generated by a system of parameters such that $e(Q', A) = e(Q, A)$.*

Proof. Induction on $d := \dim A$. The case $d = 0$ is trivial, since we can take $Q' = 0$. The difficult case is $d = 1$. Choose an element $a \in Q$ which is superficial of order 1 for Q. By definition we have

$$(Q^n : a) \cap Q^c = Q^{n-1} \tag{10.5.2}$$

for a fixed $c \in \mathbb{N}$. Now apply the Artin–Rees lemma to the ideal (a). The Q-adic topology on (a) and the topology induced by the Q-adic topology on A have bounded difference, so

$$Q^n \cap (a) \subset Q^{n-k} \cdot (a) \tag{10.5.3}$$

for some $k \in \mathbb{N}$.

Assume first that a is not a divisor of zero, so that the multiplication map

$$\mu_a \colon A \to a \cdot A$$

is injective. In this case, $\mu_a^{-1}(a \cdot Q^{n-k}) = Q^{n-k}$, while by definition

$$\mu_A^{-1}(Q^n \cap (a)) = (Q^n \cap (a) : a) = (Q^n : a).$$

So, by taking preimages in (10.5.3), we get

$$(Q^n : a) \subset Q^{n-k}.$$

When $n - k \geq c$, we can put this together with (10.5.2) and get

$$(Q^n : a) = Q^{n-1}.$$

Using Lemma 10.4.10 we get

$$\chi_{A/(a)}^{Q/(a)}(n) = \chi_A^Q(n) - \chi_A^Q(n-1) = e(Q, A),$$

since $\dim A = 1$. In particular, $\dim A/(a) = 0$, so the left-hand side is just $\ell(A/(a))$. Since a is not a zero divisor,

$$\ell\left(\frac{(a^n)}{(a^{n+1})}\right) = \ell\left(\frac{A}{(a)}\right) = e(Q, A),$$

which proves that $e((a), A) = e(Q, A)$ and we can take $Q' = (a)$. This ends the proof for the case $d = 1$, assuming a is not a divisor of 0.

For the general case—still with $d = 1$—we first notice that all chains of primes in A have the form $P \subset \mathcal{M}$, where P is a minimal prime of 0. Choose any superficial element $a \in Q$ of order 1 that lies outside the minimal primes, so that $\dim A/(a) = 0$. If a is not a divisor of 0 we are done, otherwise we can apply the above result to $A/\operatorname{Ann}(a)$. To this end, define $I := \operatorname{Ann}(a)$, $\overline{A} := A/I$ and $\overline{Q} := (Q + I)/I$. Then we can compute

$$\begin{aligned}\chi_{\overline{A}}^{\overline{Q}}(n) &= \ell\left(\frac{A}{Q^n + I}\right)\\ &= \ell\left(\frac{A}{Q^n}\right) - \ell\left(\frac{Q^n + I}{Q^n}\right)\\ &= \chi_A^Q(n) - \ell\left(\frac{I}{Q^n \cap I}\right).\end{aligned}$$

Now, I is a finitely generated module over $A/(a)$, which is Artinian, so it has finite length. Moreover, $\bigcap_n Q^n \cap I = 0$ by Krull's intersection theorem, so we have $Q^n \cap I = 0$ for n big enough. In this case, we can simplify the last equation to

$$\chi_{\overline{A}}^{\overline{Q}}(n) = \chi_A^Q(n) - \ell(I).$$

This implies that $\dim \overline{A} = 1$ and that $e(\overline{Q}, \overline{A}) = e(Q, A)$. The same reasoning, applied to (a), show that $e(\overline{(a)}, \overline{A}) = e((a), A)$. Since $\overline{a}$ is not a divisor of 0 in $\overline{A}$, we can apply the first part of the proof to $\overline{A}$ and conclude that $e(\overline{(a)}, \overline{A}) = e(\overline{Q}, \overline{A})$, so finally $e((a), A) = e(Q, A)$ and we can take $Q' = (a)$. This ends the proof for the case $d = 1$.

For $d > 1$, we perform an inductive step. Using Theorem 10.5.4, choose an element $a \in Q$ which is superficial of order 1 for Q. Let $\overline{A} = A/(a)$ and $\overline{Q} = Q/(a)$; then $e(\overline{Q}, \overline{A}) = e(Q, A)$ and by induction we find a system of parameters $\overline{a_2}, \ldots, \overline{a_d}$ for $\overline{A}$ such that

$$e((\overline{a_2}, \ldots, \overline{a_d}), \overline{A}) = e(Q, A).$$

Then $\{a, a_2, \ldots, a_d\} \subset Q$ is the desired system of parameters for A. In fact, let $Q' = (a, a_2, \ldots, a_d)$. Then using Lemma 10.4.10 we deduce that

$$e(Q', A) \leq e(Q'/(a), \overline{A}) = e(\overline{Q}, \overline{A}) = e(Q, A),$$

and vice versa $e(Q', A) \geq e(Q, A)$ in any case, so we must have equality. □

10.6. Cohen's structure theorem

In this section, we discuss a famous result of Cohen ([**Coh46**]) that gives a precise description of complete local rings. In particular, we will be able to exactly describe the structure of a complete, *regular* local ring. Under an additional hypothesis, which we will describe below, such a ring is isomorphic to a ring of power series over its residue field. Geometrically, this is essentially an algebraic version of the inverse function theorem.

In differential geometry, the inverse function theorem describes the local structure of the zero locus V of k smooth functions $f_1, \dots, f_k$ defined on an open set in $\mathbb{R}^n$. Namely, if p is a point in V, and the differentials of the functions $\{f_i\}$ are independent in p, then V admits a local chart $\phi\colon \mathbb{R}^{n-k} \to V$ around p, giving to V the structure of an embedded differentiable manifold.

In the algebraic setting, one does not have the luxury of a local chart. Still, let $V \subset \mathbb{A}^n(k)$ be an affine variety of dimension d, p a point of V, and $A, \mathcal{M}$ the local ring of V in p. If V is regular in p, we will be able to say that *the completion* $\widehat{A}$ of A with respect to the $\mathcal{M}$-adic topology is isomorphic to $k[[x_1, \dots, x_d]]$. In a sense, from an analytic point of view V has something that resembles a local chart, even though the Zariski topology is too coarse to express this.

Our treatment follows in part [**Katb**] and [**Mur**], but the reader can also consult tag 0323 in [**dJea20**]. The equicharacteristic case is Theorem 7.8 in [**Eis95**].

In order to state our results we need

Definition 10.6.1. Let $A, \mathcal{M}$ be a local ring with residue field $k = A/\mathcal{M}$. We say that the field $K \subset A$ is a *coefficient field* if the induced map $K \to A/\mathcal{M} = k$ is surjective (hence an isomorphism).

This condition cannot always be ensured. For a simple example, A may be of characteristic 0, while k may have characteristic $p > 0$ (as it happens with the ring $\mathbb{Z}_p$). In this case, a coefficient field cannot exist.

In fact, a few cases can arise. If $\operatorname{char}(k) = 0$, we must have $\operatorname{char}(A) = 0$ as well. If instead $\operatorname{char}(k) = p > 0$, then either $\operatorname{char}(A) = 0$ or $\operatorname{char}(A) = p^n$ for some n. In fact, A is local and for every prime $q \neq p$ we have $q \notin \mathcal{M}$, so q is invertible in A.

Definition 10.6.2. Let $A, \mathcal{M}$ be a local ring with residue field $k = A/\mathcal{M}$. We say that A is *equicharacteristic* if $\operatorname{char}(A) = \operatorname{char}(k)$.

By the above discussion, if A is not equicharacteristic then $\operatorname{char}(k) = p > 0$ and either $\operatorname{char}(A) = 0$ or $\operatorname{char}(A) = p^n$ for some $n > 1$. If A admits a coefficient field, then A is equicharacteristic. To handle the other cases, we need a more subtle definition.

Definition 10.6.3. Let $A, \mathcal{M}$ be a local ring with residue field $k = A/\mathcal{M}$. We say that the ring $C \subset A$ is a *coefficient ring* if

(1) C is a complete, Hausdorff local ring
(2) the ideal $C \cap \mathcal{M}$ is generated by $p = \operatorname{char}(k)$
(3) the induced map $C \to A/\mathcal{M} = k$ is surjective.

Remark 10.6.4. From 3), it follows that the maximal ideal of C is $C \cap \mathcal{M} = p \cdot C$. Moreover, since C is Hausdorff in the topology defined by its maximal ideal, $\bigcap p^k \cdot C = 0$. If $I \subset C$ is any ideal, we can find a biggest k such that $I \subset p^k \cdot C$. In particular there exists $x \in I \setminus (p^{k+1} \cdot C)$, and we can write $x = p^k \cdot y$ for some $y \in C$. But $y \notin p \cdot C$, hence y is invertible in C and $I = p^k \cdot C$. It follows that C is a principal ideal domain, and all ideals of C are powers of its maximal ideal.

If moreover, $p \notin (C \cap \mathcal{M})^2$, it follows by Nakayama's lemma that all ideals $p^k \cdot C$ are distinct. In this case, one can define a valuation v on $C \setminus \{0\}$ by declaring that $v(x) = n$ if $x \in p^n \cdot C$ but $x \notin p^{n+1} \cdot C$ and extend it to the fraction field of C. In this case, C is a discrete valuation ring.

The following proposition allows us to get some structure from the mere existence of a coefficient field or ring.

Proposition 10.6.5. *Let $A, \mathcal{M}$ be a Noetherian complete local ring, $C \subset A$ a coefficient field or ring. Then there is a surjective map*

$$\phi\colon C[[x_1, \ldots, x_t]] \to A.$$

More precisely if $\mathcal{M} = (a_1, \ldots, a_t)$, the map defined by $\phi(x_i) = a_i$ is surjective, hence one can take $t = \operatorname{embdim} A$.

Proof. For $n \geq 0$, take an element $a \in \mathcal{M}^n$, and write

$$a = \sum b_i m_i,$$

where $b_i \in A$ and m_i is a monomial of degree n in $a_1, \ldots, a_t$. By definition of a coefficient field or ring, we can write $b_i = c_i + d_i$, where $c_i \in C$ and $d_i \in \mathcal{M}$. Hence we obtain a decomposition $a = c + d$, where $d \in \mathcal{M}^{n+1}$ and c is a form of degree n in $a_1, \ldots, a_t$ with coefficients in C.

Now given any $a \in A$ we apply this construction repeatedly, starting from

$$a = c_1 + d_1,$$

with c_1 a linear term in $a_1, \ldots, a_t$ with coefficients in C and $d_1 \in \mathcal{M}$. We repeat this writing

$$d_1 = c_2 + d_2,$$

and so on. After n steps, we have

$$a = (c_1 + c_2 + \cdot + c_n) + d_n,$$

where the term in parenthesis is a polynomial in $a_1, \ldots, a_t$ with coefficients in C. Taking the limit (which exists and is unique since A is complete and Noetherian) we get

$$a = \sum_{i=1}^{\infty} c_i,$$

which is a power series in $a_1, \ldots, a_t$ with coefficients in C.

We can define ϕ by sending x_i to a_i—then ϕ is well defined since A is complete, and is surjective by the above argument. □

In view of this result, it makes sense to ask when a complete local ring admits a coefficient field or a coefficient ring. This beautiful result of Cohen gives a pretty complete answer.

Theorem 10.6.6 (Cohen). *Let A be a complete Noetherian local ring.*

(i) *If A is equicharacteristic, then A admits a coefficient field.*

(ii) *If A is not equicharacteristic, then A admits a coefficient ring that is the image of a DVR.*

Before going into the proof of the theorem, we will state and prove some important consequences.

Corollary 10.6.7. *Let $A, \mathcal{M}$ be a complete regular local ring of dimension d, with residue field k.*

(i) *If A is equicharacteristic, then*

$$A \cong k[[x_1, \ldots, x_d]].$$

(ii) *If A is not equicharacteristic,* $\operatorname{char}(k) = p > 0$ *and* $p \notin \mathcal{M}^2$, *then there exists complete DVR C such that*

$$A \cong C[[x_1, \ldots, x_{d-1}]].$$

Proof. Let us first consider the equicharacteristic case. By Cohen's theorem, there is a coefficient field $k \subset A$. If $a_1, \ldots, a_d$ is a regular system of parameters, using Proposition 10.6.5, we find a surjection

$$\phi\colon k[[x_1, \ldots, x_d]] \to A$$

such that $\phi(x_i) = a_i$. It remains to prove that ϕ is injective. There are two ways to see this. For one thing, Proposition 10.1.8 implies that ϕ induces an isomorphism between the associated graded rings, and then one can use Remark 10.4.2. Alternatively, one sees that

$$\dim k[[x_1, \ldots, x_d]] = \dim A = d.$$

If $I := \ker \phi$ was not trivial, the dimension of A would be less than d since $k[[x_1, \ldots, x_d]]$ is integral, a contradiction.

The same proof works in the mixed characteristic case with slight modifications. In this case, we can take a regular system of parameters of the form $p, a_1, \ldots, a_{d-1}$. By Cohen's theorem, we find a coefficient ring C, and using Proposition 10.6.5 a surjection

$$\phi\colon C[[x_1, \ldots, x_{d-1}]] \to A$$

such that $\phi(x_i) = a_i$. By Remark 10.6.4, C is in fact a DVR, so the dimension on the two sides are equal, and ϕ is injective as in the previous case. □

Corollary 10.6.8. *Let A be a complete Noetherian local ring. Then A is the image of a complete* regular *local ring.*

Proof. This follows immediately by Proposition 10.6.5 and Theorem 10.6.6, since the rings $k[[x_1, \ldots, x_d]]$, where k is a field, and $C[[x_1, \ldots, x_{d-1}]]$, where C is a DVR, are complete and regular (see Exercise 10). □

Corollary 10.6.9. *Let $A, \mathcal{M}$ be a complete, equicharacteristic local ring of dimension d, with residue field k. Then A is a finitely generated module over a subring $B \subset A$ such that*

$$B \cong k[[x_1, \ldots, x_d]].$$

Proof. Let $a_1, \ldots, a_d$ be a system of parameters for A, and $Q = (a_1, \ldots, a_d)$. Since $\sqrt{Q} = \mathcal{M}$ and A is complete with respect to $\mathcal{M}$, it is also complete with respect to Q. It follows that it is a well-defined homomorphism

$$\phi\colon k[[x_1, \ldots, x_d]] \to A$$

such that $\phi(x_i) = a_i$. Let $B := \operatorname{im} \phi$. Since A/Q is finitely generated over B/Q, by Corollary 7.6.4 A is finitely generated as a B-module.

In particular, $\dim B = d$ by Proposition 9.5.10. Since $k[[x_1, \ldots, x_d]]$ is an integral domain, each chain of primes ends in (0), so ϕ must be injective, otherwise $\dim B < d$. It follows that $B \cong k[[x_1, \ldots, x_d]]$ as desired. □

We can also derive from Cohen's theorem a proof of the multiplicity one criterion of Nagata, albeit with some additional assumptions.

Partial Proof of Theorem 10.2.7. Let A be a local ring with maximal ideal $\mathcal{M}$. If A is regular, then $e(A) = 1$ by Remark 10.2.3. Moreover, the completion $\widehat{A}$ with respect to the $\mathcal{M}$-adic topology is regular as well, hence an integral domain by Theorem 10.1.9. So, the only associated prime of $\widehat{A}$ is 0, and A is unmixed.

Conversely, assume that $e(A) = 1$ and A is unmixed. To prove that A is regular, it is not restrictive to assume that A is $\mathcal{M}$-adically complete. To prove this implication, we make the additional assumption that A is equicharacteristic and the residue field $k = A/\mathcal{M}$ is infinite (for instance, both are true if $\operatorname{char} k = 0$). Since k is infinite, by Proposition 10.5.5 there is an ideal $Q \subset \mathcal{M}$ generated by a system of parameters such that $e(Q, A) = 1$.

If $Q = (a_1, \ldots, a_d)$, where $\dim A = d$, using Corollary 10.6.9, we know that A is finitely generated over the power series ring $B = k[[a_1, \ldots, a_d]]$. Using the additivity formula (Proposition 10.3.3) and the hypothesis that A

is unmixed, we conclude that A has only one associated prime of 0. If P is this associated prime, by primary decomposition it follows that $P = \mathcal{N}(A)$. Moreover, by the additivity formula, $\ell(A_P) = 1$, which means that A_P is a field. Since P is nilpotent, this can only happen if $P = 0$, so in fact A is an integral domain.

We can now use Corollary 10.3.15 to deduce that A has the same fraction field as B. Since B is integrally closed and A is an integral extension of B, we must have $A = B$, so A is regular. □

We now prove Cohen's theorem. We split the proof in various cases, which we treat differently. We recall the notation that A is complete local ring, $\mathcal{M}$ its maximal ideal and $k = A/\mathcal{M}$ its residue field.

Proof of Cohen's theorem when $\operatorname{char}(A) = \operatorname{char}(k) = 0$**.**

By Zorn's lemma, there exists a maximal subfield $K \subset A$. We will prove that the induced map $K \to k$ is surjective, hence an isomorphism. Let

$$\pi\colon A \to k$$

be the projection, and let $\overline{K} = \pi(K)$ be the image of K. If $\overline{K} \subsetneq k$, we find $a \in A$ such that $\overline{a} = \pi(a) \notin \overline{K}$.

If $\overline{a}$ is transcendental over $\overline{K}$, then a is transcendental over K. In this case, $K[a] \cap \mathcal{M} = 0$, so every nonzero element of $K[a]$ has an inverse in A. This means that $K(a) \subset A$, contradicting the maximality of K.

If $\overline{a}$ is algebraic over $\overline{K}$, let $\overline{f}$ be its minimal polynomial over $\overline{K}$, where $f \in K[x]$. Then, since $\operatorname{char}(k) = 0$, $\overline{a}$ is a *simple* root of $\overline{f}$, hence it can be lifted to a simple root of f in A by Hensel's lemma 7.6.2, call it a'. Again we find that $K(a') \subset A$, contradicting the maximality of K. □

We now pass to the equicharacteristic case where $\operatorname{char}(A) = \operatorname{char}(k) = p > 0$. In this case, the last step of the proof can fail, so we need a slightly adapted argument. The argument is similar to the proof of Hensel's lemma, but we need an additional result.

Lemma 10.6.10. *Let $A, \mathcal{M}$ be a local ring with residue field k and assume that* $\operatorname{char}(A) = \operatorname{char}(k) = p > 0$. *If $\mathcal{M}^p = 0$, then A admits a coefficient field.*

Proof. Since $\operatorname{char}(A) = p$, the set $A^p := \{a^p \mid a \in A\}$ is a subring of A. Moreover let $b = a^p \in A^p$ be any nonzero element. Then $a \notin \mathcal{M}$ (as $\mathcal{M}^p = 0$), so a has an inverse in A, and b has one in A^p. It follows that $A^p \subset A$ is a *subfield.*

By Zorn's lemma, there exists a maximal subfield $K \subset A$ containing A^p. We want to prove that the induced map $K \to k$ is surjective. Let

$$\pi : A \to k$$

be the projection, $\overline{K} = \pi(K)$ the image of K. If this is not the case, we find $a \in A$ such that $\overline{a} = \pi(a) \notin \overline{K}$.

The situation is similar to the previous proof, but this time we know that $\overline{a}^p \in \overline{K}$, so the minimal polynomial of $\overline{a}$ is $x^p - \overline{a}^p$. It follow that a satisfies the same polynomial over K, hence $K(a)$ is a subfield of A, contradicting the maximality of K. □

We can now easily conclude the proof of Cohen's theorem in the equi-characteristic case.

Proof of Cohen's theorem when $\operatorname{char}(A) = \operatorname{char}(k) = p > 0$**.**

Denote

$$\pi_n : \frac{A}{\mathcal{M}^{n+1}} \to \frac{A}{\mathcal{M}^n}$$

the projection. We will recursively find a coefficient field $K_n \subset A/\mathcal{M}^n$ such that $\pi_n(K_{n+1}) = K_n$. Then the inverse limit of the fields $\{K_n\}$ is the desired coefficient field.

We start with $n = 2$. In this case, Lemma 10.6.10 applies, as $p \geq 2$, and gives a coefficient field $K_2 \subset A/\mathcal{M}^2$.

For the induction step, assume that we have found K_n and let $B := \pi_n^{-1}(K_n)$. Also, denote $P := \ker \pi_n = \mathcal{M}^n/\mathcal{M}^{n+1}$.

Given $b \in B \setminus P$, let $\overline{b} := \pi_n(b) \neq 0$. Since $\overline{b} \in K_n$, we have $\overline{b} \notin \mathcal{M}/\mathcal{M}^n$, which implies that $b \notin \mathcal{M}/\mathcal{M}^{n+1}$. It follows that b is invertible in $A/\mathcal{M}^{n+1}$, and in fact the inverse of b lies in B. (Why?) This proves that B is local with maximal ideal P.

We can then apply Lemma 10.6.10 to the ring B having residue field $B/P \cong K_n$, since $P^p = 0$. This gives us a coefficient field in B, which we can take as K_{n+1}. □

For the proof of Cohen's theorem in the mixed characteristic case, we are going to rely on the theory of Witt vectors developed in Section 7.7.

Proof of Cohen's theorem when $\operatorname{char}(A) \neq \operatorname{char}(k)$**.**

First, assume that k is a perfect field, and let $p = \operatorname{char} k$. In this case, Theorem 7.7.11 tells us that the ring $W_p(k)$ is a DVR, having maximal ideal $\mathcal{M} = p \cdot W_p(k)$, and complete with respect to the $\mathcal{M}$-adic topology. Moreover, $k \cong W_p(k)/\mathcal{M}$.

By Theorem 7.7.12, there exists a homomorphism $W_p(k) \to A$. The image of $W_p(k)$ inside A is the desired coefficient ring C. Notice that in this case if $\operatorname{char} A = 0$, the map is injective, hence C is itself a DVR.

The case where k is not perfect is done by reduction to the previous case, but we are only giving a brief sketch. The steps are:

(i) Construct a DVR V complete with respect to its maximal ideal $\mathcal{M}$ and such that $k \cong V/\mathcal{M}$, even when k is not perfect

(ii) Consider the perfect closure $\overline{k}^{per}$ of k—this is a construction similar to the algebraic closure of k, but done by recusively adding p-th roots of elements of k (see Definition A.3.30).

(iii) By a similar procedure, starting from A, construct another complete local ring $\overline{A}^{per}$ with residue field $\overline{k}^{per}$.

(iv) Apply Theorem 7.7.12 to find a lift $\phi\colon W_p(\overline{k}^{per}) \to \overline{A}^{per}$

(v) Prove that $V \subset W_p(\overline{k}^{per})$, and in fact $\phi(V) \subset A$—this is the desired coefficient ring C.

Some more details can be found in [**Katb**], or the original paper [**Coh46**]. □

10.7. Exercises

1. Prove directly that a regular local ring of dimension 1 is a discrete valuation ring.

2. Let A be a regular ring of dimension 1. Prove that A is a Dedekind domain.

3. Prove the properties stated in Proposition 10.3.1.

4. Compute the multiplicity of n lines meeting at the origin in $\mathbb{A}^2$.

5. Use Example 10.3.7 to compute the multiplicity $e(A)$, where A is the local ring of the node $y^2 = x^2 + x^3$ in a simpler way.

6. Verify the final computation (10.2.3) in Proposition 10.2.14.

7. Let A be an integral domain of dimension 1, and assume that the order function v_A is a valuation. Prove that A is a discrete valuation ring.

8. Let $B = k[x, y]/(x^2, xy)$ and A the localization of B at 0. Compute $\mu_A(y)$ and $\mu_A(y^2)$ and show that $\mu_A(y^2) \neq 2\mu_A(y)$—in particular the hypothesis that a is a regular element is necessary in Proposition 10.4.6.

9. Let A be a Noetherian ring, $a \in A$ a nonzero divisor. Assume that for $a \in \mathcal{M} \setminus \mathcal{M}^2$ for every maximal ideal $\mathcal{M} \subset A$. If $A/(a)$ is regular, prove that A is regular.

10. Let A be a regular local ring. Prove that the ring $A[[x]]$ is regular (use the previous exercise).

11. Let $A, \mathcal{M}$ be a complete local ring of dimension d, of mixed characteristic, and let $p = \operatorname{char} A/\mathcal{M}$. Assume that $\operatorname{ht}(p \cdot A) = 1$. Prove that A is a finitely generated module over a subring $B \subset A$ such that

$$B \cong C[[x_1, \ldots, x_{d-1}]],$$

where C is a DVR.

12. Prove Corollary 10.4.14—that is, $e(A[[x]]) = e(A)$ for a local Noetherian ring A—by a direct computation (it is easier to write the Hilbert polynomial of A as a sum of binomial coefficients, instead of powers of n).

The following exercises discuss the notion of *reduction* of ideals as a means to compute multiplicities. Given a ring A with two ideals $J \subset I$, we say that J is a *reduction* of I if $J \cdot I^n = I^{n+1}$ for some $n \in \mathbb{N}$.

13. Let A be a semilocal Noetherian ring, $J \subset I$ two ideals of definition. Prove that if J is a reduction of I, then $e(J) = e(I)$.

A sort of converse was proved by Rees in [**Ree61**]:

Theorem (Rees). *Let $A, \mathcal{M}$ be a local Noetherian ring. Assume that A is unmixed, and let $J \subset I$ be two $\mathcal{M}$-primary ideals such that $e(J) = e(I)$. Then J is a reduction of I.*

14. Let A be a Noetherian ring, $J \subset I \subset A$ two ideals. Let B be the integral closure of A in its total fraction ring. Prove that J is a reduction of I if and only if I and J have the same integral closure inside B.

15. Let $A = \mathbb{C}[[x, y]]$ and $I = (x^3, x^2y, y^2)$. Find a reduction of I and use it to compute $e(I)$.

16. Use Corollary 10.3.15 to give an alternative proof of the inertia-ramification formula (Theorem 6.3.2).

The following exercises, up to Exercise 21, discuss a structure result for principal ideal *rings*, due to Hungerford [**Hun68**], building on Cohen's theorem and [**ZS76a**, Theorem 33, Part IV]. A *principal ideal ring* is just a ring (not necessarily an integral domain) whose ideals are all principal. Such a ring is called *special* if it has a single prime ideal.

17. Let A be a principal ideal ring, $P_1, P_2 \subset A$ prime ideals. Prove that either P_1 and P_2 are coprime, or $P_1 \subset P_2$, or $P_2 \subset P_1$.

18. Let A be a principal ideal ring, $P_1 \subset P_2 \subset A$ prime ideals. If Q is a P_2-primary ideal, then $P_1 \subset Q$.

19. Let A be a principal ideal ring. Prove that A is a finite direct sum of principal ideal *domains* and special principal ideal rings. (Use primary decomposition for 0 and the previous exercises.)

20. Let A be a special principal ideal ring. Prove that A is an image of a PID. Deduce the theorem of Hungerford: every principal ideal ring is a finite direct sum of images of principal ideal domains. (A is a complete local ring, so we can apply the structure theorem of Cohen. If A has a coefficient field, we are done. Assume that A is the image of $C[[x]]$, where C is a coefficient ring. Find a quotient of $C[[x]]$ which is a PID and still surjects onto A.)

21. One may want a stronger form of Hungerford's theorem, but it is not the case that every principal ideal ring is a quotient of a principal ideal domain. To see this, take $A = \mathbb{R} \oplus \mathbb{Z}$, and show that $\mathbb{Q}$, seen as an A-module with a trivial $\mathbb{R}$-action, is not a direct sum of cyclic A-modules. Conclude by the classification of modules over a PID. (Compare this with Exercise 32 in Chapter 7.)

Appendix A

Fields

In this appendix, we investigate the properties of fields. The theory has a different flavor from the study of general rings, for at least two reasons. For one thing, all field homomorphisms are injective, so the study of morphisms between fields is the same as the study of *field extensions*. Second, fields have no proper ideals, and modules over fields are vector spaces, hence they are fully characterized by their dimension alone. As a consequence, many interesting questions of ring theory become moot in this setting. On the other hand, we will be able to prove much more about field extensions than about general ring homomorphisms.

This appendix goes through the standard results of field theory, up to the Galois correspondence and the Kummer theory of Abelian extensions. First, we study algebraic and transcendental extensions. Next, we specialize to the algebraic setting and introduce the notion of separability. This is in order to investigate the phenomenon—which appears in positive characteristic—that irreducible polynomials can have multiple roots. In the next section, we study normal extensions, which are maximally symmetric. Extensions that are both normal and separable admit a Galois correspondence which relates intermediate fields and subgroups of the automorphism group. We explain the Galois correspondence both in the finite and infinite case. In the last section, we specialize the theory to the case of Abelian extensions, where a more precise description can be obtained.

This appendix covers more than what is needed in the book, which only uses the notion of transcendence degree, basic results on separability, and the Galois correspondence in the finite case. Even so, many important facts about fields are not covered here. In particular, many classical applications of Galois theory to the feasibility of geometric constructions with ruler and compass are only presented in the exercises, and we just hint at the problem of inverse Galois theory.

A.1. Algebraic elements

Let K be a field. If $L \supset K$ is a bigger field and $S \subset L$ is any set, we denote by $K(S)$ the smallest subfield of L containing S—this is the field *generated* by S over K. If $\{L_i\}$ is a collection of subfields of L, we denote by $\prod_i L_i$ their *composite*, that is, the smallest subfield of L containing all the L_i. In other words,

$$\prod_i L_i = K\left(\bigcup_i L_i\right).$$

The composite of two fields L_1 and L_2 will be simply denoted L_1L_2.

As we have remarked, any ring morphism $K \to L$ between fields is injective. For this reason, we are going to be concerned with field *extensions*, that is, given a field K we will investigate the fields L such that $K \subset L$. We usually denote such an extension by the notation L/K. The first observation is that in this case L is a vector space over K, by restriction of the multiplication map.

Let us start with the case where $L = K(\alpha)$ is obtained by adding a single element to K. The natural evaluation map

$$\begin{aligned} ev_\alpha\colon\ K[x] &\longrightarrow L \\ f(x) &\longmapsto f(\alpha) \end{aligned}$$

is injective if and only if α does not satisfy any algebraic equation with coefficients in K. In this case, we can extend ev_α to an isomorphism $K(x) \cong L$ between L and the field of rational functions in one variable over K. In this case, the structure of L is easy to understand. We capture this distinction as follows:

Definition A.1.1. Let $K \subset L$ be two fields, $\alpha \in L$. We say that α is *algebraic* over K if there exists $f \in K[x]$ such that $f(\alpha) = 0$; otherwise we say that α is *transcendental* over K.

If all elements of L are algebraic over K, we say that L is an *algebraic* extension of K, otherwise that L is a *transcendental* extension of K.

Let α be an element algebraic over K. Then the set of polynomials that vanish on α is a nonzero ideal of $K[x]$. Since $K[x]$ is a UFD, this ideal is generated by a polynomial of minimal degree.

Definition A.1.2. Let α be an element algebraic over K, I the ideal of $K[x]$ consisting of polynomials vanishing at α. Any generator of I is called a *minimal polynomial* of α over K. We usually normalize such a polynomial by requiring that its leading coefficient is 1, in which case we speak of *the* minimal polynomial of α, and denote it by μ_α.

Remark A.1.3. Let $K \subset L$ be two fields, $\alpha \in L$. Then saying that α is algebraic over K amounts to a relation of linear dependence between the powers $1, \alpha, \alpha^2, \ldots$. Hence, α is algebraic over K if and only if $K(\alpha)$ is a vector space of finite dimension over K.

To restate this, we introduce another piece of terminology.

Definition A.1.4. Let $K \subset L$ be a field extension. Then we say that L is a *finite* extension of K if L is finite dimensional as a vector space over K. The dimension of this vector space is called the *degree* of L over K, denoted $[L : K]$.

With this terminology, α is algebraic over K if an only if $K(\alpha)/K$ is a finite extension.

Remark A.1.5. Let $K \subset L \subset M$ be three fields. If L/K and M/L are finite, then M/K is finite as well, and $[M : K] = [M : L] \cdot [L : K]$. We say that the degree is multiplicative in towers of field extensions.

The above remark has an important consequence:

Proposition A.1.6. *Let L/K be an extension of fields, $\alpha, \beta \in L$ elements algebraic over K. Then $\alpha + \beta$, $\alpha \cdot \beta$ and α/β (when $\beta \neq 0$) are algebraic over K.*

Proof. All such elements are contained in $K(\alpha, \beta)$, which is a finite extension of K—since $K(\alpha, \beta)/K(\alpha)$ and $K(\alpha)/K$ are finite. □

Corollary A.1.7. *Let $K \subset M$ be fields, and let $\{L_i\}$ be a set of intermediate fields. If all L_i/K are algebraic extensions, then the composite of the L_i is also algebraic over K.*

Proof. Every element in the composite is a rational function in finitely many algebraic elements. □

Corollary A.1.8. *Let L/K be an extension of fields. The set of elements of L that are algebraic over K is a subfield of L, called the* algebraic closure *of K in L.*

Usually, when we talk about the algebraic closure, we do so in an absolute sense: the algebraic closure of K is a field obtained from K by the process of adding all possible algebraic elements, that is, all possible roots of polynomials with coefficients in K. To make this precise, we need some more language.

Definition A.1.9. Let K be a field. We say that K is *algebraically closed* if every polynomial $f \in K[x]$ has at least one root in K.

The field $L \supset K$ is called an *algebraic closure* of K if it is algebraically closed and algebraic over K.

Shortly we will prove the existence and uniqueness of the algebraic closure. Still, before doing so, we first need to understand what it means to add to K the root of an irreducible polynomial $f \in K[x]$. We need to find a field $L \supset K$ such that f admits a root in L. This is easy, since $K[x]$ is a PID: the ideal generated by f is prime, because f is irreducible, hence maximal. It follows that $K[x]/(f)$ is a field, and by construction f has a root in this field (the image of x). By generalizing this construction we can prove:

Proposition A.1.10. *Let K be a field. Then K admits an algebraic closure.*

Proof. Let F be the set of irreducible polynomials of positive degree in $K[x]$, and for each $f \in F$ take a formal variable x_f. In the ring $K[x_f]_{f \in F}$ the ideal $(x_f)_{f \in F}$ is not the whole ring, hence it is contained in a maximal ideal $\mathcal{M}$. Then $K_1 := K[x_f]_{f \in F}/\mathcal{M}$ is an algebraic extension of K because of Corollary A.1.7, and each irreducible polynomial in K has a root in K_1.

Now K_1 may not be algebraically closed, but we can repeat the process to produce a field $K_2 \supset K_1$ such that all irreducible polynomials with coefficients in K_1 have a root in K_2, and so on. The union of the ascending chain $K \subset K_1 \subset K_2 \subset \cdots$ is an algebraic closure of K. □

The algebraic closure of a field enjoys a uniqueness property, although noncanonical. We can state a slightly stronger result.

Proposition A.1.11. *Let L/K be an algebraic extension, T an algebraically closed field with a morphism $\phi\colon K \to T$. Then ϕ can be extended to a morphism $\phi_L\colon L \to T$.*

Proof. By applying Zorn's lemma to pairs (M, ϕ_M)—where $K \subset M \subset L$ is a field and $\phi_M\colon M \to T$ is a morphism extending ϕ—one finds a maximal pair (H, ϕ_H). We claim that $H = L$. If this was not the case, take any $\alpha \in L \setminus H$, and let f be its minimal polynomial. By the embedding ϕ_H, we can see H as a subfield of T. The morphism ϕ_H can be extended to $H(\alpha)$ by sending α to any root of f inside T (check this!). □

Corollary A.1.12. *If L_1, L_2 are two algebraic closures of the field K, then there is an isomorphism $L_1 \cong L_2$.*

Proof. The above proposition allows us to extend the inclusion $K \subset L_2$ to a morphism $f\colon L_1 \subset L_2$. Since L_1 is a field, f is injective. It is also surjective, otherwise the inverse morphism $f(L_1) \to L_1$ could not be extended to the whole L_2. □

The isomorphism between different algebraic closures is noncanonical, but usually we are not concerned with this. Most of the time, we will just stick with one choice for the algebraic closure of a field K, which we will call *the* algebraic closure of K and denote by $\overline{K}$.

We now move to the general setting of an extension that may not necessarily be algebraic. To understand such an extension, it is useful to factor it in two stages: first a maximally transcendental extension, and then an algebraic one.

Definition A.1.13. Let L/K be an extension of fields. The set $A \subset L$ is said to be *algebraically independent* over K if there is no nonzero polynomial $f \in K[x_1, \ldots, x_n]$ such that $f(a_1, \ldots, a_n) = 0$ for some $a_1, \ldots, a_n \in A$.

A is called a *transcendence basis* of L over K if it is algebraically independent and the extension $L/K(A)$ is algebraic.

Algebraically independent sets are ordered by inclusion, and by Zorn's lemma, there is such a maximal set. Clearly, a maximal algebraically independent set is a transcendence basis, so transcendence bases always exist. Notice that if A is an algebraically independent set over K, we can form a set of indeterminates $\{x_a\}_{a \in A}$ and define a homomorphism $K(\{x_a\}) \to L$ that sends x_a to a. Hence, every field extension factors as an extension isomorphic to a rational function field, followed by an algebraic extension.

What is less obvious is the following result.

Theorem A.1.14. *Let L/K be a field extension. Any two transcendence bases for L/K have the same cardinality.*

Definition A.1.15. Given a field extension L/K, the cardinality of any transcendence bases for it is called the *transcendence degree* of L over K, denoted $\operatorname{trdeg}_K L$.

Proof. Let A, B be two transcendence bases, and assume at first that at least one of them is finite, say $A = \{a_1, \ldots, a_r\}$. Each $b \in B$ is algebraic over $K(A)$, so we can find a polynomial $p_b \in K(A)[x]$ such that $p_b(b) = 0$. Choose an element of A, say a_1. There exists a $b \in B$ such that a_1 appears in p_b, otherwise all elements of B would be algebraic over $K(a_2, \ldots, a_r)$. Since a_1 is algebraic over $K(B)$, it would follow that a_1 is algebraic over $K(a_2, \ldots, a_r)$, a contradiction.

Choose one such element b_1, and let p_1 be the associated polynomial relation. Then, we claim that $C := \{b_1, a_2, \ldots, a_r\}$ is another transcendence basis. In fact, a_1 is algebraic over $K(C)$ (using p_1), so L is algebraic over

$K(C)$ as well. Moreover, a polynomial relation between elements of C would show that b_1 is algebraic over $K(a_2, \ldots, a_r)$, which we have already excluded.

By iterating this procedure, we find elements $b_1, \ldots, b_r \in B$ that constitute a transcendence bases, hence B is finite and $|B| = r$.

It remains to prove the theorem when both A and B are infinite. In this case, for each $b \in B$ choose a finite subset $A_b \subset A$ such that b is algebraic over $K(A_b)$, and let

$$A^* := \bigcup_{b \in B} A_b.$$

Then clearly A^* is a subset of A of the same cardinality of B. Moreover, every $a \in A$ is algebraic over B, hence over A^*. Since A is algebraically independent, it follows that $A = A^*$, and so the theorem follows. □

This proof should be compared to the result that every two bases of a vector space have the same cardinality. The basic idea is similar, but in the vector space case things are made simpler by the fact that one can use a linear equation to express one variable in terms of the other ones—something that one cannot do with polynomial relations of higher degree.

Remark A.1.16. Let $K \subset L \subset M$ be three fields. If A is a transcendence basis for L/K and B one for M/L, it is immediate to check that $A \cup B$ is a transcendence basis for M/K. In particular,

$$\operatorname{trdeg}_K M = \operatorname{trdeg}_K L + \operatorname{trdeg}_L M.$$

A.2. Finite fields

Let K be a finite field. Then K has positive characteristic p, a prime, hence it contains a copy of $\mathbb{Z}/p\mathbb{Z}$, which we will call the *prime field* of K. We can regard K as a vector space over $\mathbb{Z}/p\mathbb{Z}$, of some finite dimension d—in particular K has $q := p^d$ elements.

We are now going to turn things around. We fix such prime power $q = p^d$, and let K be a field with q elements. We will show that K is uniquely determined up to isomorphism. First, all elements of K are algebraic over $\mathbb{Z}/p\mathbb{Z}$, so we can assume that K is a subfield of the algebraic closure $\overline{\mathbb{Z}/p\mathbb{Z}}$. We are going to identify K by looking at the multiplicative group K^*. This has $q-1$ element, each of which satisfies the equation $x^{q-1} = 1$. Adding 0, we get that all elements of K satisfy the equation

$$x^q = x. \tag{A.2.1}$$

Since this equation has at most q roots, this determines K as a set, that is,

$$K = \left\{ x \in \overline{\mathbb{Z}/p\mathbb{Z}} \mid x^q = x \right\}. \tag{A.2.2}$$

This is enough to prove the uniqueness of the field with q elements.

Moreover, the set of solutions to (A.2.1) is easily checked to be closed under sum, multiplication, and inversion, since K has characteristic p. It follows that we can use (A.2.2) as a *definition* of K, and this is in fact a field. The roots of (A.2.1) are distinct because the derivative of $x^q - x$ in characteristic p is -1, which never vanishes (see Proposition A.3.3). This implies that (A.2.1) has *exactly* q distinct roots. Hence the field K defined by (A.2.1) has exactly q elements, and we have an existence result.

We summarize the previous discussion:

Theorem A.2.1. *Let p be a prime, $q = p^d$. Inside $\overline{\mathbb{Z}/p\mathbb{Z}}$ there exists a unique field with q elements, which is the set of solutions to $x^q = x$.*

Since the field with q elements is essentially unique, we are going to give it a name, and denote it $\mathbb{F}_q$. In particular, $\mathbb{F}_p$ is just $\mathbb{Z}/p\mathbb{Z}$, seen as a field.

In the rest of the section, we are going to analyze the structure of $\mathbb{F}_q$ in more detail. First, $\mathbb{F}_q^*$ is cyclic. Actually, this holds in slightly greater generality:

Proposition A.2.2. *Let K be a field, $G \subset K^*$ a finite multiplicative subgroup. Then G is cyclic.*

Proof. By Corollary 1.5.7, G is a product of finite cyclic groups

$$G \cong G_1 \times \cdots \times G_r.$$

If G was not cyclic, the lowest common multiple of the cardinalities of the G_i would be strictly less than their product. If m is such a lowest common multiple, this means that $|G| > m$, and all elements $g \in G$ satisfy the equation $g^m = 1$. This cannot happen in a field, since this equation has at most m roots. □

In particular, inside $\mathbb{F}_q$, there exists a *primitive* $q-1$-th root of 1, which generates the whole field.

Proposition A.2.3. *Let p be a prime, $q = p^d$. Then there exists $\alpha \in \mathbb{F}_q$ such that α generates $\mathbb{F}_q^*$ as a multiplicative group—in particular, $\mathbb{F}_q = \mathbb{F}_p(\alpha)$.*

Since the fields $\mathbb{F}_{p^d}$ all live inside $\overline{\mathbb{F}_p}$, we can also consider whether there is any inclusion among them.

Proposition A.2.4. *Let a, b be natural numbers, p a prime. Then $\mathbb{F}_{p^a} \subset \mathbb{F}_{p^b}$ if and only if a divides b.*

Proof. One direction is obvious: if b is multiple of a, then all solutions of $x^a = x$ are also solutions of $x^b = x$.

Conversely, assume $\mathbb{F}_{p^a} \subset \mathbb{F}_{p^b}$. The inclusion between the multiplicative groups implies that $p^a - 1$ divides $p^b - 1$. In other words,

$$p^b \equiv 1 \pmod{p^a - 1}.$$

The order of p modulo $p^a - 1$ is clearly a, hence b is a multiple of a. $\square$

The fields $\mathbb{F}_q$ are also endowed with automorphisms.

Definition A.2.5. Let K be a field of characteristic p. The map $x \to x^p$ is a field automorphism of K, which is called the *Frobenius* endomorphism of K.

For a general field, the Frobenius endomorphism need not be surjective: for instance in $\mathbb{F}_p(x)$, the indeterminate x is not the p-th power of any element. However, for finite fields, every injective endomorphism is also surjective, hence an automorphism.

Let $\phi\colon \mathbb{F}_q \to \mathbb{F}_q$ be the Frobenius automorphism on $\mathbb{F}_q$, with $q = p^d$. Then $\phi^d(x) = x^q = x$ for all $x \in \mathbb{F}_q$, so ϕ^d is the identity. Also, ϕ^c is not the identity for any $c < d$, since ϕ^c only fixes elements in the field $\mathbb{F}_{p^c}$, which does not contain $\mathbb{F}_q$. We conclude:

Proposition A.2.6. *Let p be a prime, $q = p^d$. The Frobenius element is an automorphism of $\mathbb{F}_q$, fixing each elements of $\mathbb{F}_p$, and it has order d.*

A.3. Separability

Before delving into Galois theory, we need to understand a behavior which is typical of extensions of fields in positive characteristic. In many situations, it would be tempting to assume that an irreducible polynomial must have distinct roots, but this is not always the case.

Example A.3.1. Let K be a field of characteristic $p > 0$, $L = K(t)$ for some indeterminate t. The polynomial $f(x) = x^p - t \in L[x]$ is irreducible. In fact, if it was reducible, we would have a nontrivial factorization $f(x) = a(x,t)b(x,t)$ for some $a, b \in K[x,t]$ by Gauss's lemma, and this is easily excluded since f has degree 1 in t.

On the other hand, let $\alpha \in \overline{L}$ be any root of f, so that $\alpha^p = t$. Then

$$(x - \alpha)^p = x^p - \alpha^p = x^p - t = f(x),$$

since $\operatorname{char} L = p$. It follows that α has multiplicity p as a root of f, and in particular f does not have distinct roots.

The above is the prototypical example of an inseparable extension, which we are now going to define. As it turns out, this is not only a nuisance, but the manifestation of a fundamental phenomenon typical of characteristic p extensions.

Definition A.3.2. Let K be a field.

(1) The irreducible polynomial $f \in K[x]$ is called *separable* if it has distinct roots in $\overline{K}$—that is, it factorizes over $\overline{K}$ as a product of distinct linear factors.

(2) The element $\alpha \in \overline{K}$ is called *separable* over K if the minimal polynomial μ_α is separable.

(3) The algebraic extension L/K is called *separable* if every element of L is separable over K.

In all this cases, we will use the word *inseparable* to mean not separable.

The main tool to undestand this notion is the following standard criterion using derivatives.

Proposition A.3.3. *Let K be a field, $f \in K[x]$ a polynomial. Then f has distinct roots in $\overline{K}$ if and only if f and f' do not share common factors.*

Proof. If f has a multiple root a, we can write $f(x) = (x-a)^2 g(x)$ over $\overline{K}$. Since

$$f'(x) = 2 \cdot (x-a)g(x) + (x-a)^2 g'(x),$$

$(x-a)$ is a common factor of f and f'.

Conversely, if f has distinct roots $a_1, \ldots, a_n$, we can write

$$f(x) = \lambda(x-a_1)\cdots(x-a_n),$$

and then for each i we have

$$f'(a_i) = \lambda(a_i - a_1)\cdots\widehat{(a_i - a_i)}\cdots(a_i - a_n),$$

where the term $a_i - a_i$ is omitted, so $f'(a_i) \neq 0$. □

Corollary A.3.4. *Let K be a field, $f \in K[x]$ an irreducible polynomial. Then f has distinct roots if and only if $f' \neq 0$.*

Proof. Since both f and f' are defined over K, their greatest common divisor is also with coefficients in K. Assuming f is irreducible, it cannot share a nontrivial factor with f', unless $f' = 0$. □

A simple consequence is that a field of characteristic 0 is always separable: in fact, the derivative of a nonzero polynomial cannot be 0. In characteristic $p > 0$, the derivative of f can be identically 0 exactly when f contains only powers of x^p, so we conclude:

Corollary A.3.5. *Let K be a field of characteristic $p > 0$. Then $f \in K[x]$ is inseparable if and only if $f(x) = g(x^p)$ for some other polynomial $g \in K[x]$.*

Separable finite extensions are easier to study thanks to the the *primitive element theorem*, which ensures that they are generated by a single element.

Theorem A.3.6. *Let L/K be a finite separable extension. Then there exists $\alpha \in L$ such that $L = K(\alpha)$.*

Proof. By induction, it is enough to prove that if $L = K(\alpha, \beta)$, we can find a single $\gamma \in L$ such that $L = K(\gamma)$. Let μ_α, μ_β be the minimal polynomials of α and β.

Take any combination $\gamma = \alpha + t\beta$ for some fixed $t \in K$. The key observation is that the polynomial $f(x) := \mu_\alpha(\gamma - tx)$ is defined over $K(\gamma)$ and vanishes for $x = \beta$. Hence, f and μ_β have β as a common root. If we can ensure that this is the *only* common root, then the gcd of f and μ_β is $x - \beta$, and in particular $\beta \in K(\gamma)$ (and so also $\alpha \in K(\gamma)$, which means we are done).

Say μ_α, μ_β have roots $\alpha_1 = \alpha, \alpha_2, \ldots, \alpha_r$ and $\beta_1 = \beta, \beta_2, \ldots, \beta_s$. These sets of roots are distinct because L/K is separable. Then β is the only common root of f and μ_β, unless

$$\alpha_i = \gamma - t\beta_j = \alpha + t\beta - t\beta_j,$$

for some $i, j \neq 1, 1$. This can be solved for t, giving

$$t = -\frac{\alpha - \alpha_j}{\beta - \beta_j}.$$

In other words, it is enough to choose t outside this finite set of values to prove the thesis. This prove the theorem assuming that K is infinite—the finite case has been already proved as Proposition A.2.3. □

In order to measure exactly the failure of an extension to be separable, we define a *separable degree*.

Definition A.3.7. Let L/K be an algebraic field extension, $\overline{K}$ an algebraic closure of K. The *separable degree* of L over K, denoted $[L : K]_s$, is the cardinality of $\mathrm{Hom}_K(L, \overline{K})$.

We can assume that in fact $L \subset \overline{K}$, in which case, this is the same as the cardinality of the set

$$\left\{\phi \colon L \to \overline{K} \mid \phi|_K = \mathrm{id}\right\}.$$

Using this observation, it is easy to check that this notion behaves nicely in towers of extensions.

Proposition A.3.8. *Let $K \subset L \subset M$ be algebraic extensions of fields. Then*

$$[M : K]_s = [L : K]_s \cdot [M : L]_s.$$

Proof. Let $\sigma \in \mathrm{Hom}_K(L, \overline{K})$ and $\tau \in \mathrm{Hom}_L(M, \overline{K})$. We can extend σ to a homomorphism $\overline{\sigma}\colon \overline{K} \to \overline{K}$, and form the composition

$$(\overline{\sigma} \circ \tau)\big|_M \in \mathrm{Hom}_K(M, \overline{K}).$$

It is a simple verification that this gives a bijection

$$\mathrm{Hom}_K(L, \overline{K}) \times \mathrm{Hom}_L(M, \overline{K}) \to \mathrm{Hom}_K(M, \overline{K}). \qquad \square$$

The connection between this degree and the notion of separability is as follows. Let α be algebraic over K, and take $L = K(\alpha)$. Let μ_α be the minimal polynomial of α over K, and $\alpha_1, \dots, \alpha_r$ the *distinct* roots of μ_α. Then a homomorphism $\phi\colon L \to \overline{K}$ as K-algebras is uniquely defined by choosing $\phi(\alpha)$, which must be one of the α_i. Moreover, each α_i appears as $\psi(\alpha)$ for a suitable homomorphism ψ. (Why?) In other words, $[L : K]_s$ is exactly the number of distinct roots of μ_α. On the other hand, $[L : K] = \deg \mu_\alpha$ is the number of roots of μ_α, *counted with multiplicities.* We conclude:

Proposition A.3.9. *Let $L = K(\alpha)$, with α algebraic over K. Then $[L : K] = [L : K]_s$ if and only if α is separable over K.*

Corollary A.3.10. *Let L/K be a finite, separable extension. Then L admits exactly $[L : K]$ distinct embeddings into $\overline{K}$.*

Proof. By Theorem A.3.6, we can write $L = K(\alpha)$ for a separable element α, so $[L : K] = [L : K]_s$. Then use the definition of separable degree. $\square$

In general, all roots of μ_α have the same multiplicity m, so that $[L : K] = m[L : K]_s$. We can use this to prove the following important corollary, that in particular implies that separability of a field extension L/K is determined by the separability of *generators* of L over K. This makes checking whether an extension is separable much more manageable.

Corollary A.3.11. *Let L/K be a finite algebraic extension, generated by the finite set $\{\alpha_1, \dots, \alpha_s\}$. Then the following are equivalent:*

(i) *L is separable over K,*

(ii) *Each α_i is separable over K,*

(iii) *$[L : K] = [L : K]_s$.*

Proof. Implications i) to ii) to iii) are immediate, so we only need to prove that iii) implies i).

To do this, take any $\alpha \in L$, so that $[K(\alpha):K] = m[K(\alpha):K]_s$ for some $m \geq 1$. Then

$$\begin{aligned}[L:K] &= [L:K(\alpha)][K(\alpha):K] = [L:K(\alpha)] \cdot m[K(\alpha):K]_s \\ &\geq [L:K(\alpha)]_s \cdot m[K(\alpha):K]_s = m[L:K]_s = m[L:K],\end{aligned}$$

which implies that $m = 1$, and so that α is separable over K. □

This result is easily generalizable to the infinite case. Since any element algebraic over K lives in a finite extension of K, we have

Proposition A.3.12. *Let L/K be an algebraic extension, where $L = K(S)$ for some set S. Then L is separable over K if and only if each $s \in S$ is separable over K.*

We can also use the above results to guarantee that separability is preserved in a tower of extensions.

Proposition A.3.13. *Let $K \subset L \subset M$ be algebraic extensions. Then M/K is separable if and only if both L/K and M/L are separable.*

Proof. One direction is obvious, so assume that L/K and M/L are separable. If the extensions are finite, the result follows from Corollary A.3.11 and the multiplicativity of separable degree. In the infinite case, take any $\alpha \in M$, and let μ_α be its minimal polynomial over L. Then we can consider the finite extension F of K generated by the coefficients of μ_α. Both extensions $F(\alpha)/F$ and F/K are finite and separable, hence $F(\alpha)/K$ is separable as well. □

Corollary A.3.14. *Let $L, M \subset \overline{K}$ be fields separable over K. Then the composite $L \cdot M = L(M)$ is itself separable.*

The above result allows us to define the biggest separable extension of a field. In fact, let K be a field, $\overline{K}$ a fixed algebraic closure. Then the composite field

$$\overline{K}^{sep} := \prod_{L \subset \overline{K} \text{ separable}} L \tag{A.3.1}$$

is itself separable. In fact, every element of $\overline{K}^{sep}$ lives in a finite product of separable extensions of K, which is itself separable by repeated applications of Corollary A.3.14.

Definition A.3.15. The field $\overline{K}^{sep}$ defined by (A.3.1) is called the *separable closure* of K.

Now that we have a pretty clear picture of separable extensions, we are going to investigate the opposite situation. Let K be a field of characteristic $p > 0$, $f \in K[x]$ an irreducible polynomial. By Corollary A.3.5, f is inseparable if and only if we can write $f(x) = g(x^p)$ for some $g \in K[x]$. If g is itself inseparable, we can repeat the process—after a finite number of steps we end up writing $f(x) = h(x^{p^r})$ for some $r > 0$ and some separable irreducible polynomial $h \in K[x]$.

Definition A.3.16. Let L/K be a finite extension. We define its *inseparable degree*, denoted $[L : K]_i$, as the quotient

$$[L : K]_i := \frac{[L : K]}{[L : K]_s}.$$

By induction of the number of generators of L over K and the above discussion, we have:

Proposition A.3.17. *Let L/K be a finite extension of characteristic $p > 0$. Then $[L : K]_i$ is a power of p.*

Remark A.3.18. We can rephrase Corollary A.3.11 to conclude that L/K is separable if and only if $[L : K]_i = 1$. Moreover, if $K \subset L \subset M$ are finite extensions of fields, we have

$$[M : K]_i = [L : K]_i \cdot [M : L]_i$$

by Proposition A.3.8.

We can mimic Definition A.3.2 to describe the opposite phenomenon.

Definition A.3.19. Let K be a field.

(1) The polynomial $f \in K[x]$ is called *purely inseparable* if it has a single distinct root in $\overline{K}$.

(2) The element $\alpha \in \overline{K}$ is called *purely inseparable* over K if the minimal polynomial μ_α is purely inseparable.

(3) The algebraic extension L/K is called *purely inseparable* if every element of L is purely inseparable over K.

By definition, if $\alpha \in L$ is purely inseparable over K, μ_α has a single distinct root in $\overline{K}$, which means that $[K(\alpha) : K]_s = 1$. We can say a little more:

Proposition A.3.20. *Let L/K be an algebraic extension. The following are equivalent:*

(i) *L/K is purely inseparable.*

(ii) *$L = K(S)$ for a set S of purely inseparable elements.*

(iii) $[L:K]_s = 1$

(iv) *For every* $\alpha \in L$ *there exists* r *such that* $\alpha^{p^r} \in K$.

Proof. That i) implies ii) is obvious. Assume ii) and take any $\alpha \in L$. Then $\alpha \in K(S')$ for a finite set $S' \subset S$, so

$$[K(\alpha):K]_s \leq [K(S'):K]_s = 1.$$

Assume by contraction that $[L:K]_s > 1$. Then there is an embedding $\sigma\colon L \to \overline{K}$ over K which is not the identity on L. Taking any $\alpha \in L$ such that $\sigma(\alpha) \neq \alpha$, we also have $[K(\alpha):K]_s > 1$, which we have excluded. Hence, ii) implies iii).

Now assume iii) and let $\alpha \in L$. Then $[K(\alpha):K]_s = 1$, so the minimal polynomial μ_α has a single root over $\overline{K}$, which means that $\mu_\alpha(x) = x^{p^r} - c$ for some $c \in K$, so iv) holds.

Finally, assume iv). Then every $\alpha \in L$ satisfies the polynomial $x^{p^r} - \alpha^{p^r} \in K[x]$, for some $r > 0$. The minimal polynomial μ_α is a divisor of this polynomial, hence it is purely inseparable. □

Remark A.3.21. Clearly, in the finite case, all of the above are also equivalent to the condition $[L:K]_i = [L:K]$.

As a consequence of Proposition A.3.20, we prove the analogue of Proposition A.3.13.

Proposition A.3.22. *Let* $K \subset L \subset M$ *be algebraic extensions. Then* M/K *is purely inseparable if and only if both* L/K *and* M/L *are purely inseparable.*

Proof. Use the fact that L/K is purely inseparable if and only if $[L:K]_s = 1$, together with the multiplicativity of the separable degree in towers. □

Corollary A.3.23. *Let* $L, M \subset \overline{K}$ *be fields purely inseparable over* K. *Then the composite* $L \cdot M = L(M)$ *is itself purely inseparable.*

The above theory allows us to factorize any extensions as a tower of a separable and a purely inseparable one, thus cleanly splitting the two phenomena. To see this, take any algebraic extension L/K, and define the field

$$(L/K)_s := \{\alpha \in L \mid \alpha \text{ is separable over } K\} = L \cap \overline{K}^{sep}. \tag{A.3.2}$$

By Proposition A.3.14, $(L/K)_s$ is a field, and we can factorize the extension as $K \subset (L/K)_s \subset L$.

Theorem A.3.24. *Let* L/K *be an algebraic extension. Then* $(L/K)_s$ *is the only subfield of* L *such that* $L/(L/K)_s$ *is purely inseparable extension and* $(L/K)_s/K$ *is separable. Moreover,*

$$[L:K]_s = [(L/K)_s : K],$$

and if L/K is finite,

$$[L : K]_i = [L : (L/K)_s].$$

Proof. First, $(L/K)_s$ is separable over K, by construction. Take any $\alpha \in L$ and let $\mu_\alpha \in K[x]$ be its minimal polynomial. Then we can write $\mu_\alpha(x) = h(x^{p^r})$ for some $h \in K[x]$ separable. This means that $\alpha^{p^r} \in (L/K)_s$, and if $r > 1$, α is purely inseparable over $(L/K)_s$.

The equality $[L : K]_s = [(L/K)_s : K]$ then follows from the fact that $[L : (L/K)_s]_s = 1$, and the claim about $[L : K]_i$ follows by multiplicativity assuming L/K is finite.

Finally, let $K' \subset L$ be any other field such that K'/K is separable and L/K' is purely inseparable. Then $K' \subset (L/K)_s$, and $(L/K)_s$ is at the same time a separable and a purely inseparable extension of K' (because L/K' is purely inseparable, while $(L/K)_s/K$ is separable). It follows that $K' = (L/K)_s$. □

We end this section by considering fields whose finite extensions are always separable.

Definition A.3.25. Let K be a field. We say that K is *perfect* if all finite extensions L/K are separable.

Example A.3.26.

(a) Any field of characteristic 0 is perfect.

(b) Any finite field is perfect. In fact, any finite field extension $\mathbb{F}_{p^b}/\mathbb{F}_{p^a}$ is generated by the p^b-th roots of 1, and the p^b-th cyclotomic polynomial has exactly $\phi(p^b) = (p-1)p^{b-1}$ distinct roots in $\overline{\mathbb{F}_p}$.

Remark A.3.27. If K is a perfect field, any *algebraic* extension of K is separable, since it is generated by finite, separable extensions.

There is a convenient characterization of perfect fields.

Proposition A.3.28. *Let K be a field of characteristic $p > 0$. Then K is perfect if and only if $K^p = K$.*

Proof. Assume that K is perfect and let $\alpha \in K$. Let $\beta \in \overline{K}$ be any p-th root of α. Then

$$(x - \beta)^p = x^p - \beta^p = x^p - \alpha,$$

so β is the only root of $x^p - \alpha$. This means that $\beta \in K$, otherwise $K(\beta)/K$ would be inseparable. It follows that $\alpha \in K^p$.

Conversely, assume that $K = K^p$, and take any $\alpha \in \overline{K}$. Then we can write its minimal polynomial as $\mu_\alpha(x) = h(x^{p^r})$ for some h separable over K. Taking p^r-th roots of the coefficients of h, we can write $h(x^{p^r}) = g(x)^{p^r}$

for some $g \in K[x]$, which implies that $p^r = 1$, since μ_α is irreducible. It follows that $\mu_\alpha = h$, so α is separable, hence all finite extensions of K are separable. □

Remark A.3.29. A finite extension of a perfect field is perfect as well. In fact, let $K \subset L \subset M$ be finite extensions of fields, with K perfect. Then M/K is separable, and so M/L is separable as well (for instance, by computing the separable degree).

With the above criterion, we can rephrase this by saying that if K is a field of characteristic p such that $K^p = K$ and L is a finite extension of K, then $L^p = L$ as well, something that is much less obvious to prove directly.

Using this characterization, we can always enlarge a field to make it become perfect. Starting from a field $K_0 = K$ of characteristic $p > 0$, we let

$$K_1 = K^{1/p} = K(\{\alpha \in \overline{K} \mid \alpha^p \in K\}).$$

Iterating this procedure, we define $K_2 = K_1^{1/p}$ and so on. The union

$$\overline{K}^{per} := K^{1/p^\infty} = \bigcup_{i \geq 0} K_i \tag{A.3.3}$$

satisfies $\overline{K}^{per} = (\overline{K}^{per})^p$, so it is perfect. In fact, it is the smallest perfect field containing K.

Definition A.3.30. Let K be a field of characteristic $p > 0$. The field $\overline{K}^{per}$ defined by (A.3.3) is called the *perfect closure* of K.

By construction, $\overline{K}^{per}$ is the union of a tower of purely inseparable extensions of K, so the extension $\overline{K}^{per}/K$ is purely inseparable as well. Given an algebraic extension L/K, we define

$$(L/K)_i := \{\alpha \in L \mid \alpha^{p^r} \in K \text{ for some } r\} = L \cap \overline{K}^{per}. \tag{A.3.4}$$

Clearly, $(L/K)_i$ is a purely inseparable extension of K. The analogue of Theorem A.3.24 does not necessarily hold, though. In other words, $L/(L/K)_i$ can fail to be separable.

Example A.3.31. Let k be a field of characteristic $p > 2$, and $K = k(x, y)$. Consider the polynomial

$$f(t) = t^{2p} - xt^p - y,$$

and let α be a root of f in $\overline{K}$. We take $L := K(\alpha)$. By construction, f is not separable, hence L/K is an inseparable extension. On the other hand, no element of L is purely inseparable over K, that is, $(L/K)_i = K$. To see this, assume that some element $\beta \in L$ satisfies $\beta^p \in K$. Write

$$\beta = g(\alpha) = \sum_{i=0}^{2p-1} g_i \alpha^i$$

for some $g \in K[t]$. Then

$$\beta^p = g(\alpha)^p = \sum_{i=0}^{2p-1} g_i^p (\alpha^p)^i = \sum_{i=0}^{2p-1} g_i^p \gamma^i,$$

where $\gamma = \alpha^p$. Using the fact that $f(\alpha) = 0$, we can simplify this equation to eliminate all powers γ^d for $d \geq 2$. Namely, $\gamma^2 = x\gamma + y$ so we end up with

$$\beta^p = a\alpha^p + b,$$

with $a, b \in K$. The condition $\beta^p \in K$ amounts to $a = 0$.

We now choose $p = 3$ and carry out the computation explicitly, to find

$$a = g_1^3 + xg_2^3 + (x^2 + y)g_3^3 + (x^3 + 2y)g_4^3 + (x^4 + y^2)g_5^3 = 0. \tag{A.3.5}$$

This is a linear equation in the powers g_i^3 with coefficients in K. We can derive (A.3.5) with respect to either x or y to find new equations. Using the fact that the derivatives of g_i^3 are 0 in characteristic 3, these are new equations in the powers g_i^3. It is not difficult to five independent equations, which means that $\beta^p \in K$ implies that $g_i = 0$ for all $i \geq 1$, or equivalently $\beta \in K$.

In fact, we can tell exactly when an algebraic extension admits such a splitting.

Proposition A.3.32. *Let L/K be an algebraic extension. Then L is separable over $(L/K)_i$ if and only if $L = (L/K)_s(L/K)_i$.*

Proof. Assume that $L = (L/K)_s(L/K)_i$. Then L is generated over $(L/K)_i$ by elements that are separable over K, and a fortiori over $(L/K)_i$. This means that L is a separable extension of $(L/K)_i$.

Conversely, assume that L/K is separable, and consider the composite $M = (L/K)_s(L/K)_i$. Then L is at the same time separable and purely inseparable over M, hence $L = M$. □

A.4. Normal extensions

In this section, we study those field extensions that satisfy as many symmetries as possible.

Definition A.4.1. Let L/K be an algebraic extension, and regard L as a subfield of $\overline{K}$. We say that L is *normal* if for every field morphism $\sigma\colon L \to \overline{K}$ such that $\sigma|_K = \mathrm{id}$, we have $\sigma L \subset L$.

Remark A.4.2. In fact, in the above definition we can equivalently say that $\sigma(L) = L$. Otherwise, $\sigma(L) = L' \subsetneq L$, and we can extend $\sigma^{-1}|_{L'}\colon L' \to \overline{K}$ to the whole of L (by Proposition A.1.11), thereby showing that L is not normal.

Let L/K be a normal extension, $\alpha \in L$ and let μ_α be its minimal polynomial. Then every other root of μ_α lies in L. Otherwise we could define a morphism $K(\alpha) \to \overline{K}$ which sends α to a root outside L, and extend this to a morphism $L \to \overline{K}$. This prompts the next definition.

Definition A.4.3. Let K be a field, F a set of polynomials over K. The *splitting field* of F is the smallest subfield of $\overline{K}$ containing all roots of all polynomials in F.

The two concepts are clearly linked, and we make this precise in the next result.

Proposition A.4.4. *Let F be a set of polynomials defined over K. Then the splitting field of F is a normal extension of K. Conversely, if L/K is a normal extension, there is a set F of polynomials over K such that L is the splitting field of F.*

Before proving this, it is useful to introduce a last piece of terminology.

Definition A.4.5. Let K be a field $\alpha, \beta \in \overline{K}$. We say that α and β are conjugate if they have the same minimal polynomial. Equivalently, there is a morphism $\sigma\colon K(\alpha) \to \overline{K}$ such that $\sigma(\alpha) = \beta$.

Proof. Let L be the splitting field of F, and let S be the set of roots of all polynomials in F. Then each morphism $L \to \overline{K}$ must permute the set S, and so preserve L. Conversely, assume that L/K is normal. If S is any set of generators of L, all conjugates of elements in S also lie in L. Hence L is the splitting field of the set of the minimal polynomials of elements in S. □

We now give some examples.

Example A.4.6.

(a) Let α be purely inseparable over K. Then $K(\alpha)$ is the splitting field of μ_α, hence $K(\alpha)/K$ is normal. By the same reasoning, any purely inseparable extension is normal.

(b) Let K be a field of characteristic different from 2, L/K an extension of degree 2. Then $L = K(\alpha)$ for some α that satisfies an equation of degree 2. By the quadratic formula, $L = K(\sqrt{\Delta})$ for some $\Delta \in K$. There is only one nontrivial morphism $L \to \overline{K}$, sending $\sqrt{\Delta}$ to $-\sqrt{\Delta}$, hence L/K is normal.

(c) Every finite field extension is normal, for cardinality reasons: $\mathbb{F}_q$ is the only subfield of $\overline{\mathbb{F}_q}$ with q elements, so every embedding of $\mathbb{F}_q$ into its algebraic closure must actually map $\mathbb{F}_q$ into itself.

(d) The extension $\mathbb{Q}(\sqrt[3]{2})/\mathbb{Q}$ is not normal. In fact, the conjugates of $\sqrt[3]{2}$ are $\zeta_3 \sqrt[3]{2}$ and $\zeta_3^2 \sqrt[3]{2}$, where ζ_3 is a third root of 1, and $\zeta_3 \sqrt[3]{2} \notin \mathbb{Q}(\sqrt[3]{2})$ since it is not a real number.

(e) Unlike other concepts we have introduced, normality is not preserved in towers of extensions. That is, if $K \subset L \subset M$ are fields, with M/L and L/K normal, it may be the case that M/K is not normal. A simple example is the composition of two degree 2 extensions, such as $\mathbb{Q} \subset \mathbb{Q}(\sqrt{2}) \subset \mathbb{Q}(\sqrt[4]{2})$. The two extensions of degree 2 are normal by b), but the minimal polynomial of $\sqrt[4]{2}$ over $\mathbb{Q}$ is $x^4 - 2$, so $\sqrt[4]{2}$ is conjugate to $i \cdot \sqrt[4]{2}$, which is not real.

We can single out some easy properties of normal extensions, whose proof is a simple verification.

Proposition A.4.7. (i) *Let L/K and M/K be two normal extensions. Then the composite LM is normal over K.*

(ii) *Let $K \subset L \subset M$ be extensions. If M/K is normal, then M/L is normal.*

(iii) *Let L/K be a normal extension. Then $(L/K)_s$ is also normal over K.*

With the concept of normal extension we can also revisit Example A.3.31.

Example A.4.8. Let L/K the extension of example A.3.31. Notice that in that example $[L : K]_s = 2$, hence $(L/K)_s$ is a normal extension of K. If $(L/K)_i$ was not trivial, it would have degree p over K, which implies that the composite of $(L/K)_s$ and $(L/K)_i$ is the whole L. Since both are normal, L/K would be normal as well, by the previous proposition. But it is easy to check that L/K is not normal.

In fact, keeping the notation of that example, let $\gamma = \alpha^p$ be one solution of $t^2 - xt - y = 0$, and let γ_* be the other one. Also, choose α_* such that $\alpha_*^p = \gamma_*$. Then we have the equations

$$\alpha^p \alpha_*^p = x$$
$$\alpha^p \alpha_*^p = -y.$$

If L/K is normal, we have $\alpha_* \in L$, which means that $\sqrt[p]{x}, \sqrt[p]{y} \in L$. However, this is impossible, since $K(\sqrt[p]{x}, \sqrt[p]{y})/K$ is an extension of degree p^2.

It is not a coincidence that normality can shed a light on Example A.3.31. In fact, under normality assumptions, we can prove an analogue of Proposition A.3.24.

Proposition A.4.9. *Let L/K be a normal extension. Then L is separable over $(L/K)_i$, and in fact $(L/K)_i$ is the only intermediate field K' such that L/K' is separable and K'/K is purely inseparable.*

Proof. Let $\alpha \in L$, and let $\alpha_1 = \alpha, \alpha_2, \ldots, \alpha_r$ be its conjugates over $(L/K)_i$. Then the polynomial

$$f(x) = (x - \alpha_1) \cdots (x - \alpha_r)$$

lies in $(L/K)_i$, and so must be the minimal polynomial of α over $(L/K)_i$. To see this, notice that all embeddings $L \hookrightarrow \overline{K}$ send L into itself, hence they form a group $G = \mathrm{Aut}_K(L)$. The elements of L fixed by G are by definition purely inseparable, so the fixed field of G is $(L/K)_i$. But every elements of G permutes the set $\{\alpha_i\}$, so the polynomial f is unchanged under the action of G, which means that $f \in (L/K)_i[x]$. Since f is separable by construction, α is separable over $(L/K)_i$.

To prove the second assertion, every field K' purely inseparable over K is contained in $(L/K)_i$, so assuming L/K' separable, we must have $K' = (L/K)_i$. □

A.5. The Galois correspondence

Let L/K be a finite extension, where we assume that $L \subset \overline{K}$ is a fixed algebraic closure of K. We have seen that the number of distinct embeddings $L \hookrightarrow \overline{K}$ that fix K equals the separable degree $[L : K]_s$. Moreover, for such an embedding σ, we do not have necessarily $\sigma(L) \subset L$, unless L/K is a normal extension. If $\sigma(L) \subset L$, then one has equality $\sigma(L) = L$, and σ can be regarded as an automorphism of L fixing K. Thus, L/K has the maximal possible number of automorphisms precisely when L/K is separable and normal.

Definition A.5.1. Let L/K be an algebraic extension. We say that L/K is *Galois* if L/K is separable and normal. In this case, the automorphism group $\mathrm{Aut}_K(L)$ will be called the *Galois group* of L over K, and denoted $Gal(L/K)$. If $f \in K[x]$ is a separable polynomial, the splitting field of f is a Galois extension, and we define the Galois group of f as the Galois group of this extension.

We remark again that if L/K is a Galois extension, every embedding $\sigma\colon L \hookrightarrow \overline{K}$ that fixes K actually satisfies $\sigma(L) = L$, and so can be regarded as an element of $Gal(L/K)$.

Remark A.5.2. When L/K is finite and Galois, by the above discussion the Galois group $Gal(L/K)$ has exactly $[L : K]$ elements.

The Galois group can be used to study subfields of L, through the *Galois correspondence*, which relates subgroups of $Gal(L/K)$ and subfields of L that contain K. In one direction, let $L' \subset L$ be a subfield such that $K \subset L'$. Then L is Galois over L', and one can consider the Galois group $Gal(L/L')$.

By construction, this is a subgroup of $Gal(K/L)$, since the automorphisms of L that fix L' a fortiori fix all elements of K.

Going in the other direction, let $H < Gal(L/K)$ be a subgroup. Then the set

$$L^H := \{a \in L \mid \sigma(a) = a \text{ for all } \sigma \in H\}$$

is a subfield of L that contains K. The Galois correspondence is given by the two maps

$$\text{(A.5.1)} \quad \{\text{ subfields of } L \text{ containing } K\} \longrightarrow \{\text{subgroups of } Gal(L/K)\}$$

$$L' \longmapsto Gal(L/L')$$

$$L^H \longleftarrow H.$$

There are trivial inclusions $H < Gal(L/L^H)$ and $L' \subset L^{Gal(L/L')}$. In fact, much more is true. We first state the main theorem of Galois theory for the case of finite extensions.

Theorem A.5.3 (Main theorem of Galois theory, finite case)**.** *Let L/K be a finite Galois extension. Then the maps of Galois correspondence* (A.5.1) *are inverse to each other.*

Proof. Let $H < Gal(L/K)$ be a subgroup, L^H its fixed field. Using Theorem A.3.6, write $L = L^H(\alpha)$, and let

$$f(x) := \prod_{\sigma \in H} (x - \sigma(\alpha)).$$

Then f is invariant under H, hence $f \in L^H[x]$. Since f has degree $n := |H|$, we have $[L : L^H] \leq n$. But L is a Galois extension of L^H, and its Galois group has exactly $[L : L^H]$ elements. Since $H < Gal(L/L^H)$, we must have equality, so $H = Gal(L/L^H)$.

In the other direction, let $L' \subset L$ be an intermediate field, and let $H = Gal(L/L')$. Consider the field $L^H \supset L'$. The extension L/L^H is Galois, and by the first part of the proof its Galois group is H. This implies that $[L : L^H] = [L : L']$, which means that $L^H = L'$. □

Corollary A.5.4. *Let L/K be a finite separable extension. Then there are only finitely many subfield $L' \subset L$ containing K.*

Proof. Let M be the normal closure of L/K, that is, the composite of the extensions $\sigma(L)$ for all embeddings $\sigma\colon L \hookrightarrow \overline{K}$. Then M is a finite extension of K, and it is normal and separable by construction. The main theorem of Galois theory applied to M/K shows that the intermediate fields between

K and M are in bijection with the subgroups of $Gal(M/K)$, hence they are finite in number. A fortiori, this is true of the extension L/K. □

Let L/K be a Galois extension, L' an intermediate extension that is *also* Galois over K. Then one can consider the Galois group $Gal(L'/K)$. Any automorphism of L that fixes K will send L' to itself, hence restriction gives a natural map $Gal(L/K) \to Gal(L'/K)$. This map is surjective, since every automorphism of L' can be extended to a map $L \to \overline{K}$, that in turn will send L to itself. The main theorem of Galois theory can be strengthened to also describe this picture, by connecting Galois groups of the form $Gal(L'/K)$ to *quotients* of $Gal(L/K)$.

Theorem A.5.5. *Let L/K be a finite Galois extension. Then a subgroup $H < Gal(L/K)$ is normal if and only if L^H is normal over K. In this case, H is the kernel of the natural surjective homomorphism $Gal(L/K) \to Gal(L^H/K)$.*

Proof. Let $\sigma \in Gal(L/K)$, and $H' = \sigma H \sigma^{-1}$ a conjugate of H. Then the fixed field of H' is exactly $\sigma(L^H)$, so L^H is invariant under $Gal(L/K)$ if and only if H is stable under conjugation, which proves the first half of the theorem. The kernel of the restriction map $Gal(L/K) \to Gal(L^H/K)$ consists of those elements of $Gal(L/K)$ that fix L^H, and so is exactly H by the main theorem of Galois theory. □

The main theorem of Galois theory can be extended to the case of infinite extensions, but it requires some subtlety. In one direction, the correspondence works flawlessly even in the infinite case.

Proposition A.5.6. *Let L/K be a (possibly infinite) Galois extension, $L' \subset L$ an intermediate field, and let $H := Gal(L/L')$. Then $L^H = L'$.*

Proof. To simplify the notation, we can assume without loss of generality that $L' = K$. Let $\alpha \in L \setminus K$, and let M be the normal closure of $K(\alpha)$ over K. Then M/K is Galois, so using the Galois correspondence in the finite case we find $\sigma\colon M \to M$ such that $\sigma(\alpha) \neq \alpha$. We can extend σ to a map $\overline{\sigma}\colon L \to \overline{K}$, and since L/K is Galois, $\overline{\sigma} \in Gal(L/K)$. It follows that $\alpha \notin L^H$, and since α is arbitrary, $L^H = K$. □

The other half of Galois correspondence, though, cannot be extended literally, for the following reason. An algebraic extension is the composite of its finite subextensions. Dually, this should entail that the Galois group is determined by its finite quotients. However, not all groups have this property. To make this observation precise, we introduce some topological language.

Let G be a topological group. We recall from Section 7.5 that this is a group endowed with the structure of a topological space in such a way that the group operations (multiplication and inverse) are continuous. In this case, since translation in the group are continuous homeomorphisms, the topology is determined by the set of neighborhoods of the identity element.

Definition A.5.7. The topological group G is called *profinite* if the subgroups of G of finite index form a fundamental system of neighborhoods of the identity.

In other words, for every homomorphism $\phi\colon G \to H$, where H is a finite group, we require that the map ϕ is continuous, where H is given the discrete topology. In the language of Section 7.4, a profinite group is the inverse limit (as topological groups) of a family of finite groups (Exercise 22).

Remark A.5.8. A *finite* profinite group necessarily has the discrete topology, so the notion is only meaningful for infinite groups.

If L/K is a Galois extension, we can always endow $Gal(L/K)$ with a profinite topology. Namely, let $\{L_i\}$ be the family of *finite* Galois extensions of K contained in L. Each element of $Gal(L/K)$ is determined by its action on the finite extensions L_i/K, which shows that there is a natural injection

$$Gal(L/K) \hookrightarrow \prod_{L_i} Gal(L_i/K). \tag{A.5.2}$$

Inside this product, the image of $Gal(L/K)$ can be identified as the set of *compatible* sequences—that is, sequences (σ_i) where $\sigma_i \in Gal(L_i/K)$ such that σ_i and σ_j agree on the intersection $L_i \cap L_j$. This shows that $Gal(L/K) = \varprojlim Gal(L_i/K)$. If we give each $Gal(L_i/K)$ the discrete topology, then $Gal(L/K)$ can be given the smallest topology that makes all restriction maps

$$Gal(L/K) \to Gal(L_i/K)$$

continuous. In other words, $Gal(L/K)$ inherits the subspace topology from the injection (A.5.2), where the right-hand side is endowed with the product topology. This description exhibits the group $Gal(L/K)$ as the inverse limit (as topological groups) of a family of discrete finite groups, hence $Gal(L/K)$ is a profinite group.

We can use this language to make the remark after Proposition A.5.6 precise:

Proposition A.5.9. *Let L/K be a Galois extension, $H < Gal(L/K)$ a subgroup, L^H its fixed field. Then $Gal(L/L^H) = \overline{H}$ is the topological closure of H.*

Proof. Let $\{L_i\}$ be the family of *finite* Galois extensions of K contained in L, so that L is the composite of the L_i. If we let H_i be the image of H inside $Gal(L^i/K)$, then $L_i \cap L^H = L_i^{H_i}$, hence

$$L^H = \bigcup_i L_i^{H_i}.$$

Let $\phi_i\colon Gal(L/K) \to Gal(L_i/K)$ be the restriction map. Then, by definition of the profinite topology,

$$\overline{H} = \bigcap_i \phi_i^{-1}(H_i).$$

On the other hand, $\overline{H}$ is the biggest subgroup of $Gal(L/K)$ that fixes all fields $L_i^{H_i}$ (here we are using the Galois correspondence inside L_i). Equivalently, $\overline{H}$ is the biggest subgroup that fixes L^H, and this is $Gal(L/L^H)$ by definition. □

By putting together Propositions A.5.6 and A.5.9, we obtain the main theorem of Galois theory in its general form.

Theorem A.5.10 (Main theorem of Galois theory)**.** *Let L/K be a Galois extension. Then the maps of Galois correspondence*

(A.5.3)

$$\{\ \textit{subfields of } L \textit{ containing } K\} \longrightarrow \{\text{closed } \textit{subgroups of } Gal(L/K)\}$$

$$L' \longmapsto Gal(L/L')$$

$$L^H \longleftarrow\!\shortmid\ H.$$

are inverse to each other.

We remark again that in (A.5.3) we are only considering *closed* subgroups with respect to the natural profinite topology on the Galois group. Once the Galois correspondence is established, the following refinement can be proved exactly as in the finite case.

Theorem A.5.11. *Let L/K be a Galois extension. Then a closed subgroup $H < Gal(L/K)$ is normal if and only if L^H is normal over K. In this case, H is the kernel of the natural surjective homomorphism $Gal(L/K) \to Gal(L^H/K)$.*

Given this result, all Galois groups of the form $Gal(L/K)$, where L/K is a Galois extension, can be regarded as quotients of a fixed group:

Definition A.5.12. Let K be a field. The *absolute Galois group* of K is $Gal(\overline{K}^{sep}/K)$, endowed with its profinite topology.

The absolute Galois group is the biggest Galois group that can be taken over K, and encodes in a single (albeit complicated) object, the structure of all possible (separable) algebraic extensions of K.

A.6. Some computations

In the previous section, we have exposed the Galois correspondence, without actually computing even a single Galois group. In this section, we are giving some examples.

Example A.6.1.

(a) Let L/K be a finite Galois extension *of prime* degree p. Then $Gal(L/K)$ is a group with p elements, which is necessarily isomorphic to $\mathbb{Z}/p\mathbb{Z}$.

(b) Take an extension of finite fields $\mathbb{F}_{p^b}/\mathbb{F}_{p^a}$. Then we know that a divides b by Proposition A.2.4. The extension is normal because $\mathbb{F}_{p^b}$ is the only subfield of $\overline{\mathbb{F}_p}$ of cardinality p^b. Let $\phi\colon \overline{F_p} \to \overline{F_p}$ be the Frobenius homomorphism given by $\phi(x) = x^p$. Then ϕ^a leaves each element of $\mathbb{F}_{p^a}$ fixed, hence it is an element of $Gal(\mathbb{F}_{p^b}/\mathbb{F}_{p^a})$. Letting $b = ka$, we see that the power $(\phi^a)^k = \phi^b$ acts as the identity on $\mathbb{F}_{p^b}$, while no smaller power of ϕ^a is the identity. It follows that ϕ^a generates a subgroup of $Gal(\mathbb{F}_{p^b}/\mathbb{F}_{p^a})$ of cardinality k. Since the extension itself has degree k, we conclude that $Gal(\mathbb{F}_{p^b}/\mathbb{F}_{p^a})$ is cyclic, generated by ϕ^a.

(c) We can extend the previous example to compute the absolute Galois group $Gal(\overline{\mathbb{F}_p}/\mathbb{F}_p)$. Namely, each finite quotient $Gal(\mathbb{F}_{p^n}/\mathbb{F}_p)$ is isomorphic to $\mathbb{Z}/n\mathbb{Z}$, and these form an inverse systems with the natural maps $\mathbb{Z}/n\mathbb{Z} \to \mathbb{Z}/m\mathbb{Z}$ each time m divides n. The absolute Galois groups is then $\widehat{\mathbb{Z}}$, the inverse limit of the finite cyclic groups ordered by divisilibity. This group naturally contains $\mathbb{Z}$ (it is its completion with respect to the topology generated by arithmetic progressions), and the Frobenius element ϕ can be identified with 1 under this map. In particular, the subgroup generated by ϕ is dense in $Gal(\overline{\mathbb{F}_p}/\mathbb{F}_p)$.

(d) Let $\mathbb{Q}(\zeta_m)/\mathbb{Q}$ be a cyclotomic extension, where m is not twice an odd integer. Then we compute in Proposition 6.8.4 that $Gal(\mathbb{Q}(\zeta_m)/\mathbb{Q})$ is isomorphic to $(\mathbb{Z}/m\mathbb{Z})^*$, where the element $a \in (\mathbb{Z}/m\mathbb{Z})^*$ acts by sending ζ_m to ζ_m^a.

(e) Let L/K be a finite Galois extension of degree n. By the primitive element theorem, $L = K(\alpha)$ for some $\alpha \in L$. Let $\alpha_1 = \alpha, \alpha_2, \ldots, \alpha_n$ be the conjugates of α. Then the Galois group $Gal(L/K)$ acts on the elements $\{\alpha_1, \alpha_2, \ldots, \alpha_n\}$ by permutations. Since an element of $Gal(L/K)$ is determined by its action on α, it follows that the restriction homomorphism $Gal(L/K) \to \mathfrak{S}_n$ it injective. This exhibits the Galois group as a subgroup of a group of permutations.

(f) As a partial converse, let K be a field and let $\mathfrak{S}_n$ act on $K(x_1, ..., x_n)$ by permuting the indeterminates. If we let $L = K(x_1, \ldots, x_n)^{\mathfrak{S}_n}$ be the subfield of symmetric rational function, then

$$Gal(K(x_1, \ldots, x_n)/L) = \mathfrak{S}_n.$$

In fact, in this case *every* permutation of $\{1, \ldots, n\}$ defines a field automorphism. Define the polynomial

$$f(T) = \prod_{i=1}^{n} (T - x_i) = T^n - \sigma_1 T^{n-1} + \cdots \pm \sigma_n$$

over the field $K(x_1, \ldots, x_n)$, where the polynomials $\{\sigma_i\}$ (defined by this equality) are the elementary symmetric functions. By constructions, the σ_i are symmetric, hence $f \in L[T]$. This proves that $K(x_1, \ldots, x_n)$ is the splitting field of f, so the Galois group of f is $\mathfrak{S}_n$.

(g) Let G be any finite group. By letting G act on itself, we find an injective homomorphism $G \hookrightarrow \mathfrak{S}_n$ into some symmmetric group. Let L/K be any Galois extension with Galois group $\mathfrak{S}_n$—for instance the one in the previous example. Then G corresponds to a subfield $L^G \subset L$ such that $Gal(L/L^G) \cong G$. This allows us to realize every finite group as a Galois group of *some* Galois extension. Unfortunately, this does not allow us to control L^G, in particular to construct—for example—Galois extensions with given Galois group over $\mathbb{Q}$. The problem of realizing a given finite group as a Galois group over a given field is called the *inverse Galois problem*, and is in general open even over $\mathbb{Q}$ [**MM18**].

To give some more examples, we quote Hilbert's irreducibility theorem.

Theorem (Hilbert). *Let K be a number field, $f[x_1, \ldots, x_n, y]$ an irreducible polynomial. Then there exist $(a_1, \ldots, a_n) \in K^n$ such that $f(a_1, \ldots, a_n, y)$ remains irreducible as a polynomial in y.*

In fact, much more is known: the same can be done for finitely many polynomials at once, and the set of specializations that leave the polynomials irreducible is "big" in an appropriate sense (see [**Ser88**] for the notion of thin set, or [**CD16**] for quantitative results).

Example A.6.2.

(h) Let K be a number field, and consider the general polynomial of degree n

$$f(T) = \prod_{i=1}^{n}(T - x_i) = T^n - \sigma_1 T^{n-1} + \cdots \pm \sigma_n,$$

which is defined over $K(\sigma_1, \ldots, \sigma_n)$. By Hilbert's irreducibility theorem, we can find $(a_1, \ldots, a_n) \in K^n$ such that $g(T) = f(a_1, \ldots, a_n)(T)$ remains irreducible as a polynomial in T. This entails that the splitting field of G has degree $n!$ over K. Since its Galois groups is a subgroup of $\mathfrak{S}_n$, we must have equality, which shows that one can always realize $\mathfrak{S}_n$ as a Galois group over K.

A.7. The trace and norm

Let L/K be a finite extension of separable degree $n = [L : K]_s$ and inseparable degree $f = [L : K]_i$. Denote by $\sigma_1, \ldots, \sigma_n$ the distinct embeddings $L \hookrightarrow \overline{K}$. We define two important group homomorphisms.

Definition A.7.1. The *trace* is the additive homomorphism

$$\mathrm{Tr}_{L/K} \colon L \to K$$

defined by

$$\mathrm{Tr}_{L/K}(\alpha) = f \cdot (\sigma_1(\alpha) + \cdots + \sigma_n(\alpha)).$$

Similarly we define a multiplicative homomorphism called *norm*

$$\mathrm{N}_{L/K} \colon L^* \to K^*$$

by

$$\mathrm{N}_{L/K}(\alpha) = (\sigma_1(\alpha) \cdots \sigma_n(\alpha))^f.$$

It is not a priori clear that these homomorphisms actually take value in K. However, this follows since the values $\mathrm{Tr}_{L/K}(\alpha)$ and $\mathrm{N}_{L/K}(\alpha)$ are fixed by every element of the Galois group of the normal closure of L.

Remark A.7.2. We do not ask in the definition that L is a normal extension of K, so the individual values $\sigma_i(\alpha)$ may lie outside L.

Remark A.7.3. The coefficient f appears in the above formulas because we want to count each embedding σ_i with multiplicity f. This ensures that the number of (repeated) embeddings considered in these formulas is $[L : K]$.

Example A.7.4.

(a) The simplest example is a quadratic extension of K, say $L = K(\sqrt{a})$. In this case the extension is Galois, and the only non-trivial automoprhism of L over K sends $\sqrt{a}$ to $-\sqrt{a}$. It follows that

$$\mathrm{Tr}_{L/K}(x + y\sqrt{a}) = 2x$$
$$\mathrm{N}_{L/K}(x + y\sqrt{a}) = x^2 + ay^2.$$

(b) More generally assume that $L = K(\alpha)$. Let $\alpha_i = \sigma_i(\alpha)$ be the conjugates of α. Then the minimal polynomial of α is

$$\mu_\alpha(x) = x^n + a_{n-1}x^{n-1} + \cdots + a_0 = (x - \alpha_1) \cdots (x - \alpha_n).$$

In particular we see that $\mathrm{Tr}(\alpha) = -a_{n-1}$ and $\mathrm{N}(\alpha) = \pm a_0$ are, up to sign, coefficients of the minimal polynomial of α.

(c) Even more generally, take any $\alpha \in L$, which does not necessarily generate the field. Multiplication by α gives a K-linear map

$$m_\alpha \colon L \to L,$$

which sends β to $\alpha\beta$. Let

$$f_\alpha(x) = x^n + a_{n-1}x^{n-1} + \cdots + a_0$$

be the characteristic polynomial of m_α. Then $\mathrm{Tr}(\alpha) = -a_{n-1}$ and $\mathrm{N}(\alpha) = \pm a_0$. This relation is the reason why we considered the factor $[L : K]_i$ in the definition of the trace and norm.

For a tower of extensions, trace and norm behave nicely.

Proposition A.7.5. *Let $K \subset L \subset M$ be extensions of fields. Then*

$$\mathrm{Tr}_{M/K} = \mathrm{Tr}_{L/K} \circ \mathrm{Tr}_{M/L}$$
$$\mathrm{N}_{M/K} = \mathrm{N}_{L/K} \circ \mathrm{N}_{M/L}\,.$$

Proof. Let $\{\sigma_i\}$ be the embeddings of M in $\overline{K}$, $\{\tau_i\}$ be those of M in $\overline{L} = \overline{K}$ (fixing L) and $\{\eta_i\}$ those of L in $\overline{K}$. Since M is algebraic over L, we can extend the homomorphisms τ_i to homomorphisms

$$M \to \overline{K},$$

which we still denote τ_i.

Note that every composition $\tau_i \circ \eta_j$ is an embedding of M into $\overline{K}$, hence it is equal to some σ_k. Moreover, since

$$[M : K]_s = [M : L]_s \cdot [L : K]_s,$$

the two sets $\{\sigma_i\}$ and $\{\tau_j \circ \eta_k\}$ have the same number of elements. If we prove that they are equal, the desired relations follow immediately.

So we only need to prove that there are no repetitions in the set $\{\tau_j \circ \eta_k\}$. If we have

$$\tau_j \circ \eta_k = \tau_l \circ \eta_m$$

we obtain $\tau_l^{-1} \circ \tau_j = \eta_m \circ \eta_k^{-1}$, so $\eta_m \circ \eta_k^{-1}$ fixes L. It follows that $m = k$, and in turn $l = j$. □

The trace can be used as a measure of separability.

Proposition A.7.6. *Let L/K be a finite extension. Then L/K is separable if, and only if,* $\mathrm{Tr}_{L/K}$ *is not identically* 0.

We only prove half of this result here; the last part of the proof will be given as a consequence of Artin's theorem on the independence of characters.

Proof of Proposition A.7.6, first half. If L/K is inseparable, we can factor the extension through $F := (L/K)_s$. The extension L/F is purely inseparable by Proposition A.3.24. Using the property of composition of the trace in Proposition A.7.5, it is enough to show that $\mathrm{Tr}_{L/F} = 0$—in other words, we only need to consider a purely inseparable extension. But if $\alpha \in L$ is purely inseparable over F, its minimal polynomial is a divisor of $x^r - \alpha^r$ for some $r > 0$, hence its second coefficient is 0, that is, $\mathrm{Tr}_{L/F}(\alpha) = 0$. □

A.8. Abelian extensions

Let L/K be a Galois extension with Galois group G. According to the nature of G, we can distinguish some classes of extensions.

Definition A.8.1. Let L/K be a Galois extension with Galois group G

(1) We say that L/K is *cyclic* if G is (finite) cyclic.
(2) We say that L/K is *Abelian* if G is Abelian.
(3) We say that L/K is *cyclic* if G is solvable.

We recall that a group G is called *solvable* if there exists a finite chain of subgroups

$$\{e\} = G_0 < G_1 < \cdots < G_r = G,$$

with G_k normal inside G_{k+1}, such that the quotiens G_{k+1}/G_k are Abelian. If G is finite, it is equivalent to require that each quotient is cyclic (up to refining the chain).

Remark A.8.2. Certain authors call an extension L/K solvable if there exists a field $M \supset L$ such that M/K is a Galois extension with solvable Galois group. In our terminology, a solvable extension will always be implicitly Galois (and the same holds for cyclic and Abelian).

In this section, we want to show that these conditions on the Galois group have a natural interpretation in terms of extensions. In particular, we will be able to characterize the properties of Abelian extensions. The prototypical result is the following.

Proposition A.8.3. *Let L/K be a Galois extension of degree d, where $L = K(\alpha)$ for some α such that $\alpha^d \in K$. Assume that* $\operatorname{char} K$ *does not divide d, and that K contains the d-th roots of 1. Then L/K is cyclic.*

Proof. Let $\zeta_d \in K$ be a primitive d-th root of 1, and let $\beta = \alpha^d \in K$. Then α is a root of $f(x) = x^d - \beta$. Since $[L : K] = d$, f is irreducible, so α is conjugate to $\zeta_d\alpha$, which is another root of f. Let $\sigma \in Gal(L/K)$ such that $\sigma(\alpha) = \zeta_d\alpha$. Then $\sigma^k(\alpha) = \zeta_d^k\alpha$. In other words, σ acts by cyclically permuting the conjugates of α. Since an element of $Gal(L/K)$ is determined by its action on α, it follows that the Galois group is cyclic, generated by σ. □

Actually, a kind of converse holds: a cyclic extension L/K (under suitable conditions) is generated by adding roots of elements of K. Let us try to invert the argument in the above proof. Assume that $Gal(L/K) = \langle\sigma\rangle$ is cyclic of order d, and that a primitive d-th root of 1, ζ_d, lies in K. If we are able to find an element $\alpha \in L^*$ such that $\sigma(\alpha) = \zeta_d\alpha$, then

$$\alpha^d = \prod_{i=1}^{d} \zeta_d^i\alpha = \prod_{i=1}^{d} \sigma^i(\alpha) = \mathrm{N}_{L/K}(\alpha) \in K^*.$$

This suggests that we are able to find d-th roots of elements of K inside L, *provided* $\zeta_d \in K$ and we are able to write

$$\zeta_d = \frac{\sigma(\alpha)}{\alpha}$$

for some $\alpha \in L^*$.

Remark A.8.4. Let L/K be a Galois extension and $\sigma \in G = Gal(L/K)$. Then for any $\alpha \in L^*$ we have

$$\mathrm{N}_{L/K}\left(\frac{\sigma(\alpha)}{\alpha}\right) = \frac{\prod_{\tau\in G}\tau(\sigma(\alpha))}{\prod_{\tau\in G}\tau(\alpha)} = 1.$$

This observation is consistent with our setting, since we assumed that $\zeta_d \in K$, hence $\mathrm{N}_{L/K}(\zeta^d) = \zeta_d^d = 1$. The key result is the following:

Theorem A.8.5 (Hilbert's Theorem 90)**.** *Let L/K be a finite cyclic extension with Galois group $G = \langle\sigma\rangle$.*

(i) *If $\beta \in L^*$ satisfies $\mathrm{N}_{L/K}(\beta) = 1$, there exists $\alpha \in L^*$ such that $\beta = \sigma(\alpha)/\alpha$.*

(ii) *If* $\beta \in L$ *satisfies* $\operatorname{Tr}_{L/K}(\beta) = 0$, *there exists* $\alpha \in L$ *such that* $\beta = \sigma(\alpha) - \alpha$.

Before being able to prove Hilbert's Theorem 90, we need some terminology. If G is a group and L is a field, a homomorphism $G \to L^*$ will be called a *character* of G (with values in L).

Theorem A.8.6 (Artin). *Let* $\sigma_1, \ldots, \sigma_n$ *be character of* G *with values in* L. *Then the characters are linearly independent over* L.

The following proof is taken from [**Kata**].

Proof. Let A be the group algebra $L[G]$. Its elements are formal finite linear combinations

$$\sum_{i=1}^{n} a_i g_i$$

with $a_i \in L$ and $g_i \in G$. The sum is made formally, while multiplication is lifted by linearity from the multiplication of G (all we need here is that G is a monoid). By construction, A is a (noncommutative) algebra with unit over L. Every character of G gives rise to a L-linear homomorphism $A \to L$—we will still denote these homomorphisms by $\sigma_1, \ldots, \sigma_n$.

We claim that—more generally—such homomorphisms of algebras are linearly independent for every L-algebra A. In fact, let $\sigma\colon A \to L^n$ be the map of L-algebras given by $\sigma = (\sigma_1, \ldots, \sigma_n)$. We claim that σ is surjective—this is easily done by adapting the proof of the Chinese remainder theorem to the case of noncommutative rings. In particular, we can choose elements $x_1, \ldots, x_n \in A$ such that $\sigma_i(x_j) = \delta_{ij}$. Assume a linear relation

$$\sum_{i=1}^{n} a_i \sigma_i = 0$$

with $a_i \in L$. Evaluating at x_j gives $a_j = 0$, hence the characters are linearly independent. □

With Artin's theorem on the independence of characters at hand, we can prove the second half of Proposition A.7.6.

Proof of Proposition A.7.6, second half. Assume that L/K is separable. The embeddings $\sigma_i\colon L \to \overline{K}$ can be seen as characters of the group L^* with values in $\overline{K}$. Hence they are linearly independent, and in particular the element $\operatorname{Tr}_{L/K} = \sigma_1 + \cdots + \sigma_n$ is not zero. □

We can now prove Hilbert's Theorem 90.

Proof of Theorem A.8.5. To prove i), take $\alpha \in L^*$ with $\mathrm{N}_{L/K}(\alpha) = 1$. Define a function $f\colon G \to L$ given by $f(e) = 1$, $f(\sigma) = \alpha$, $f(\sigma^2) = \alpha\sigma(\alpha)$ and so on, up to $f(\sigma^{d-1}) = \alpha\sigma(\alpha)\cdots\sigma^{d-2}(\alpha)$. This is done to ensure the relation $f(\tau_1\tau_2) = f(\tau_1)\cdot\tau_1(f(\tau_2))$ for all $\tau_1, \tau_2 \in G$.

By Artin's theorem, the linear combination

$$\chi := \sum_{\tau\in G} f(\tau)\tau$$

is not 0, hence we find $\gamma \in L$ such that $\beta := \chi(\gamma) \neq 0$. We can compute

$$\sigma(\beta) = \sum_{\tau\in G} \sigma(f(\tau))\sigma(\tau(\gamma)) = \sum_{\tau\in G} \frac{f(\sigma\tau)}{f(\sigma)}\sigma(\tau(\gamma)) = \frac{\beta}{f(\sigma)},$$

which we can rewrite as

$$\frac{\sigma(\beta)}{\beta} = 1/f(\sigma) = 1/\alpha.$$

Since $\mathrm{N}_{L/K}(\alpha) = 1$ if and only if $\mathrm{N}_{L/K}(1/\alpha) = 1$, we have proved the first part.

The proof of ii) is similar. Define $f\colon G \to L$ by $f(e) = 0$, $f(\sigma) = \alpha$ and so on, up to $f(\sigma^d) = \alpha + \sigma(\alpha) + \cdots + \sigma^{d-2}(\alpha)$. As above, consider

$$\chi := \sum_{\tau\in G} f(\tau)\tau.$$

This time, just choose any $\gamma \in L$ having $\mathrm{Tr}_{L/K}(\gamma) \neq 0$ (here we use Proposition A.7.6), and let $\beta := \chi(\gamma)/\mathrm{Tr}_{L/K}(\gamma)$. A computation similar as before (do it!) shows that $\beta - \sigma(\beta) = f(\sigma) = \alpha$. □

We can now prove a converse to Proposition A.8.3:

Proposition A.8.7. *Let L/K be a cyclic extension of degree d. Assume that* char K *does not divide d, and that K contains the d-th roots of* 1. *Then $L = K(\alpha)$ for some α such that $\alpha^d \in K$.*

Proof. Let $\zeta_d \in K$ be a primitive d-th root of 1. Then $\mathrm{N}_{L/K}(\zeta_d) = 1$, so by Hilbert's Theorem 90 we can write $\zeta_d = \sigma(\alpha)/\alpha$, where σ is a generator of $Gal(L/K)$ and $\alpha \in K$. Raising to the d-th power we deduce that α^d is invariant under σ, so $\alpha^d \in K$. Moreover, d is the smallest exponent k such that $\alpha^k \in K$, so α has degree d over K, and it generates the whole of L. □

This fact, together with Proposition A.8.3, is enough to derive the classical results of Galois theory about the solvability of equations using roots. To state the result precisely, let us first agree what it means to solve a polynomial with a formula involving roots.

Definition A.8.8. Let L/K be a finite extension. We say that L is obtained from K by adding roots if there is a tower of fields

$$K = L_0 \subset L_1 \subset \cdots \subset L_r = L$$

such that $L_{i+1} = L_i(\alpha_i)$ for some α_i such that $\alpha_i^{d_i} \in L_i$, for all $i = 0, \ldots, r-1$.

The polynomial $f \in K[x]$ is said *solvable by radicals* if the splitting field of f is obtained from K by adding roots.

We have already proved that extensions that are obtained by adding a single root are cyclic (under some additional conditions), so the following should not come unexpected.

Theorem A.8.9. *Let L/K be a Galois extension of degree d, and assume* $\operatorname{char} K = 0$ *or is a prime not dividing $[L : K]$. Then L is obtained from K by adding roots if and only if L/K is a solvable extension.*

Proof. Let $d = [L : K]$. It is not restrictive to assume that K contains the d-th roots of 1. In fact, if this is not the case, one can consider a primitive d-th root of 1 ζ_d and apply the result to the extension $L(\zeta_d)/K(\zeta_d)$. Since $K(\zeta_d)/K$ is both a cyclic extension and is obtained from K by adding roots, the result for L/K follows.

By the Galois correspondence, L/K is solvable if and only if there is a tower of fields

$$K = L_0 \subset L_1 \subset \cdots \subset L_r = L$$

such that each extension L_{i+1}/L_i has cyclic Galois group. Using Propositions A.8.3 and A.8.7, this is the same as saying that L is obtained by K adding roots. □

Remark A.8.10. Of course, we introduced the terminology in a way that is backward with respect to historical usage. First, the theorem about solvability of equations was proved, then it became customary to call groups obtained as a tower of cyclic extensions solvable.

Corollary A.8.11 (Abel). *Let K be a field of characteristic* 0. *The generic equation of degree* ≥ 5 *does not have a solution in radicals—that is, the generic polynomial*

$$f(t) = \sum_{i=0}^{d} a_i x^i$$

is not solvable by radicals over $K(a_0, \ldots, a_d)$ if $d \geq 5$.

Proof. By Theorem A.8.9, this amounts to saying that the Galois group of f is not solvable. We computed in Example A.6.1 that this Galois group is the full symmetric group $\mathfrak{S}_d$, and this is not solvable for $d \geq 5$ (Exercise 13). □

Remark A.8.12. Using Hilbert's irreducibility theorem, one can give concrete polynomials over $\mathbb{Q}$ of degree 5 that are not solvable by radicals. This means that not only is there no generic formula to solve the quintic equation, there are explicit quintic equations whose roots are not writable in terms of radicals (see Exercise 16 for an explicit example).

Propositions A.8.3 and A.8.7 admit a generalization to Abelian extensions. In this case, one needs to add simultaneously many roots of elements of the base field at once. To simplify the statement of the result, we introduce some terminology.

Definition A.8.13. Let L/K be a Galois extension. We say that L/K is a *Kummer extension* of exponent n if

(1) The Galois group $Gal(L/K)$ is Abelian and has exponent n, that is, $\sigma^n = e$ for all $\sigma \in Gal(L/K)$
(2) the field K contains a primitive nth root of 1
(3) the characteristic of K does not divide n.

Remark A.8.14. If $\operatorname{char} K$ is not multiple of n and K contains a primitive nth root of 1, K^* contains the subgroup U_n of nth roots of 1. Moreover, if $a \in K$ and $b, c \in \overline{K}$ are two elements such that $b^n = c^n = a$, the fields $K(b)$ and $K(c)$ are the same, since b and c differ by a nth root of 1. We will simply denote this field by $K\left(\sqrt[n]{a}\right)$. Similarly, if $\Delta \subset K^*$ is any set, we will denote by $K\left(\sqrt[n]{\Delta}\right)$ the composite of $K\left(\sqrt[n]{a}\right)$ for all $a \in \Delta$.

Propositions A.8.3 generalizes as follows.

Proposition A.8.15. *Let K be a field such that* $\operatorname{char} K$ *does not divide n and K contains all nth roots of* 1. *Let $\Delta \subset K^*$ and denote $L = K\left(\sqrt[n]{\Delta}\right)$. Then L/K is a Kummer extension of exponent n.*

Proof. By definition, L is the splitting field of a family of separable polynomials, hence L/K is Galois. For each $a \in K^*$, the extension $K\left(\sqrt[n]{a}\right)$ is cyclic of exponent n, by Proposition A.8.3. Putting together all restriction homomorphisms gives an embedding

$$Gal(L/K) \hookrightarrow \prod_{a \in \Delta} Gal(K\left(\sqrt[n]{a}\right)/K),$$

which proves that $Gal(L/K)$ is Abelian of exponent n. □

In the other direction, we can generalize Proposition A.8.7.

Proposition A.8.16. *Let L/K be a Kummer extension of exponent n. Then $L = K\left(\sqrt[n]{\Delta}\right)$, where $\Delta := (L^*)^n \cap K^*$.*

Proof. By definition, $K\left(\sqrt[n]{\Delta}\right) \subset L$. As we have already observed, by the Galois correspondence, L/K is the composite of cyclic extensions. If M/K is a cyclic extension with $M \subset L$, we have $M = K(\sqrt[n]{a})$ for some $a \in K$, by Proposition A.8.7. It follows that $M \subset K\left(\sqrt[n]{\Delta}\right)$, and since L is the composite of all such extensions, $L \subset K\left(\sqrt[n]{\Delta}\right)$. □

The statement of Proposition A.8.16 tells more than just the converse of Proposition A.8.15. In fact, it tells how to reconstruct L from the subgroup $\Delta := (L^*)^n \cap K^*$. This hints at a correspondence between subgroups of K^* and Abelian extensions. In fact, a famous result of Kummer describes precisely such a link.

Let K be a field such that $\operatorname{char} K$ does not divide n and K contains all nth roots of 1. The Kummer correspondence is given by the two maps (A.8.1)

$$\{L/K \text{ Kummer of exponent } n\} \longrightarrow \{\text{subgroups } \Delta \text{ of } K^*/(K^*)^n\}$$

$$L \xmapsto{\quad\alpha\quad} \frac{(L^*)^n \cap K^*}{(K^*)^n}$$

$$K\left(\sqrt[n]{\Delta}\right) \xleftarrow{\quad\beta\quad} \Delta.$$

Proposition A.8.16 already proves that $\beta \circ \alpha$ is the identity. The converse is more subtle, and needs a little detour on characters of finite groups.

Definition A.8.17. Let G be a finite Abelian group. The *dual group* of G is defined as

$$\widehat{G} := \operatorname{Hom}(G, \mathbb{C}^*).$$

Remark A.8.18. Let $U_n < \mathbb{C}^*$ be the subgroup of nth roots of 1. If G has exponent n, then $\widehat{G} := \operatorname{Hom}(G, U_n)$, since the image of each element of g by a homomorphism is a nth root of 1.

As in the case of vector spaces, there is a natural homomorphism f from G to its double dual $\widehat{\widehat{G}}$, given by

$$f(g)(\alpha) = \alpha(g)$$

for $g \in G$ and $\alpha \in \widehat{G}$. The following result, known as *Pontryagin duality*, holds in greater generality (see [**Fol94**, Section 4.3]), but we will only need the finite case.

Theorem A.8.19 (Pontryagin). *Let G be a finite Abelian group. The natural homomorphism $f\colon G \to \widehat{\widehat{G}}$ is an isomorphism.*

Proof. Since every finite Abelian group is a product of cyclic ones (for instance by Corollary 1.5.9), it is enough to prove the result for cyclic groups. If $G = \mathbb{Z}/m\mathbb{Z}$, every character ϕ of G sends 1 to a m-th root of 1—call it ω—and moreover ϕ is determined by ω, since $\phi(k) = \omega^k$.

It follows that the dual of $\mathbb{Z}/m\mathbb{Z}$ can be identified with U_m, and in particular G and $\widehat{G}$ are isomorphic, albeit not canonically. Still, this implies that G and $\widehat{\widehat{G}}$ have the same cardinality, so it is sufficient to check that f is injective.

If $f(g) = e$, then $\phi(g) = 1$ for all $\phi \in \widehat{G}$. This means that every character on G descends to a character on $G/\langle g\rangle$—in other words there is an injection $\widehat{G} \hookrightarrow \widehat{G/\langle g\rangle}$. This can only happen when $g = e$, for cardinality reasons. □

If $H < G$ is a subgroup, we can define its orthogonal

$$H^{\perp} := \left\{\phi \in \widehat{G} \mid \phi(h) = 1 \text{ for all } h \in H\right\}.$$

Remark A.8.20. There is a natural identification

$$H^{\perp} \cong \widehat{G/H},$$

since characters in $H^{\perp}$ descend to the quotient modulo H. With the natural identification $G \cong \widehat{\widehat{G}}$, we have the inclusion $H \subset \left(H^{\perp}\right)^{\perp}$, and by comparing cardinalities we conclude that this is in fact an equality $H = \left(H^{\perp}\right)^{\perp}$.

With these remarks on group duality out of the way, we can now state a converse to the stronger statement in Proposition A.8.16.

Theorem A.8.21 (Kummer). *Let K be a field such that $\operatorname{char} K$ does not divide n and K contains a primitive nth root of 1. The Kummer correspondence in* (A.8.1) *is a bijection. Moreover, for each $\Delta < K^*/(K^*)^n$, there is an isomorphism*

$$\begin{aligned} \psi\colon\ \Delta &\longrightarrow \operatorname{Hom}_c(Gal(L/K), U_n),\\ a &\longmapsto \chi_a \end{aligned}$$

where $L = K\left(\sqrt[n]{\Delta}\right)$ is the field corresponding to Δ, χ_a is the character defined by

$$\chi_a(\sigma) = \frac{\sigma(\sqrt[n]{a})}{\sqrt[n]{a}},$$

and Hom_c denotes continuous homomorphism with respect to the natural profinite topology on $Gal(L/K)$ (on U_n we just use the discrete topology).

Proof. We already know from Proposition A.8.16 that $\beta \circ \alpha$ is the identity. To prove that $\alpha \circ \beta$ is the identity requires the second part of the statement, so we proceed to that first.

Fix a Kummer extension L/K and let $\Delta := (L^*)^n \cap K^*$, $G := Gal(L/K)$. First, we prove that ψ is well defined. To start, notice that $\sigma(\sqrt[n]{a})/\sqrt[n]{a}$ is a nth root of 1, and does not depend on the choice of an nth root of a, as we assumed that $U_n \subset K$. It is immediate that χ_a is a homomorphism, and it is continuous since

$$\ker \chi_a = \{\sigma \in G \mid \sigma(\sqrt[n]{a}) = \sqrt[n]{a}\} = Gal(L/K\left(\sqrt[n]{a}\right))$$

is a subgroup of finite index. Finally, χ_a only depends on the class of a modulo $(K^*)^n$, so ψ is well defined. Also,

$$\ker \psi = \{a \in \Delta \mid \sigma(\sqrt[n]{a}) = \sqrt[n]{a} \text{ for all } \sigma \in G\} = (K^*)^n/(K^*)^n = \{1\},$$

so ψ is injective.

To check that ψ is surjective, assume first that L/K is finite. Given a homomorphism $\chi\colon G \to U_n$, as in the proof of Hilbert's Theorem 90, we find an element $b \in K^*$ such that

$$\chi(\sigma) = \frac{\sigma(b)}{b},$$

for all $\sigma \in G$. But then b^n is fixed by all elements of G, so $b^n \in K$. Writing $b = \sqrt[n]{a}$ for some $a \in K^* \cap (L^*)^n = \Delta$, we recover that $b = \chi_a$.

In the infinite case, let $\chi\colon G \to U_n$ be a *continuous* homomorphism. Then $\ker \chi$ has finite index, and there exists a subfield $M \subset L$ such that $\ker \chi = Gal(L/M)$. By the finite case, we find $a \in (L^*)^n \cap M^*$ such that $\chi = \chi_a$. But actually, given $\sigma \in G$ we have

$$\frac{\sigma(a)}{a} = \left(\frac{\sigma(\sqrt[n]{a})}{\sqrt[n]{a}}\right)^n = 1,$$

so $\sigma(a) = a$. Since this holds for all $\sigma \in G$, $a \in (L^*)^n \cap K^* = \Delta$.

At this point, we have the second part of the theorem. To check that $\alpha \circ \beta$ is the identity, fix $\Delta < K^*/(K^*)^n$, and let $L = \beta(\Delta) = K\left(\sqrt[n]{\Delta}\right)$ and $\Delta' = \alpha(L) = (L^*)^n \cap K^*$. Clearly, $\Delta \subset \Delta'$, and by the previous part of the proof, we have an isomorphism

$$\Delta' \cong \mathrm{Hom}_c(G, U_n),$$

where $G = Gal(L/K)$. Our goal is to show that $\Delta = \Delta'$.

Assume first that L/K is finite. In this case, we can identify $\mathrm{Hom}_c(G, U_n)$ with $\widehat{G}$. By Pontryagin duality, it is enough to check that the orthogonal of $\psi(\Delta)$ is trivial. But

$$\psi(\Delta)^\perp = \{\sigma \in G \mid \chi_a(\sigma) = 1 \text{ for all } a \in \Delta\}.$$

Saying that $\chi_a(\sigma) = 1$ amounts to $\sigma(\sqrt[n]{a}) = \sqrt[n]{a}$; if this happens for all $a \in \Delta$, σ is the identity on L. So $\phi(\Delta)^{\perp}$ is trivial, which implies that $\Delta = \Delta'$.

In the infinite case, take any $\Delta_* < \Delta$ which is finite. Correspondingly, construct L_* and Δ'_*. By the finite case, we have $\Delta_* = \Delta'_*$. Notice that Δ is the union of all such finite subgroups Δ_*, since it has finite exponent. Correspondingly, L is the union of all such subfields L_*. It follows that Δ' is the union of all such Δ'_*, hence $\Delta = \Delta'$. $\square$

A.9. Exercises

1. Show that the infinite recursion in the proof of Theorem A.1.10 is not necessary: namely, the field K_1 constructed in the proof is already algebraically closed.

The following exercises, up to Exercise 9, apply Galois theory to the classical problem of geometric constructions with ruler and compass. Let S be a set of points in the plane. Two distinct points $p, q \in S$ determine the line $\overline{pq}$, as well as the circle having p as center and passing through q. In turn, each pair of these curves (lines and circles) is either disjoint or determines 1 or 2 intersection points. We can add these new points to the set S and repeat the construction. We say that a point s is *constructible with ruler and compass* from S if it can be obtained by a finite number of steps as above.

2. Assume we start with $S = \{(0,0),(1,0)\}$. Prove that a point s is constructible with ruler and compass if and only if both its coordinates (x, y) are algebraic of degree 2^n for some n.

More generally, if S is a finite set of points—regarded as complex numbers $a_1, \ldots, a_k$—prove that $s \in \mathbb{C}$ is constructible if and only if it is algebraic of degree 2^n for some n over $\mathbb{Q}(a_1, \ldots, a_n)$.

3. Use Exercise 2 to prove that doubling a cube cannot be done with ruler and compass. The problem of doubling a cube consists of starting from a segment—which is the side of a cube—and producing a segment which is the side of a cube having twice the volume.

4. Use Exercise 2 to prove that angle trisection cannot be done with ruler and compass. The problem of trisecting the angle consists of starting from an angle α (determined by its vertex and two points on its sides) and constructing the angle $\alpha/3$.

5. Determine for which $n \geq 3$ the regular n-gon is constructible with ruler and compass. Your answer should predict that the regular 9-gon is not constructible, but the regular 17-gon is.

The last classical problem with ruler and compass is the squaring of the circle. This requires understanding the algebraic properties of π. We offer some exercises proving the classical theorem of Lindemann that π is transcendental. The approach we take is from [**Fil11**], where the transcendence of e is proved as well, using similar methods.

6. Let f be a real polynomial of degree n. Define

$$I(t) := \int_0^t e^{t-u} f(u)du.$$

Prove the equality

$$I(t) = e^t \sum_{j=0}^{n} f^{(j)}(0) - \sum_{j=0}^{n} f^{(j)}(t),$$

where $f^{(j)}$ denotes the j-th derivative of f.

7. Let f be a real polynomial of degree n, say

$$f(x) = \sum_{j=0}^{n} a_j x^j,$$

and define the polynomial $\overline{f}$ by

$$\overline{f}(x) = \sum_{j=0}^{n} |a_j| \, x^j.$$

If $I(t)$ is the quantity defined in the previous exercise, prove the bound

$$|I(t)| \leq |t| \, e^{|t|} \overline{f}(|t|).$$

8. Prove the π is transcendental, as follows. Assume that $\theta = i\pi$ is algebraic, and let g be its minimal polynomial, with roots $\theta = \theta_1, \theta_2, \ldots, \theta_r$. Denote by b the leading coefficient of g. Expand the identity

$$(1 + e^{\theta_1}) \cdots (1 + e^{\theta_r}) = 0$$

to get

$$q + e^{\phi_1} + \cdots + e^{\phi_n} = 0,$$

where each $\phi_i \neq 0$ has the form $\epsilon_1\theta_1 + \cdots + \epsilon_r\theta_r$ for some set of $\epsilon_i \in \{0, 1\}$, and $q = 2^r - n \in \mathbb{N}$.

Fix a large prime p, and introduce the polynomial

$$f(x) := b^{np} x^{p-1} (x - \phi_1)^p \cdots (x - \phi_r)^p,$$

and prove that $f \in \mathbb{Z}[x]$. With $I(t)$ defined in Exercise 6, let

$$J = I(\phi_1) + \cdots + I(\phi_r).$$

Prove that J is an integer, and in fact J is the sum of two terms J_1 and J_2, where $p!$ divides J_2 but not J_1 (for p large enough), while $(p-1)!$ divides both. Conclude that $|J| \geq (p-1)!$.

On the other hand, derive from the previous exercise a bound of the kind $|J| \leq c_1 c_2^p$ for suitable constants c_1, c_2, and observe that this is a contradiction.

9. The problem of squaring the circle starts from a circle (determined by its center and a point on the circumference) and requires to construct a square having the same area. Use the previous exercises to show that this is not doable with ruler and compass.

The theory of cyclic extensions from Propositions A.8.3 and A.8.7 generalizes to the case where characteristic of the field is not prime to the degree of the extension, but with a twist. The results of the following two exercises go under the name of *Artin–Schreier theory.*

10. Let K be a field of positive characteristic p, L/K a cyclic extension of degree p. Prove that $L = K(\alpha)$, where α satisfies a polynomial $x^p - x + c \in K[x]$.

11. Let K be a field of positive characteristic p, $\alpha \in \overline{K}$ a root of the polynomial $f(x) = x^p - x + c \in K[x]$, and let $L := K(\alpha)$. Prove that either $L = K$ or L/K is cyclic of degree p. In particular, f is either irreducible or a product of linear factors.

12. Given a group G, denote G' the subgroup of G generated by the commutators (i.e., elements of the form $ghg^{-1}h^{-1}$ for $g, h \in G$). The group G' is called the *derived group* of G, and the sequence of groups defined by $G^{(0)} = G$ and $G^{n+1} = (G^{(n)})'$ is called the *derived series* of G. Prove that a group G is solvable if and only if $G^{(n)} = \{e\}$ for n big enough.

13. Prove that the symmetric group $\mathfrak{S}_n$ is solvable if and only if $n \leq 4$ (look at the conjugacy classes in the alternating group A_5).

14. Let $G < \mathfrak{S}_n$ be a subgroup of the nth symmetric group. Assuming that G contains a transposition and n is prime, prove that G is the whole $\mathfrak{S}_n$.

15. Let $f \in \mathbb{Q}[x]$ be an irreducible polynomial of prime degree n. Assume that f has exactly two nonreal roots. Then the Galois group of f is the whole $\mathfrak{S}_n$.

16. Prove that the quintic polynomial $x^5 + 3x + 3$ is not solvable in radicals over $\mathbb{Q}$. More generally, do this for $x^5 + px + p$, where p is a positive prime number.

17. Construct a quartic Galois extension of $\mathbb{Q}$ having Galois group $\mathbb{Z}/4\mathbb{Z}$.

18. Construct a quartic Galois extension of $\mathbb{Q}$ having Galois group $\mathbb{Z}/2\mathbb{Z} \times \mathbb{Z}/2\mathbb{Z}$.

19. Use Galois theory to find the formula for the solution to a general equation of degree 3

$$f(x) = x^3 + px + q = 0$$

over a field K of characteristic different from $2, 3$. (You can assume that $\zeta_3 \in K$. Let a_1, a_2, a_3 be the roots of f, $L = K(a_1, a_2, a_3)$. Corresponding to the subgroup $A_3 < \mathfrak{S}_3$ there is a field C such that $[L : C] = 3$ and $[C : K] = 2$—prove that $L = C(a_1)$. Let σ be a generator of A_3, acting as $\sigma(a_1) = a_2$, $\sigma(a_2) = a_3$, $\sigma(a_3) = a_1$. Follow the proof of Hilbert Theorem 90 and find that $\zeta_3 = \beta/\sigma(\beta)$, where

$$\beta = a_1 + \zeta_3 a_2 + \zeta_3^2 a_3.$$

Conclude that $\beta^3 \in C$; then compute β^3 explicitly and express it in terms of p and q. Use the relations you find to solve for a_1, a_2, and a_3 in terms of β and β^2.)

20. State and prove a form of the Chinese remainder theorem for rings that are not necessarily commutative, as we used in the proof of Theorem A.8.6.

21. Give an alternative proof of Artin's theorem A.8.6 by induction on the number of characters (given a dependence relation between n characters, you can get more using the fact that characters are multiplicative).

22. Prove that a topological group G is profinite if and only if it is the inverse limit (as topological groups) of a family of discrete finite groups.

23. A topological group G is profinite if and only if it is Hausdorff, compact and totally disconnected.

24. Use Corollary A.5.4 to give an alternative proof of the primitive element theorem.

25. Prove the *fundamental theorem of algebra*, namely show that $\mathbb{C}$ is algebraically closed. (Prove that $\mathbb{R}$ does not have finite extensions of odd degree, and that $\mathbb{C}$ does not have quadratic extensions. If $K/\mathbb{C}$ is a finite extension, use Galois theory to show that $K = \mathbb{C}$.)

Bibliography

[Aki35] Yasno Akizuki, *Einige Bemerkungen über primäre Integritätsbereiche mit Teilerkettensatz*, Proc. Phys.-Math. Soc. Japan. **17** (1935), 327–336.

[Alo99] Noga Alon, *Combinatorial nullstellensatz: Recent trends in combinatorics (Mátraháza, 1995)*, Combin. Probab. Comput. **8** (1999), no. 1–2, 7–29, DOI 10.1017/S0963548398003411. MR1684621

[Alo99] Noga Alon, *Combinatorial Nullstellensatz*, Combin. Probab. Comput. **8** (1999), nos. 1–2, 7–29, DOI 10.1017/S0963548398003411. Recent trends in combinatorics (Mátraháza, 1995). MR1684621

[AM69] M. F. Atiyah and I. G. Macdonald, *Introduction to commutative algebra*, Addison-Wesley Publishing Co., Reading, Mass.-London-Don Mills, Ont., 1969. MR0242802

[Art71] M. Artin, *On the joins of Hensel rings*, Advances in Math. **7** (1971), 282–296 (1971), DOI 10.1016/S0001-8708(71)80007-5. MR289501

[AT51] Emil Artin and John T. Tate, *A note on finite ring extensions*, J. Math. Soc. Japan **3** (1951), 74–77, DOI 10.2969/jmsj/00310074. MR44509

[Ax68] James Ax, *The elementary theory of finite fields*, Ann. of Math. (2) **88** (1968), 239–271, DOI 10.2307/1970573. MR229613

[Bau15] Oswald Baumgart, *The quadratic reciprocity law*. A collection of classical proofs; Edited, translated from the German, and with contributions by Franz Lemmermeyer, Birkhäuser/Springer, Cham, 2015. DOI 10.1007/978-3-319-16283-7. MR3361935

[BDRH$^+$09] Thomas Bauer, Sandra Di Rocco, Brian Harbourne, Michał Kapustka, Andreas Knutsen, Wioletta Syzdek, and Tomasz Szemberg, *A primer on Seshadri constants*, Interactions of classical and numerical algebraic geometry, Contemp. Math., vol. 496, Amer. Math. Soc., Providence, RI, 2009, pp. 33–70, DOI 10.1090/conm/496/09718. MR2555949

[Bel] J. Bellaïche, *Answer on MathOverflow*, available at `https://mathoverflow.net/questions/162030/`.

[Buc65] Bruno Buchberger, *Ein Algorithmus zum Auffinden der Basiselemente des Restklassenringes nach einem nulldimensionalen Polynomideal*, Ph.D. thesis, University of Innsbruck, 1965.

[Car] Dustin Cartwright, *Answer on MathOverflow*, available at `http://mathoverflow.net/questions/71699/`.

[Car80] A. Caruth, *A short proof of the principal ideal theorem*, Quart. J. Math. Oxford Ser. (2) **31** (1980), no. 124, 401, DOI 10.1093/qmath/31.4.401. MR596975

[CD16] Abel Castillo and Rainer Dietmann, *On Hilbert's irreducibility theorem*, Acta Arith. **180** (2017), no. 1, 1–14, DOI 10.4064/aa8380-2-2017. MR3693563

[Ces] Kestutis Cesnavicius, *Answer on MathOverflow*, available at `http://mathoverflow.net/questions/64399/`.

[Che45] Claude Chevalley, *Intersections of algebraic and algebroid varieties*, Trans. Amer. Math. Soc. **57** (1945), 1–85, DOI 10.2307/1990167. MR12458

[CK90] C. C. Chang and H. J. Keisler, *Model theory*, 3rd ed., Studies in Logic and the Foundations of Mathematics, vol. 73, North-Holland Publishing Co., Amsterdam, 1990. MR1059055

[CL] Antoine Chambert-Loir, *Answer on MathOverflow*, available at `https://mathoverflow.net/questions/15611/`.

[Cla65] Luther Claborn, *Dedekind domains and rings of quotients*, Pacific J. Math. **15** (1965), 59–64. MR178005

[Cla15] Pete L. Clark, *Commutative algebra*, available at `http://math.uga.edu/~pete/integral.pdf`, 2015.

[CLO96] David Cox, John Little, and Donal O'Shea, *Ideals, varieties, and algorithms: An introduction to computational algebraic geometry and commutative algebra*, 2nd ed., Undergraduate Texts in Mathematics, Springer-Verlag, New York, 1997. MR1417938

[Coh46] I. S. Cohen, *On the structure and ideal theory of complete local rings*, Trans. Amer. Math. Soc. **59** (1946), 54–106, DOI 10.2307/1990313. MR16094

[Con] Keith Conrad, *Ideal factorization*, Lecture notes, available at `http://www.math.uconn.edu/~kconrad/blurbs/gradnumthy/idealfactor.pdf`.

[CS46] I. S. Cohen and A. Seidenberg, *Prime ideals and integral dependence*, Bull. Amer. Math. Soc. **52** (1946), 252–261, DOI 10.1090/S0002-9904-1946-08552-3. MR15379

[CSS97] Gary Cornell, Joseph H. Silverman, and Glenn Stevens (eds.), *Modular forms and Fermat's last theorem*, Springer-Verlag, New York, 1997. Papers from the Instructional Conference on Number Theory and Arithmetic Geometry held at Boston University, Boston, MA, August 9–18, 1995, DOI 10.1007/978-1-4612-1974-3. MR1638473

[dJea20] Aise Johan de Jong and et al., *The stack project*, 2020, available at `https://stacks.math.columbia.edu/`.

[Dol03] Igor Dolgachev, *Lectures on invariant theory*, London Mathematical Society Lecture Note Series, 296, Cambridge University Press, Cambridge, 2003, DOI 10.1017/CBO9780511615436. MR2004511

[Eak68] Paul M. Eakin Jr., *The converse to a well known theorem on Noetherian rings*, Math. Ann. **177** (1968), 278–282, DOI 10.1007/BF01350720. MR225767

[Eis50] G. Eisenstein, *Über die Irreductibilität und einige andere Eigenschaften der Gleichung, von welcher die Theilung der ganzen Lemniscate abhängt* (German), J. Reine Angew. Math. **39** (1850), 160–179, DOI 10.1515/crll.1850.39.160. MR1578663

[Eis95] David Eisenbud, *Commutative algebra: With a view toward algebraic geometry*, Graduate Texts in Mathematics, vol. 150, Springer-Verlag, New York, 1995, DOI 10.1007/978-1-4612-5350-1. MR1322960

[Ele] George Elencwajg, *Answer on MathOverflow*, available at `http://mathoverflow.net/questions/16869/`.

[Fer] Andrea Ferretti, *Homological methods in commutative algebra*, Graduate Studies in Mathematics, vol. 234, American Mathematical Society, Providence, RI, 2023.

[Fil11] Michael Filaseta, *The transcendence of e and π*, available at `http://people.math.sc.edu/filaseta/gradcourses/Math785/Math785Notes6.pdf`, 2011.

[Fol94] Gerald B. Folland, *A course in abstract harmonic analysis*, Studies in Advanced Mathematics, CRC Press, Boca Raton, FL, 1995. MR1397028

[For73] Edward Formanek, *Faithful Noetherian modules*, Proc. Amer. Math. Soc. **41** (1973), 381–383, DOI 10.2307/2039099. MR379477

[Ful84] William Fulton, *Intersection theory*, Ergebnisse der Mathematik und ihrer Grenzgebiete (3) [Results in Mathematics and Related Areas (3)], vol. 2, Springer-Verlag, Berlin, 1984, DOI 10.1007/978-3-662-02421-8. MR732620

[gne] gnesis, *Answer on Math Stack Exchange*, available at `https://math.stackexchange.com/questions/1109732/`.

[Gra74] Anne Grams, *Atomic rings and the ascending chain condition for principal ideals*, Proc. Cambridge Philos. Soc. **75** (1974), 321–329, DOI 10.1017/s0305004100048532. MR340249

[Gro66] A. Grothendieck, *Éléments de géométrie algébrique. IV. Étude locale des schémas et des morphismes de schémas. III*, Inst. Hautes Études Sci. Publ. Math. **28** (1966), 255. MR217086

[GT02] Andrew Granville and Thomas J. Tucker, *It's as easy as abc*, Notices Amer. Math. Soc. **49** (2002), no. 10, 1224–1231. MR1930670

[Har77] Robin Hartshorne, *Algebraic geometry*, Graduate Texts in Mathematics, No. 52, Springer-Verlag, New York–Heidelberg, 1977. MR0463157

[Haz09] Michiel Hazewinkel, *Witt vectors. Part 1*, vol. 6, pp. 319–472, Elsevier, 2009, Available at `https://arxiv.org/abs/0804.3888`.

[Hes05] Lars Hesselholt, *Lecture notes on Witt vectors*, Lecture notes, available at `https://www.math.nagoya-u.ac.jp/~larsh/papers/s03/wittsurvey.pdf`, 2005.

[HH92] Melvin Hochster and Craig Huneke, *Infinite integral extensions and big Cohen-Macaulay algebras*, Ann. of Math. (2) **135** (1992), no. 1, 53–89, DOI 10.2307/2946563. MR1147957

[HHO18] Jürgen Herzog, Takayuki Hibi, and Hidefumi Ohsugi, *Binomial ideals*, Graduate Texts in Mathematics, vol. 279, Springer, Cham, 2018, DOI 10.1007/978-3-319-95349-6. MR3838370

[Hil93] David Hilbert, *Ueber die vollen Invariantensysteme* (German), Math. Ann. **42** (1893), no. 3, 313–373, DOI 10.1007/BF01444162. MR1510781

[Hir97] Morris W. Hirsch, *Differential topology*, Graduate Texts in Mathematics, vol. 33, Springer-Verlag, New York, 1994. Corrected reprint of the 1976 original. MR1336822

[Hor76] James Hornell, *The multiplicity function of a local ring*, Trans. Amer. Math. Soc. **220** (1976), 321–341, DOI 10.2307/1997648. MR409491

[Huc76] James A. Huckaba, *The integral closure of a Noetherian ring*, Trans. Amer. Math. Soc. **220** (1976), 159–166, DOI 10.2307/1997638. MR401734

[Hun68] Thomas W. Hungerford, *On the structure of principal ideal rings*, Pacific J. Math. **25** (1968), 543–547. MR227159

[Jen80] C. U. Jensen, *Peano rings of arbitrary global dimension*, J. London Math. Soc. (2) **21** (1980), no. 1, 39–44, DOI 10.1112/jlms/s2-21.1.39. MR576181

[Jot00] P. Jothilingam, *Cohen's theorem and Eakin–Nagata theorem revisited*, Comm. Algebra **28** (2000), no. 10, 4861–4866, DOI 10.1080/00927870008827126. MR1779878

[Kap70] Irving Kaplansky, *Commutative rings*, Revised edition, University of Chicago Press, Chicago, Ill.-London, 1974. MR0345945

[Kata] Makoto Kato, *Question on Math Stack Exchange*, available at `https://math.stackexchange.com/questions/131892/`.

[Katb] Damien Katz, *A guide to Cohen's structure theorem for complete local rings*, Lecture notes, available at `https://pdfs.semanticscholar.org/ae9d/954d57285883acd94e55b45ed4c349f50ffc.pdf`.

[Knaa] Knaf Hagen, *Answer on Math Stack Exchange*, available at https://math.stackexchange.com/questions/24406/.

[Knab] Knaf Hagen, *Answer on Math Stack Exchange*, available at https://math.stackexchange.com/questions/363166/.

[KO10] Keith A. Kearnes and Greg Oman, *Cardinalities of residue fields of Noetherian integral domains*, Comm. Algebra **38** (2010), no. 10, 3580–3588, DOI 10.1080/00927870903200893. MR2760675

[KR00] Martin Kreuzer and Lorenzo Robbiano, *Computational commutative algebra. 1*, Springer-Verlag, Berlin, 2000, DOI 10.1007/978-3-540-70628-1. MR1790326

[Lec57] Christer Lech, *On the associativity formula for multiplicities*, Ark. Mat. **3** (1957), 301–314, DOI 10.1007/BF02589424. MR85245

[Leq85] Yves Lequain, *A local characterization of Noetherian and Dedekind rings*, Proc. Amer. Math. Soc. **94** (1985), no. 3, 369–370, DOI 10.2307/2045215. MR787874

[LLL82] A. K. Lenstra, H. W. Lenstra Jr., and L. Lovász, *Factoring polynomials with rational coefficients*, Math. Ann. **261** (1982), no. 4, 515–534, DOI 10.1007/BF01457454. MR682664

[Mar77] Daniel A. Marcus, *Number fields*, Universitext, Springer-Verlag, New York-Heidelberg, 1977. MR0457396

[Mat70] Hideyuki Matsumura, *Commutative algebra*, W. A. Benjamin, Inc., New York, 1970. MR0266911

[Mat86] Hideyuki Matsumura, *Commutative ring theory*, Cambridge University Press, 1986.

[MG02] Daniele Micciancio and Shafi Goldwasser, *Complexity of lattice problems: A cryptographic perspective*, The Kluwer International Series in Engineering and Computer Science, vol. 671, Kluwer Academic Publishers, Boston, MA, 2002, DOI 10.1007/978-1-4615-0897-7. MR2042139

[Mir11] Diego Mirandola, *Some basic ideas about Groebner bases and their application to Euclidean geometry*, Postgraduate students Seminar, available at http://math.sun.ac.za/wp-content/uploads/2011/04/Diego-Mirandola-Groebner-bases.pdf, 2011.

[MM18] Gunter Malle and B. Heinrich Matzat, *Inverse Galois theory*, Springer Monographs in Mathematics, Springer, Berlin, 2018. Second edition [MR1711577]. MR3822366

[Mor08] Kaitlyn Moran, *Gröbner bases and their applications*, Lecture notes, available at http://home.sandiego.edu/~aboocher/writings/kaitlyn.pdf, 2008.

[Mur] Daniel Murfet, *Cohen's theorem*, Lecture notes, available at http://therisingsea.org/notes/CohensTheorem.pdf.

[Nag62] Masayoshi Nagata, *Local rings*, Interscience Tracts in Pure and Applied Mathematics, No. 13, Interscience Publishers (a division of John Wiley & Sons, Inc.), New York-London, 1962. MR0155856

[Nag68] Masayoshi Nagata, *A type of subrings of a noetherian ring*, J. Math. Kyoto Univ. **8** (1968), 465–467, DOI 10.1215/kjm/1250524062. MR236162

[Nag05] A. R. Naghipour, *A simple proof of Cohen's theorem*, Amer. Math. Monthly **112** (2005), no. 9, 825–826, DOI 10.2307/30037605. MR2179863

[Now97] Krzysztof Jan Nowak, *A proof of the criterion for multiplicity one*, Univ. Iagel. Acta Math. **35** (1997), 247–249. MR1458061

[Pol] Jason Polak, *A non-Noetherian subring of a polynomial ring*, available at http://blog.jpolak.org/?p=1427.

[Rab29] J. L. Rabinowitsch, *Zum Hilbertschen Nullstellensatz* (German), Math. Ann. **102** (1930), no. 1, 520, DOI 10.1007/BF01782361. MR1512592

[Rab14] Joseph Rabinoff, *The theory of Witt vectors*, 2014, Available at https://arxiv.org/abs/1409.7445.

[Ray70] Michel Raynaud, *Anneaux locaux henséliens* (French), Lecture Notes in Mathematics, Vol. 169, Springer-Verlag, Berlin-New York, 1970. MR0277519

[Ree61] D. Rees, *$\mathfrak{a}$-transforms of local rings and a theorem on multiplicities of ideals*, Proc. Cambridge Philos. Soc. **57** (1961), 8–17, DOI 10.1017/s0305004100034800. MR118750

[Rot79] Joseph J. Rotman, *An introduction to homological algebra*, Pure and Applied Mathematics, vol. 85, Academic Press, Inc. [Harcourt Brace Jovanovich, Publishers], New York–London, 1979. MR538169

[RZ10] Hamid Rahkooy and Zafeirakis Zafeirakopoulos, *On Computing Elimination Ideals Using Resultants with Applications to Gröbner Bases*, available at `https://static1.squarespace.com/static/559921a3e4b02c1d7480f8f4/t/585c2df3d1758e618c87fca5/1482436084200/Rahkooy.pdf`, 2010.

[Sam49] Pierre Samuel, *La notion de multiplicité en algèbre et en géométrie algébrique* (French), J. Math. Pures Appl. (9) **30** (1951), 159–205. MR48103

[Sam53] Pierre Samuel, *Algèbre locale* (French), Mémor. Sci. Math., no. 123, Gauthier-Villars, Paris, 1953. MR0054995

[Sch99] Hans Schoutens, *The use of ultraproducts in commutative algebra*, Lecture Notes in Mathematics, vol. 1999, Springer-Verlag, Berlin, 2010, DOI 10.1007/978-3-642-13368-8. MR2676525

[Sch03] René Schoof, *Number theory*, Lecture notes, available at `http://www.mat.uniroma2.it/~schoof/tn.html`, 2003.

[Sei53] A. Seidenberg, *A note on the dimension theory of rings*, Pacific J. Math. **3** (1953), 505–512. MR54571

[Ser88] Jean-Pierre Serre, *Topics in Galois theory*, 2nd ed., Research Notes in Mathematics, vol. 1, A K Peters, Ltd., Wellesley, MA, 2008. With notes by Henri Darmon. MR2363329

[Ser00] Jean-Pierre Serre, *Local algebra*, Springer Monographs in Mathematics, Springer-Verlag, Berlin, 2000. Translated from the French by CheeWhye Chin and revised by the author, DOI 10.1007/978-3-662-04203-8. MR1771925

[Sim10] Denis Simon, *Selected applications of LLL in number theory*, pp. 265–282, Springer Berlin Heidelberg, Berlin, Heidelberg, 2010.

[Sla] Slade, *Answer on MathOverflow*, available at `http://mathoverflow.net/questions/1267419/`.

[Sta69] H. M. Stark, *On the "gap" in a theorem of Heegner*, J. Number Theory **1** (1969), 16–27, DOI 10.1016/0022-314X(69)90023-7. MR241384

[Tha99] Dinesh S. Thakur, *Fermat's last theorem for regular primes*, Cyclotomic fields and related topics (Pune, 1999), Bhaskaracharya Pratishthana, Pune, 2000, pp. 165–173. MR1802383

[vDdB] Remy van Dobben de Bruyn, *Answer on MathOverflow*, available at `https://mathoverflow.net/questions/242641/`.

[Wei95] Charles A. Weibel, *An introduction to homological algebra*, Cambridge Studies in Advanced Mathematics, vol. 38, Cambridge University Press, Cambridge, 1994, DOI 10.1017/CBO9781139644136. MR1269324

[Wit36] Ernst Witt, *Zyklische Körper und Algebren der Charakteristik p vom Grad p^n. Struktur diskret bewerteter perfekter Körper mit vollkommenem Restklassenkörper der Charakteristik p* (German), J. Reine Angew. Math. **176** (1937), 126–140, DOI 10.1515/crll.1937.176.126. MR1581526

[Wof] Eric Wofsey, *Answer on Math Stack Exchange*, available at `https://math.stackexchange.com/questions/1515854/`.

[Zak82] Abraham Zaks, *Atomic rings without a.c.c. on principal ideals*, J. Algebra **74** (1982), no. 1, 223–231, DOI 10.1016/0021-8693(82)90015-1. MR644228

[ZS76a] Oscar Zariski and Pierre Samuel, *Commutative algebra. Vol. 1*, Graduate Texts in Mathematics, No. 28, Springer-Verlag, New York-Heidelberg-Berlin, 1975. With the cooperation of I. S. Cohen; Corrected reprinting of the 1958 edition. MR0384768

[ZS76b] Oscar Zariski and Pierre Samuel, *Commutative algebra II*, Graduate Texts in Mathematics, no. 29, Springer New York, 1976.

Index of Notation

Index

SELECTED PUBLISHED TITLES IN THIS SERIES

234 **Andrea Ferretti,** Homological Methods in Commutative Algebra, 2023
233 **Andrea Ferretti,** Commutative Algebra, 2023
232 **Harry Dym,** Linear Algebra in Action, Third Edition, 2023
231 **Luís Barreira and Yakov Pesin,** Introduction to Smooth Ergodic Theory, Second Edition, 2023
229 **Giovanni Leoni,** A First Course in Fractional Sobolev Spaces, 2023
228 **Henk Bruin,** Topological and Ergodic Theory of Symbolic Dynamics, 2022
227 **William M. Goldman,** Geometric Structures on Manifolds, 2022
226 **Milivoje Lukić,** A First Course in Spectral Theory, 2022
225 **Jacob Bedrossian and Vlad Vicol,** The Mathematical Analysis of the Incompressible Euler and Navier-Stokes Equations, 2022
224 **Ben Krause,** Discrete Analogues in Harmonic Analysis, 2022
223 **Volodymyr Nekrashevych,** Groups and Topological Dynamics, 2022
222 **Michael Artin,** Algebraic Geometry, 2022
221 **David Damanik and Jake Fillman,** One-Dimensional Ergodic Schrödinger Operators, 2022
220 **Isaac Goldbring,** Ultrafilters Throughout Mathematics, 2022
219 **Michael Joswig,** Essentials of Tropical Combinatorics, 2021
218 **Riccardo Benedetti,** Lectures on Differential Topology, 2021
217 **Marius Crainic, Rui Loja Fernandes, and Ioan Mărcuţ,** Lectures on Poisson Geometry, 2021
216 **Brian Osserman,** A Concise Introduction to Algebraic Varieties, 2021
215 **Tai-Ping Liu,** Shock Waves, 2021
214 **Ioannis Karatzas and Constantinos Kardaras,** Portfolio Theory and Arbitrage, 2021
213 **Hung Vinh Tran,** Hamilton–Jacobi Equations, 2021
212 **Marcelo Viana and José M. Espinar,** Differential Equations, 2021
211 **Mateusz Michałek and Bernd Sturmfels,** Invitation to Nonlinear Algebra, 2021
210 **Bruce E. Sagan,** Combinatorics: The Art of Counting, 2020
209 **Jessica S. Purcell,** Hyperbolic Knot Theory, 2020
208 **Vicente Muñoz, Ángel González-Prieto, and Juan Ángel Rojo,** Geometry and Topology of Manifolds, 2020
207 **Dmitry N. Kozlov,** Organized Collapse: An Introduction to Discrete Morse Theory, 2020
206 **Ben Andrews, Bennett Chow, Christine Guenther, and Mat Langford,** Extrinsic Geometric Flows, 2020
205 **Mikhail Shubin,** Invitation to Partial Differential Equations, 2020
204 **Sarah J. Witherspoon,** Hochschild Cohomology for Algebras, 2019
203 **Dimitris Koukoulopoulos,** The Distribution of Prime Numbers, 2019
202 **Michael E. Taylor,** Introduction to Complex Analysis, 2019
201 **Dan A. Lee,** Geometric Relativity, 2019
200 **Semyon Dyatlov and Maciej Zworski,** Mathematical Theory of Scattering Resonances, 2019
199 **Weinan E, Tiejun Li, and Eric Vanden-Eijnden,** Applied Stochastic Analysis, 2019
198 **Robert L. Benedetto,** Dynamics in One Non-Archimedean Variable, 2019
197 **Walter Craig,** A Course on Partial Differential Equations, 2018
196 **Martin Stynes and David Stynes,** Convection-Diffusion Problems, 2018
195 **Matthias Beck and Raman Sanyal,** Combinatorial Reciprocity Theorems, 2018

For a complete list of titles in this series, visit the
AMS Bookstore at **www.ams.org/bookstore/gsmseries/**.